Crystal Structure Analysis

Crystal Structure Analysis

A Primer

Second Edition

Jenny Pickworth Glusker

The Institute for Cancer Research—Fox Chase, Philadelphia

Kenneth N. Trueblood

University of California, Los Angeles

New York Oxford

Oxford University Press

1985

OXFORD UNIVERSITY PRESS

Oxford New York Toronto
Delhi Bombay Calcutta Madras Karachi
Kuala Lumpur Singapore Hong Kong Tokyo
Nairobi Dar es Salaam Cape Town
Melbourne Auckland

and associated companies in Beirut
Berlin Ibadan Mexico City Nicosia

Published by Oxford University Press, Inc. 200 Madison Avenue New York, New York 10016

Library of Congress Cataloging in Publication Data
Glusker, Jenny Pickworth.
Crystal structure analysis.
Bibliography: p.
Includes index.
I. X-ray crystallography. I. Trueblood, Kenneth N.
II. Title.
QD945.G58 1985 547′.83 84-14823
ISBN 0-19-503531-3
ISBN 0-19-503543-7 (pbk.)

Printing (last digit): 9 8 7 6 5 4 3 2 1

Printed in the United States of America

Preface to the Second Edition

In the thirteen years since the first edition of this book appeared there have been numerous advances in the practice of structural crystallography. Furthermore, many users of the first edition have suggested ways in which the book might have been improved. In this revision, we have endeavored to incorporate those suggestions and to describe the most significant advances in practice. The major changes include a considerable elaboration of the treatment of direct methods, a new chapter on anomalous dispersion and absolute configuration, a more detailed treatment of biological macromolecules, a reorganization and expansion into a separate chapter of the discussion of microcrystalline and non-crystalline materials, enlargement of the section on experimental methods to include discussion of area detectors and synchrotron radiation, and a new appendix on molecular geometry. The bibliography has been expanded by more than 50 percent, and the glossary doubled in length.

Our aim is to explain how and why the detailed three-dimensional architecture of molecules can be determined by an analysis of the diffraction patterns produced when X rays (or neutrons) are scattered by the atoms in single crystals. As with the first edition, the book is intended primarily for those who want to understand the fundamental concepts on which crystal structure determination is based without necessarily themselves becoming specialists in crystallography—an audience that includes advanced undergraduates who have studied some physics and chemistry, as well as graduate students and other research workers.

The book is divided, as before, into three parts; each has been expanded, the last two significantly. Part I, comprising the first four chapters, deals with the nature of the crystalline state, certain relevant facts about diffraction generally and diffraction by crystals in particular, and the experimental procedures used. Part II, consisting of Chapters 5 through 10, examines the problem of converting the experimentally obtained data (directions and intensities of diffracted beams) into a model of the atomic arrangement that scattered these beams—in other words, the problem of determining the approximate structure of this scattering matter, a "trial structure" suitable for refinement. Part III (Chapters 11 through 14) is concerned with techniques for refining this approximate structure to the degree warranted by the experimental data and with discussions of the structural parameters and other information that can be derived from a careful structure determination. It also includes a discussion of microcrystalline materials and glasses, and an overall summary of the various stages in structure analysis.

We wish to thank those who have helped us in this endeavor, particularly Bill Stallings and John Stezowski, who read through our manuscript and made most helpful comments, and Jack Dunitz who helped us with the glossary. We are also grateful to Margaret J. Adams, Bob Bryan, Bud Carrell, Philip Coppens, Dick Dickerson, Jose Donnay, David Eisenberg, Doris Evans, Setsuo Kashino, Henry Katz, Lisa Keefe, Bill Parrish, Eileen Pytko, Miriam Rossi, Christopher Smart, Verner Schomaker and David Zacharias for their help. One of us (J.P.G.) acknowledges financial support from the National Institutes of Health, U.S.P.H.S. (grant CA-10925).

Finally, we appreciate the help of all of you who have encouraged us through the years with your comments and constructive criticisms.

Philadelphia J.P.G.
Los Angeles K.N.T.
April 1985

Preface to the First Edition

This book, which developed from a talk to the California Association of Chemistry Teachers at Asilomar in 1966, is designed to serve as an introduction to the principles underlying structure analysis by X-ray diffraction from single crystals. It is intended both for undergraduates who have had some previous chemistry and physics and for graduate students and other research workers who do not intend to become specialists in crystallography but who want to understand the fundamental concepts on which this widely used method of structure determination is based. We have included many illustrations, with legends that form an important part of the text, a rather detailed glossary of common terms, an extensive annotated bibliography, and a list of the symbols used.

Our aim is to explain how and why the detailed three-dimensional architecture of molecules can be determined by an analysis of the diffraction patterns produced when X rays (or neutrons) are scattered by the atoms in single crystals. Part I, consisting of the first four chapters, deals with the nature of the crystalline state, certain relevant facts about diffraction generally and diffraction by crystals in particular, and, briefly, the experimental procedures that are used. Part II comprises an examination of the problem of converting the experimentally obtained data (directions and intensities of diffracted beams) into a model of the atomic arrangement that scattered these beams, that is, the problem of determining the approximate structure of this scattering matter. Part III is concerned with techniques for refining this approximate structure to the degree warranted by the experimental data, and also includes a brief discussion of some of the auxiliary information, beyond the geometric details of the structure, that can be learned from modern structure analysis. Most mathematical details have been relegated to several Appendices.

We are indebted to D. Adzei Bekoe, Helen Berman, Herbert Bernstein, Carol Ann Casciato, Anne Chomyn, Joyce Dargay, David Eisenberg, Emily Maverick, Walter Orehowsky, Jr., Joel Sussman, and David E. Zacharias for their help in suggesting revisions of earlier drafts, and to all those writers on crystallography whose ideas and illustrations we have included here.

One of us (J.P.G.) acknowledges financial support from the National Institutes of Health, U.S.P.H.S. (grants CA-10925, CA-06927 and RR-05539), and an appropriation from the Commonwealth of Pennsylvania. This book is Contribution No. 2609 from the Department of Chemistry, University of California, Los Angeles.

Finally, we want to express our gratitude to Miss Doris E. Emmott for her patient, painstaking, and precise typing of the manuscript and to Miss Leona Capeless of Oxford University Press for her help through the stages of publication.

Philadelphia J.P.G.
Los Angeles K.N.T.
December 1971

Contents

APPENDIXES

List of Figures

Symbols Used in This Book

a	An additional undetermined sign (+ or −) used in analytical phase determination in a centrosymmetric structure.
a	The width of each of a series of (or single) diffracting slits.
A	Amplitude of a wave.
A, B, $A(hkl)$, $B(hkl)$, A_j, B_j	Values of $\|F\|$ cos α and $\|F\|$ sin α respectively; that is, the components of a structure factor $F = A + iB$. The subscript j denotes the atom j.
A', B', A'', B'', A_d, B_d	Values of A and B taking into account $\Delta f'$ (to give A' and B'), $\Delta f'$ and $\Delta f''$ (to give A'' and B''), and the anomalously scattering atom alone (A_d and B_d).
a, b, c	Unit cell axial lengths.
a, **b**, **c**	Unit cell vectors of the direct lattice.
a^*, b^*, c^*	Lengths of the unit cell edges of the reciprocal lattice.
a*, **b***, **c***	Unit cell vectors in reciprocal space.
a, b, c, n, d, g	Glide planes. The row parallel to the translation is designated; it is the side of the net (a, b, or c) or its diagonal (n in a primitive net, d in a centered net). In two dimensions a glide-reflection line is represented by g.
Abs	Absorption factor.
B_{iso}, B,	Isotropic vibration parameter.
b_{11}, b_{22}, b_{33}, b_{12}, b_{23}, b_{31}, b_{ij}, b_{11j}	Six anisotropic vibration parameters representing anisotropic temperature motion; a third subscript j denotes the atom j.
C	A complex number $C = x + iy$.
C^*	The complex conjugate of C, where $C^* = x - iy$.
$\|C\|$	The magnitude of a complex number $\|C\| = (CC^*)^{1/2} = (x^2 + y^2)^{1/2}$.

c_i, c_1, c_2, c_r	Wave amplitudes (see Chapter 5)
d	The distance between two diffracting slits
d_{hkl}, d	The spacing between the lattice planes (hkl) in the crystal.
$d_{\mathrm{A-B}}$	Bond distance between atoms A and B.
E, E_{hkl}, E_{H}	Values of F corrected to remove thermal motion and scattering factor effects. These are called "normalized structure factors."
F	Face-centered lattice.
$F(hkl)$, F, $F(000)$	The structure factor for the unit cell, for the reflection hkl. It is the ratio of the amplitude of the wave scattered by the entire contents of the unit cell to that scattered by a single electron. A phase angle for the scattered wave is also involved. $F(000)$ is thus equal to the total number of electrons in the unit cell.
$\|F(hkl)\|$, $\|F\|$	The amplitude of the structure factor for hkl with no phase implied.
$\|F_o\|$, $\|F_c\|$	Amplitudes of structure factors observed ($\|F_o\|$), that is, derived from measurements of the intensity of the diffracted beam, and calculated ($\|F_c\|$) from a postulated trial structure.
F_{P}, F_{PH1}, F_{PH2}, F_{H1}, F_{H2}, F_{M}, $F_{\mathrm{M'}}$, F_{R}, F_{T}, $F_{\mathrm{T'}}$	Structure factors for a given value of hkl for a protein (P), two heavy-atom derivatives (PH1 and PH2), the parts of F due to certain atoms (M, M′, H1 and H2) and the rest of the molecule (R), and for the total structure (T and T′).
$\mathbf{F}$	Structure factor when represented as a vector.
F_{novib}	Value of F for a structure containing only non-vibrating atoms.
F_{+}, F_{-}	Values of $F(hkl)$ and $F(\bar{h}\bar{k}\bar{l})$ when anomalous dispersion effects are measurable.
$f(hkl)$, f, f_j	Atomic scattering factor, also called atomic form factor, for the hkl reflection relative to the scattering by a single electron. The subscript j denotes atom j.
G(r)	Radial distribution function.
G, H	Values of A_{M} and B_{M} with the scattering factor contribution ($f + \Delta f' + \Delta f''$) removed (Appendix 10).
$\mathbf{H}$	Reciprocal lattice vector.
H, K	Indices of two "reflecting planes."
H, K	$H = h, k, l$; $K = h', k', l'$.
hkl, $-h$, $-k$, $-l$, $\bar{h}\bar{k}\bar{l}$, $hkil$	Indices of the reflection from a set of parallel planes; also the coordinates of a reciprocal lattice point. If h, k, or l are negative they are represented as $-h$, $-k$, $-l$ or $\bar{h}\bar{k}\bar{l}$. In hexagonal systems a fourth index, $i = -(h + k)$ may be used (see Appendix 1).
(hkl)	Indices of a crystal face, or of a single plane, or of a set of parallel planes.
I	Body-centered lattice.

I	Intensity (on an arbitrary scale) for each reflection.
I_{corr}	Value of I corrected for Lp and Abs.
i	An "imaginary number," $i = \sqrt{-1}$.
i, j	Any integers.
k	(See *hkl*.)
K	A scale factor to bring F^2 values to an absolute scale (relative to the scattering by a single electron).
K	Reciprocal lattice constant (see caption to Figure 3.5).
Lp	Lorentz and polarization factors. These are factors that are used to correct values of I for the geometric conditions of their measurement.
l	(See *hkl*.)
l	The distance between two points in the unit cell.
l	A direction cosine.
M	Molecular weight of a compound.
M, M′	Atoms or groups of atoms that are interchanged during the preparation of an isomorphous pair of crystals.
M_1, M_2, M	Heavy atoms substituted in a protein, P.
m	Mirror planes.
m	The number of experimental observations.
m	Figure of merit.
N	The number of X-ray reflections observed for a structure
N	The number of atoms in the unit cell.
$N_{Avog.}$	Avogadro's number. The number of molecules in a mole, 6.02×10^{23}.
n	Used for n-fold rotation axes. Also used as a general constant.
n_r	Screw axis designations where n and r are integers (2,3,4,6 and 1, . . . , $(n - 1)$), respectively.
P, PH1, PH2	Protein (P), also heavy atom derivatives PH1 and PH2.
$P(uvw)$, P, P_s	The Patterson function, evaluated at points of u, v, w in the unit cell. The P_s function is used with anomalous dispersion data (see Chapter 10).
P, A, B, C, F, I	Lattice symbols. Primitive (P), centered on one set of faces (A, B, C), or all faces (F) of the unit cell, or body-centered (I).
P_+	Probability that a triple product is positive (see eqn. 8.3).
p, q	Path differences.
p	Total number of parameters to be refined.
Q	The quantity minimized in a least-squares calculation.
R	Discrepancy index $R = \dfrac{\Sigma\|(\|F_0\| - \|F_c\|)\|}{\|F_0\|}$.

R	Remainder of a structure.
R/S	System of Cahn and Ingold for describing the aboslute configuration of a chiral molecule.
r	The distance on a radial distribution function.
s	Number of symbolic signs used.
$s(hkl)$	The sign of the reflection hkl for a centrosymmetric structure.
t	Crystal thickness.
U_{11}, U_{ii}, U_{ij}	Anisotropic vibration parameters.
$\langle u^2 \rangle$	Mean square amplitude of atomic vibration.
u, v, w	The coordinates of any one of a series of systematically spaced points, expressed as fractions of a, b, and c, in the unit cell for a Patterson function.
V_c, V, V^*	The unit cell volume in direct and reciprocal space.
$w(hkl)$	The weight of an observation in a least squares refinement.
X, Y, Z	Cartesian coordinates for atomic positions.
x, y, z; x_j, y_j, z_j; x, y, z, u	Atomic coordinates as fractions of a, b, and c. The subscript j denotes the atom under consideration. If the system is hexagonal a fourth coordinate, u, may be added (see Appendix 1)
x_1, x_2, x_j, x_r	Displacements of a wave at a given point. The waves are each designated 1, 2, j; and r is the resultant wave from the summation of several waves.
x, y, z	Coordinates of any one of a series of systematically spaced points, expressed as fractions of a, b, c filling the unit cell at regular intervals.
Z	Number of molecules in a unit cell.
Z_i, Z_j	The atomic number of atoms i and j.
α, β, γ	Interaxial angles between **b** and **c**, **a** and **c**, and **a** and **b**, respectively (alpha, beta, gamma).
α^*, β^*, γ^*	Interaxial angles in reciprocal space.
$\alpha(hkl)$, α, α_M, α_P, α_H	Phase angle of the structure factor for the reflection hkl. $\alpha = \tan^{-1}(B/A)$.
α_1, α_2, α_j, α_r	Phases of waves 1, 2, j, and r, the resultant of the summation of waves, relative to an arbitrary origin.
$\Delta\|F\|$	The difference in the amplitudes of the observed and calculated structure factors, $\|F_o\| - \|F_c\|$ (delta $\|F\|$).
$\Delta f'$, $\Delta f''$	When an anomalous scatterer is present the value of f is replaced by $(f + \Delta f') + i\Delta f''$.
$\Delta\rho$	Difference electron density.
δ	Interbond angle.
δ_{ij}	An index that is 1 when $i = j$ and 0 elsewhere. i and j are integers (delta).

θ, θ_{hkl}	The glancing angle (complement of the angle of incidence) of the X-ray beam to the "reflecting plane." 2θ is the deviation of the diffracted beam from the direct X-ray beam (theta).
λ	Wavelength, usually that of the radiation used in the diffraction experiment (lambda).
μ/ρ	Mass absorption coefficient. μ linear absorption coefficient. ρ, density.
ϕ	An angular variable, proportional to the time, for a travelling wave. It is of the form $2\pi\nu t$, where ν is a frequency and t is the time (phi).
ϕ_H, ϕ_K, ϕ_{H-K}	The phase angle of the structure factor.
ϕ	Angle on spindle axis of goniometer head. See diffractometer (Figure 4.2).
ψ	Angle incident beam makes with lattice rows (see Appendix 3) (psi).
χ	Angle between ϕ axis and diffractometer axis (see Figure 4.2) (chi).
$\rho(xyz)$, ρ_{obs}, ρ_{calc}	Electron density, expressed as number of electrons per unit volume, at the point x, y, z in the unit cell (rho).
Σ	Summation sign (sigma).
Σ_1, Σ_2	Listing of triple products of normalized structure factors (see Chapter 8).
τ	Torsion angle.
ω	Angle between diffraction vector and plane of χ circle on diffractometer (Figure 4.2) (omega).
$\langle\ \rangle$	The mean value of a quantity.
1, 2, 3, 4, 6	Rotation-axes.
$\bar{1}$, $\bar{2}$, $\bar{3}$, $\bar{4}$, $\bar{6}$	Rotatory-inversion axes.
2_1, 4_1, 4_2, 4_3	Screw axes n_r.

To those who taught us crystallography, most especially Dorothy Hodgkin, A. L. Patterson, J. H. Sturdivant, R. B. Corey, and V. Schomaker, and to the memory of AC_3, the Advisory Council on College Chemistry.

I

Crystals and Diffraction

1

Introduction

Much of our present knowledge of the architecture of molecules has been derived from studies of the diffraction of X rays by crystals. This method was first used by W. L. Bragg in 1913. He showed that in crystalline sodium chloride each sodium is surrounded by six equidistant chlorines and each chlorine by six equidistant sodiums. No discrete molecules of NaCl are present and Bragg inferred that the crystal contained sodium ions and chloride ions; this had been predicted by Pope and Barlow in the nineteenth century but had not been previously demonstrated experimentally. A decade and a half later Kathleen Lonsdale used X-ray diffraction to show that the benzene ring is a regular hexagon rather than having alternating single and double bonds, a result that was of great significance in theoretical chemistry. Since then X-ray and neutron diffraction have served to establish detailed features of the molecular structure of every kind of stable chemical species, from the simplest to those containing many thousands of atoms.

We address ourselves here to those concerned with or interested in the structural aspects of chemistry and biology, who wish to know how crystal diffraction methods can be made to reveal structure and how the results of such structure determinations may be critically assessed. In attempting to explain why molecular structure can be determined by single-crystal diffraction of X rays, we shall try to answer several questions: Why use crystals and not liquids or gases? Why use X rays (or neutrons) and not other radiation? What experimental measurements are needed? What are the stages in a typical structure determination? How are the structures of macromolecules such as proteins and viruses determined? Why is the process of structure analysis sometimes lengthy and complex? Why is it necessary to "refine" the approximate structure that is first obtained? How can one assess the reliability of a structure analysis? This book should be regarded not as an account of "how to do it" or of practical procedural details, but rather as an effort to explain "why it is possible to do it," to give an account of the underlying physical principles and of the kinds of experiments and methods of handling the experimental data that make this approach to molecular structure determination such a powerful and fruitful one. Practitioners are urged to look elsewhere for details that are not of fundamental significance. Most such details have been omitted.

The primary aim of a crystal structure analysis by X-ray (or neutron) diffraction is to obtain a detailed picture of the contents of the crystal at the atomic level, as if one had viewed it through an extremely powerful micro-

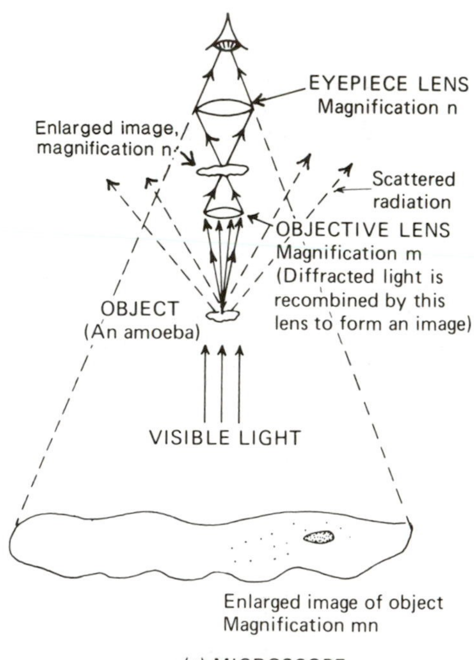

FIGURE 1.1 Analogies Between Light Microscopy and X-ray Diffraction.

Certain analogies between these two methods of using scattered radiation for determining structure are shown here. The object (sample) in both set-ups scatters some of the incident radiation into a diffraction pattern.

(a) In the ordinary microscope there is no need to record the diffraction pattern because the scattered light can be focused by the objective lens to give a magnified image of the object under study. The closer this lens is to this object, the wider the angle through which scattered radiation is caught by the lens. Thus, if this distance is small, most of the diffracted light will be caught by the objective lens and focused to form an image. The rest of the radiation is lost to the surroundings.

scope. Once this information is available, and the positions of the individual atoms are therefore known precisely, one can calculate interatomic distances, bond angles, and other features of the molecular geometry that are of interest, such as the planarity of a particular group of atoms, the angles between planes, and conformation angles around bonds. Frequently the resulting three-dimensional representation of the atomic contents of the crystal establishes a structural formula and geometrical details hitherto completely unknown. This information is of primary interest to most chemists and biochemists who are concerned with the relation of structural features to chemical properties. Precise molecular dimensions (and other information about

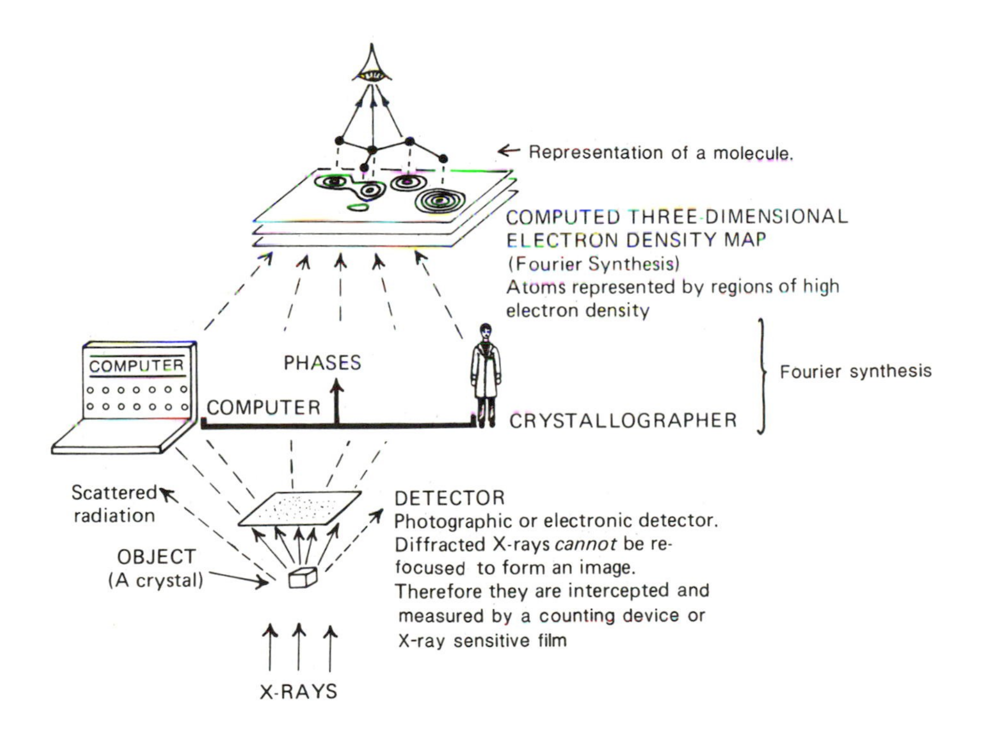

(b) X-RAY DIFFRACTION

(b) With X rays the diffraction pattern has to be recorded electronically or photographically (as indicated schematically here), because X rays cannot be focused by any known lens. Therefore the recombination of the diffracted beams that is done by a lens in the microscope must, when X rays are used, be done mathematically by a crystallographer with the aid of a computer. As stressed later (Chapter 5), this recombination cannot be done directly because the phase relations among the different diffracted beams cannot usually be measured directly. However, once these phases have been derived, deduced, guessed, or measured indirectly (which is what this book is mostly concerned with), an approximate image of the scattering matter can be formed.

molecular packing, molecular motion in the crystal, and molecular charge distribution) may be determined for a substance whose structure has been established already. Such results are more likely to be of interest to those with a theoretical bent, particularly those concerned with electronic structure, molecular strain, and the interactions between molecules.

In an ordinary optical or electron microscope the radiation scattered by the object being viewed can be recombined by the lens system to give an image of the scattering matter, appropriately magnified (Figure 1.1a) (as discussed and illustrated later in Chapter 3). In such a microscope the flow of radiation (light) through the specimen and beyond through the lens system is continuous, and therefore the relationships between the phases of the scattered waves are maintained when they are recombined by the lens. In a similar way X rays are scattered by the electrons in atoms (Figure 1.1b), but, in contrast to the situation with visible light, such radiation cannot normally be focused by presently known experimental techniques. This is because no electrical or magnetic field or material has yet been found that can refract X-rays sufficiently that an X-ray lens may be made. Consequently, an X-ray microscope cannot be used to view atoms (which have dimensions too small to permit them to be visible with an ordinary light microscope).

Since a lens system cannot be used to recombine scattered X-radiation some other technique must be used. In practice the diffracted (scattered) X-radiation is intercepted by a detecting system, but this means that the relationships between the phases of scattered waves are lost. Only the intensities (not the phases) of the diffracted waves can be measured. Therefore, information is lost. However, it *is* possible to *simulate* the recombination of scattered X-rays—just as if a lens had done it—by an appropriate, though complicated, calculation (Figure 1.1b). This mathematical calculation, the *Fourier synthesis* of the pattern of scattered or "diffracted" radiation, is a fundamental step in crystal structure determination by diffraction methods and is a central subject of our discussion (Chapters 5, 6, 8, and 9). The major complication is referred to as the "phase problem," that is, discovering the relative phases of the diffracted beams that are needed in order to compute a recognizable image of the molecule ("scattering object") by Fourier synthesis. When these relative phases are known (derived, deduced, guessed, or measured indirectly) the three-dimensional structure of the molecule* will be revealed as a result of such a Fourier synthesis.

When a crystal structure analysis by diffraction methods is successful, a wealth of information results. It can be adapted to a wide range of temperatures, pressures, and environments and has been successfully used to estab-

*We use the terms *molecule* and *molecular* in a general sense here to refer to any of the basic units that comprise a crystal—atoms, molecules or ions.

lish the molecular architecture and packing of an enormous diversity of substances, from elementary hydrogen and simple salts to proteins and nucleic acids. It has also contributed significantly to our understanding of natural and synthetic partially crystalline materials such as rubber and polyethylene. Although structure determinations of organic and biochemically significant molecules have received the most attention in recent years, the contributions of the technique to inorganic chemistry have been equally profound, initially through clarification of the chemistry of the silicates and of other chemical mysteries of minerals and inorganic solids, with recent applications to such diverse materials as the boron hydrides, alloys, hydrates, compounds of the rare gases, and novel organometallic compounds including ferrocene and metal-cluster compounds.

Why make the effort to carry out a crystal structure analysis? The reason is that when the method is successful it is unique in providing an unambiguous and complete *three-dimensional representation* of the atoms in a crystal. Other chemical and physical methods of structure determination merely provide relationships from which one can deduce the number and nature of the atoms bonded to each atom present (the topology of the molecule), or, for relatively simple molecules, provide some quantitative information from which geometrical details can be deduced. High resolution infrared and microwave spectroscopic techniques can give quantitative structural information in very simple molecules; high field nuclear magnetic resonance (NMR) can provide distances between identified atoms, in favorable circumstances. No other method can give the entire detailed picture that X-ray and neutron diffraction techniques can produce. Crystal diffraction methods have their limitations, chiefly connected with obtaining samples with the highly regular long-range three-dimensional order characteristic of the ideal crystalline state; the success of high resolution diffraction analysis requires that the sample be organized or prepared as an ordered array. It is also worthwhile to point out that a crystal structure analysis gives a fairly static or time-averaged representation of molecular structure. Although information about molecular motion in crystals—both overall and intramolecular motion *can* be appreciable—may be obtained from precise diffraction data (see Chapter 12), the freedom of motion is in general severely restricted in solids relative to that in liquids or gases.

Throughout the book you may encounter symbols or terms that are unfamiliar—such as d_{010} or d_{110} in Figure 2.3. We have included a list of symbols at the start (p. xi ff.) and a glossary (p. 221 ff.) to provide definitions of such symbols and words and a bibliography (p. 243 ff.) at the end; references listed by number in the text may be found in the bibliography. We urge you to use all of these sections frequently as you work your way through the book.

2

Crystals

The fundamental characteristic of the crystalline state is a very *high degree of internal order,* at least ideally*; in other words, the objects (the atoms, molecules, or ions) of which a crystal is composed are arranged in a regular way that is repeated over and over in all directions. The most obvious characteristic of crystals is that they usually have flat faces bounded by straight edges, but this property is not necessary or sufficient to define a crystal. It is possible to cut and polish a piece of glass or plastic, neither of which is crystalline, so that its faces are flat and its edges straight. However, it does not thereby become a crystal, for its disordered internal structure is not made regular by the polishing (even though the word "crystal" is misleadingly used for some quality glassware).

The fact that crystals have an internal structure that is periodic (regularly repeating) in three dimensions was surmised by Kepler (1611) and Hooke (1665) and logically concluded by Bergman (1773) and Haüy (1782). The precise internal order of crystalline materials can be demonstrated when a crystal is used as a three-dimensional diffraction grating for radiation that has a wavelength comparable to the interatomic distances within the crystal. It was realization of this fact that, in 1912, led von Laue to suggest that Friedrich and Knipping try diffracting X rays by crystals in order to test the

*Real crystals often exhibit a variety of imperfections—for example, short-range or long-range disorder, dislocations, irregular surfaces, twinning, and other kinds of defects—but for our present purposes, it is a good approximation to consider that in a specimen of a single crystal the order is perfect and three-dimensional. We discuss very briefly at the end of Chapter 13 the way in which our discussion must be modified when some disorder is present—for example, when the order is only one-dimensional, as in many fibers.

hypothesis that X rays are wavelike with wavelengths of the order of 1 Å (10^{-10} m). The resulting experiment was dramatically successful and led, within less than a year, to the first structure determination by X-ray methods (sodium chloride by W. L. Bragg). Even molecules as large as proteins and viruses can be crystallized and the microcrystallinity of proteins has long been used by biochemists as a measure of the purity of their preparations. Suitably sized crystals of proteins can also give good diffraction patterns, and analyses of the structures of biological macromolecules have contributed greatly to our understanding of biochemistry and molecular biology.

The elegance and beauty of crystals have always been a source of delight, and studies of their external features, known as crystal morphology, have been made since early times, particularly by those interested in minerals (for practical as well as esthetic reasons). The regularity of form of different crystals was noted, and, in the seventeenth century, it was speculated by Kepler in 1611 and Hooke in 1665 (among others) that this resulted from the ordered packing of spheroidal particles within the crystals. The Danish physician Steno noted in 1669 that although the faces of a crystalline substance, such as quartz, often varied greatly in shape and size (depending on the conditions under which the crystals were formed), the angle between pairs of similar faces on different crystals of a particular substance were always the same. Such angles can be measured approximately with a protractor or more precisely with an optical goniometer (Greek: *gonia* = angle), and a great many such measurements have been recorded over the past three centuries. This constancy of the interfacial angles for a given crystalline form of a substance reflects the regularity of its internal structure (its molecular or ionic packing) and has been used with success as an aid in characterizing and identifying compounds. Crystals tend to develop most rapidly those faces that are most densely packed in the underlying structure, although the crystal habit (overall shape) may be modified by adding soluble foreign materials.

Crystal Growth The growth of crystals from solution (or from a melt) is a fascinating experimental exercise, which almost anyone can practice successfully with a little patience. In general the growth should be very slow so that a regular arrangement of molecules or ions, leading to a well-formed crystal, may be obtained. The aim is to reach a point at which the solution is just saturated (this depends on factors such as solubility as a function of pH, temperature, and ionic strength) and then to very slowly lower the saturation point while limiting the rate of nucleation (which depends on the presence of foreign particles, including seed crystals that may have been added) so that only a few crystals can grow to a large size. Methods used to grow crystals include slow evaporation of a solution, slow cooling of a melt,

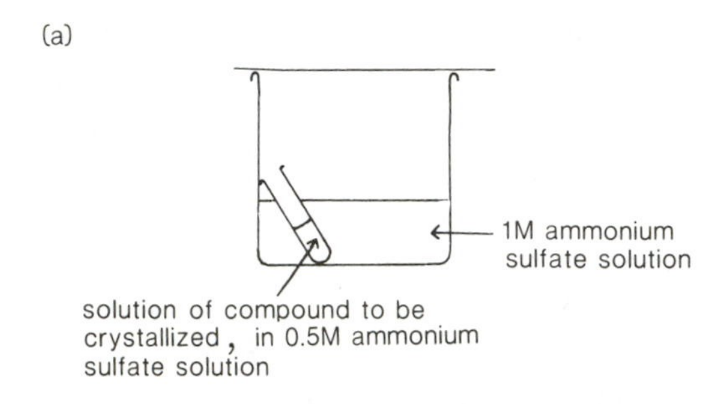

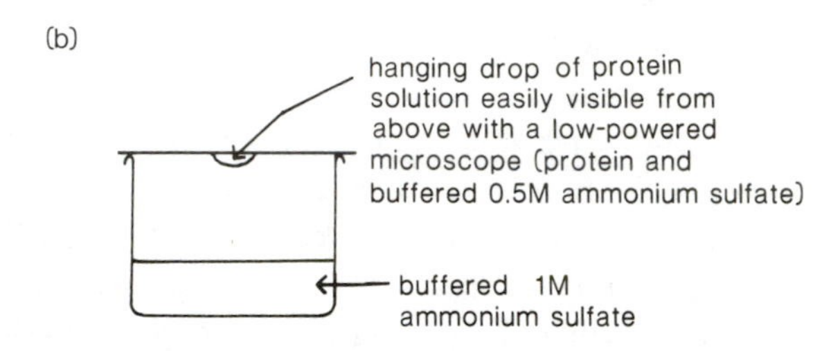

FIGURE 2.1 Crystal Being Grown by the Vapor Diffusion Method.

(a) A sample compound to be crystallized is soluble in 0.5M ammonium sulfate but not in 1M ammonium sulfate. The test tube containing the sample dissolved in 0.5M ammonium sulfate is sealed with a solution of 1M ammonium sulfate as illustrated. Vapor phase diffusion of the water molecules from the test tube to the larger reservoir will bring the concentration of ammonium sulfate in the test tube nearly to 1M; the volume of the reservoir solution is considered to be much larger than that in the test tube. If the conditions in the test tube, such as pH, temperature, ionic strength, and sample concentration, are optimized, the sample will separate from solution in the crystalline state as the ammonium sulfate concentration in the test tube approaches 1M.

(b) Scheme (a) is practical only for materials available in relatively large quantities. Pure protein is usually available only in limited quantities and this scheme (b) has been commonly adopted to circumvent this difficulty. A drop of protein solution ($\sim 10^{-2}$ ml) is placed on a cover slip, which is sealed with grease over a small well, usually one of many in a biological culture tray. Just as in scheme (a), the protein drop contains precipitant at a concentration below the protein precipitation point; the well contains precipitant at or slightly above the concentration of the protein precipitation point. Water evaporates slowly from the drop until the concentration of precipitant in the drop is the same as that in the well. The conditions in the drop (e.g., ionic strength, temperature, protein concentration) control whether the protein will crystallize, and if so at what rate.

slow precipitation, and sublimation. In the slow precipitation method a precipitant (that is, a liquid in which the substance is insoluble, such as methylpentanediol or a concentrated ammonium sulfate solution) is layered on a solution of the substance or, alternatively, some of the solvent is slowly removed from the solution by equilibration through the vapor phase in a closed system. With proteins, a solution containing precipitant salt (usually ammonium sulfate) at a concentration below that needed for precipitation, is put in a *sealed* container with another, more concentrated, precipitant salt solution. The concentration of the precipitant in the protein solution gradually increases by vapor diffusion (to equilibrium) of the volatile components, usually only water. As the precipitation point of the protein is approached, factors such as pH, temperature, ionic strength, and choice of buffer control whether the protein will separate from solution in the crystalline state rather than as an amorphous precipitate. Other precipitants such as other inorganic salts and polyethylene glycol work by similar principles, but with a number of organic precipitants—for example, ethanol—the precipitant is also quite volatile and participates in establishing the equilibrium through the vapor phase. Many variations of these techniques are possible, but most depend upon conditions for slowly bringing a solution just past saturation and adjusting the solution to optimize conditions for crystal growth. Simple examples are illustrated diagrammatically in Figure 2.1. This method is commonly used for the crystallization of biological macromolecules.

Crystals suitable for modern single-crystal diffraction need not be large. For X-ray work, specimens with dimensions 0.2–0.4 mm on an edge are usually appropriate. Such a crystal normally weighs only a small fraction of a milligram and, unless there is radiation damage or crystal deterioration during exposure, can be reclaimed intact at the end of the experiment. Larger samples (1–3 mm) are needed for neutron diffraction studies.

Sometimes a crystal is difficult to prepare or is unstable under ordinary conditions. It may react with oxygen or water vapor, or may effloresce (that is, lose solvent of crystallization and form a noncrystalline powder) or deliquesce (that is, take up water from the atmosphere and eventually form a solution). Many biologically interesting materials are unstable unless the relative humidity is extremely high. Special techniques, such as sealing the crystal in a capillary tube in a suitable amosphere, cooling the crystal, or even growing it at very low temperatures, are often used to surmount such experimental difficulties.

Symmetry of Crystals, Lattices, and Unit Cells Any crystal may be regarded as being built up by the continuing *three-dimensional translational repetition* of some basic *structural pattern,* which may comprise one or more atoms, a molecule, or a complex assembly of molecules. Figure 2.2a illustrates a schematic two-dimensional example, which might serve as a pattern for wallpaper. In three dimensions an imaginary parallelepiped that contains one unit of the translationally repeating pattern is termed the *unit cell.* The consideration of some properties of such a repetition of a structural pattern can be simplified by replacing each unit of the pattern by a point. This leads to a *crystal lattice,* which is a regular three-dimensional *arrange-*

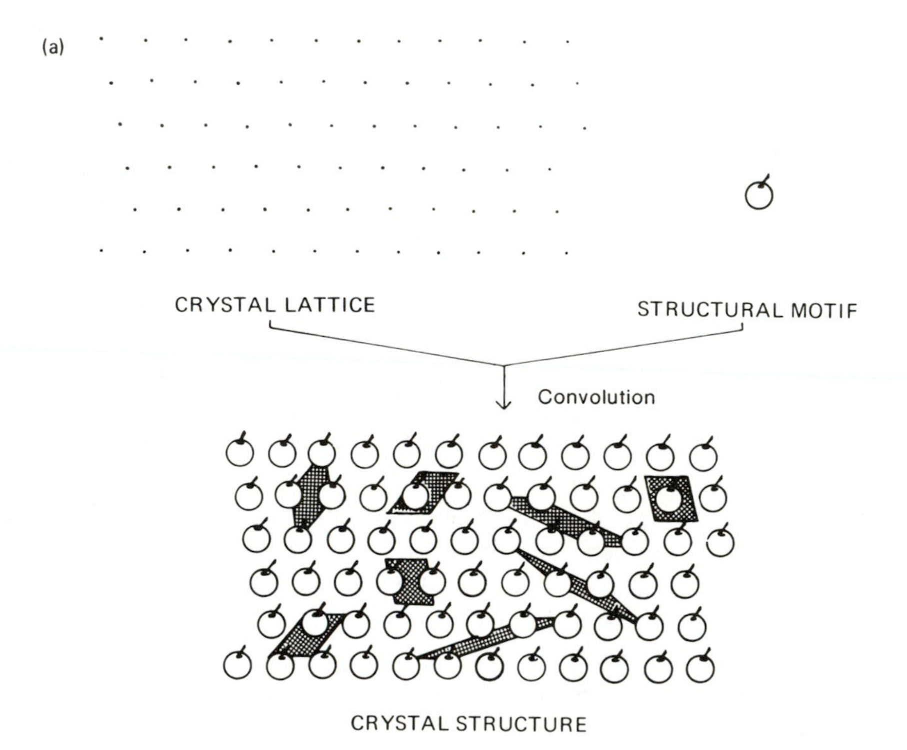

FIGURE 2.2 Some Illustrations of Choices of Unit Cells.

(a) The generation of a two-dimensional "crystal structure" from a lattice and a structural motif; the replacement of each point by an apple would lead to a two-dimensional structure. This operation may be described alternatively as the convolution of an apple and the crystal lattice. There are many ways in which unit cells may be chosen in a repeating pattern. Various possible alternative choices are shown, each having the same area despite varying shape. This can be verified by noting that the total content of any chosen unit cell is one apple. Infinite repetition in two dimensions of any one of these choices for unit cell will reproduce the entire pattern.

ment of points such that the view in a given direction from each point in this lattice is identical with the view in the same direction from any other lattice point.

The crystal lattice is, then, an infinite set of points that may be generated from a single starting point by the infinite repetition of a set of fundamental translations that characterize the lattice. These fundamental translations

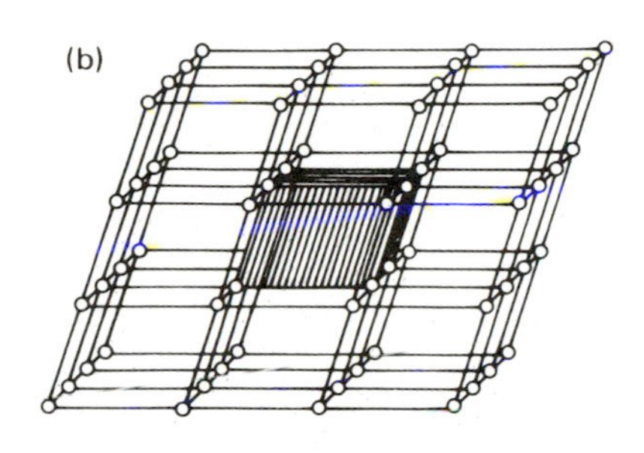

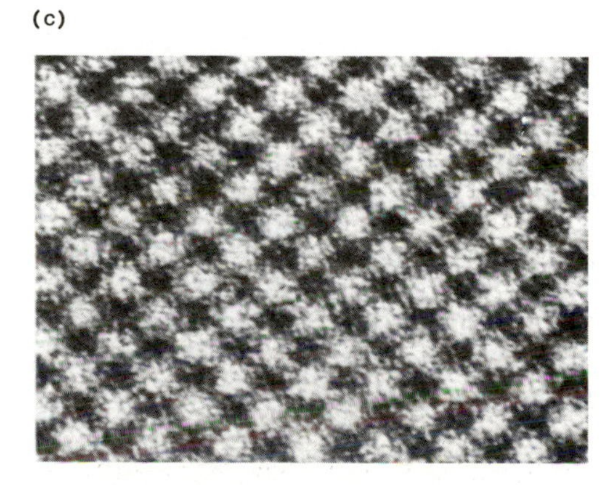

(b) Perspective view of a triclinic lattice (no translations necessarily equal in length, no interaxial angles necessarily equal or necessarily having special values, e.g., 90°). In order that the perspective may be seen by the reader, the lattice points are joined by lines that define imaginary edges of a unit cell. One unit cell is shaded; it could have been chosen with a different shape (but the same volume), as in the two-dimensional example in Figure 2.2a.

(c) An electron micrograph of a crystalline protein, fumarase, molecular weight about 200,000. The individual molecules, in white, are visible as approximately spherical structures at low resolution. Note that several choices of unit cell are possible.

(Photograph courtesy of Dr. L. D. Simon)

are, in most cases, the edges of the conventional unit cell* used in describing the lattice. The crystal lattice is *not* the same as the *structure,* which is an array of objects (atoms, molecules, ions) rather than merely of (imaginary) points. Thus *the lattice is the basic network of points on which the repeating unit (the contents of the unit cell) may be imagined to be laid down so that the regularly repeating structure of the crystal is obtained.* The combination (convolution†) of the lattice with an object gives the crystal structure. The most general kind of lattice, composed of unit cells with three unequal edges and three unequal angles, is called the triclinic lattice and is illustrated in Figure 2.2b.

The unit cell is the basic parallelepiped-shaped repeating unit in the crystal. It contains a complete representation of the contents of the repeating unit of the crystal. However, Figure 2.2a shows that several possible choices of unit cell can be made from a two-dimensional arrangement of objects. How, then, can we speak of *the* unit cell for a given crystal? The conventional choice of unit cell is made by examining the lattice of the crystal and choosing a unit cell whose shape displays the full symmetry of the lattice—rotational as well as translational—and that is most convenient (for example, the axial lengths may be the shortest possible ones that can be chosen and the interaxial angles may be as near as possible to 90°). In choosing a unit cell, it is essential to consider the rotational symmetry of the lattice. The conventional unit cell shapes in three dimensions are listed in Appendix 1; *they are classified according to their rotational symmetry.* The presence of an n-fold rotation axis, where n is any integer, means that if the unit cell is rotated $(360/n)°$ about this axis, the unit cell so obtained is indistinguishable from the first. In a triclinic lattice there is no rotational symmetry and, in general, all unit-cell axial lengths (a, b, and c) are unequal, as are all interaxial angles (α, β, and γ). A monoclinic lattice ($\alpha = \gamma = 90°$) has a two-fold rotation axis parallel to the unique axis, **b** (by convention). This means that

*Some lattices are conventionally described in terms of a "non-primitive" unit cell. These lattices have lattice points not only at the corners of the conventional unit cell but also at the centers of this cell or at the center of one pair, or all three pairs of opposite faces (see Appendix 1.) More than one lattice point is then associated with a unit cell so chosen, but the requirement that every lattice point must have identical surroundings is still fulfilled. Non-primitive unit cells are chosen because they display the full symmetry of the lattice, or are more convenient for calculation; any given lattice may always be described in terms of either primitive or non-primitive cells.

†Consider two functions: A (x, y, z) and B (x, y, z). The *convolution* of A and B at the point (u_0, v_0, w_0) is found by multiplying together the values A (x, y, z) and B $(x + u_0, y + v_0, z + w_0)$ for *each* set of possible values of x, y, and z and *summing all these products.* To find the convolution as a general function of (u, v, w), it is then necessary to perform this multiplication and summation for *all* desired values of u, v and w. A crystal structure viewed as the convolution of a lattice with the contents of a single unit cell (see Figure 2.2a) is a simple example because the lattice exists only at discrete points.

a rotation of 180° about one axis (**b**) gives a lattice indistinguishable from the original. If the lattice is orthorhombic, all interaxial angles are 90°. However it must be stressed that it is the symmetry of the contents of the unit cell that is important in defining crystal class, not the magnitude of the interaxial angles (some monoclinic structures have been found with $\beta = 90°$, some triclinic structures with all interaxial angles very close to 90°).

In choosing a unit cell it is customary to take advantage of the full symmetry. Thus when an orthorhombic unit cell can be chosen from a three-dimensional arrangement of objects, this is a preferable choice of cell to one of lower symmetry that does not show the symmetry of the lattice (*e.g.*, three mutually perpendicular two-fold rotation axes rather than one or none). These conventions then help define the appropriate unit cell.

Before the discovery of X-ray diffraction in 1912, it was possible to deduce only the *relative* lengths of the axes (and the values of the interaxial angles) from optical measurements of interfacial angles in crystalline specimens. The way this can be done is illustrated in Figure 2.3. However, as we shall see shortly, X rays provide a tool for measuring the actual lengths of these axes and thus the size, as well as the shape, of the unit cell of any crystal.

There are 14, and only 14, distinct crystal lattices. These were deduced by Frankenheimer and Bravais in the nineteenth century, and are named after Bravais. The unit cells of 9 of these 14 lattices are illustrated in Appendix 1, including one of each fundamental symmetry type and the two common non-primitive cubic lattices as well. In this Appendix all lattice points, designated by small circles, are equivalent by translational symmetry, and, as indicated, all except the triclinic lattice display rotational symmetry of various kinds. All but the triclinic lattice also have at least one mirror plane, and all lattices have centers of inversion (see Chapter 7 for details).

However, *structures* arranged on these lattices do not necessarily display all of these possible symmetry elements (rotational symmetry, mirror planes, translational symmetry, centers of inversion). When one considers the possible combinations of symmetry elements consistent with these 14 lattices, and thus the possible symmetry elements of the structures that can be arranged on the latttices, it is found that 230, and only 230, distinct combinations of the possible symmetry elements exist. These different combinations are referred to as the 230 different crystallographic space groups. They are listed in *International Tables for X-ray Crystallography* (hereafter referred to as *International Tables*). Thus the innumerable different ways of arranging atoms or ions in structures to give a regularly repeating arrangement such as must be found in a crystalline solid fall into 230, and only 230, different symmetry classes. In most structures some groups of atoms in each unit cell are related to others in the same unit cell by some symmetry operation other than translation. For example, one part of the contents of the unit cell may be related to another part by rotational symmetry or by a mirror plane. The smallest part of a crystal structure from which the complete structure can be obtained from the space group symmetry operations (including translations) is called the asymmetric unit. The operation of the symmetry elements (other than lattice translations) on it will generate the entire contents of the primitive unit cell. (Symmetry and space groups are discussed further in Chapter 7.)

As emphasized already the term lattice is sometimes misleadingly used to refer to the structure itself. The *structure* is an ordered array of objects (atoms, molecules, ions) while a *lattice* is an ordered array of imaginary points. Although lattice points are conventionally placed at the corners of the unit cell (as in Appendix 1), there is no reason why this need be done. The lattice may be imagined to be free to move in a straight line (although not to rotate) in any direction relative to the structure. Each lattice point

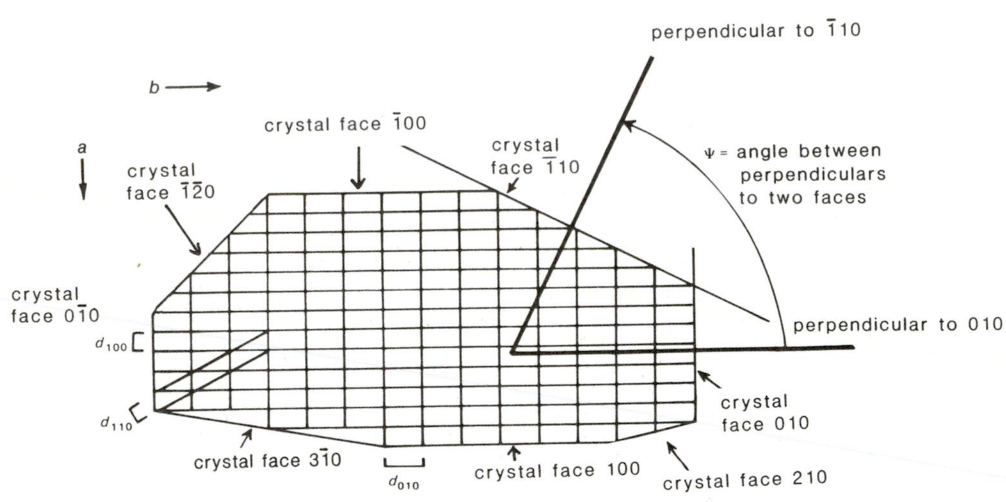

(b)

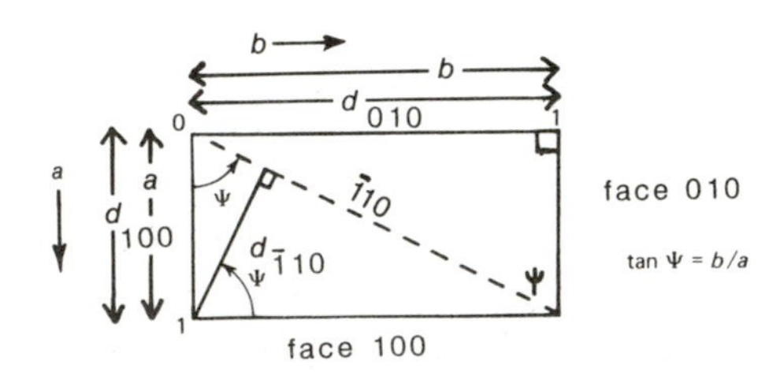

then always occupies exactly the same relative position in each unit cell, whatever this position may be. The lattice can thus be used to pick out translationally equivalent points in different unit cells of the structure.

It is a common misconception, perhaps arising from the abundance of illustrations of the simplest elementary and ionic structures in textbooks, that an atom must lie at the corner (origin) of each unit cell. It is possible to choose the origin arbitrarily and place it at the site of some atom, but in most structures the choice of origin is dictated by convenience because of its relation to symmetry elements that may be present (i.e., the appropriate space group), and in the great majority of known structures no atom is present at the origin. Another misconception is that what a chemist finds convenient to regard as a single molecule or formula unit must lie entirely within one unit cell. Portions of a single bonded aggregate may lie in two adjacent unit cells, although if this does happen, any single cell will necessarily still contain all of the independent atoms in the molecule—they simply comprise portions of different molecules. This is illustrated in Figure 2.2a, which shows that a given unit cell may contain one apple or portions of two or more apples; Figure 2.2c could equally well serve to show this possibility for a biological macromolecule.

The optical properties of crystals may reveal information about their symmetry and, in certain cases, their internal structure. Some crystals, such as

---

FIGURE 2.3 The Determination of the Probable Shape of the Unit Cell from Interfacial Angles in the Crystal.

The upper sketch, (a), shows a cross section of a crystal viewed down the **c**-axis. The faces of a crystal develop, in general, along the edges of, or diagonals across, the unit cell. A crystal face or plane is indexed (hkl), with these indices relatively prime, when it makes intercepts a/h, b/k, c/l with the edges of the unit cell of lengths a, b, and c (Law of Rational Indices). If a face is parallel to some crystal axis, its intercept on that axis is at infinity, so that the corresponding index is zero. A slice of a crystal perpendicular to **c** is illustrated in (a); the bounding lines are traces of the crystal faces perpendicular to the page and thus parallel to **c**. Hence for each face in this figure, $l = 0$. In practice, the indexing of faces of a crystal involves only small integers for h, k, and l. If the faces can be indexed and the angles between these faces measured with either a contact or an optical goniometer, it is possible to derive, as in (b), the ratio of the lengths of the unit cell edges (in this example b/a) and hence the shape (but not the absolute dimensions) of the unit cell. The indices of some faces and the relative lengths of some interplanar spacings, d_{hkl} (the spacing between the lattice planes (hkl) in the crystal), are indicated in both (a) and (b). An index with a line above it means that the value is negative. For example, $3\,\bar{1}\,0$ means $h = 3$, $k = -1$, $l = 0$ (that is, for the $3\,\bar{1}\,0$ plane the intercepts with the axes are $a/3$ and $-b$; the plane lies parallel to **c**). The same principles apply in three dimensions.

cubic crystals, are optically isotropic (like glass); the refractive index is independent of direction. Other crystals may be birefringent with different refractive indices in different directions. An analysis of the details of the anisotropy can help in the determination of crystal class. In addition, if the crystal structure contains an approximately planar group, measurements of refractive indices (by immersing the crystal in a series of liquids of known refractive index and determining when it becomes "invisible") may permit deduction of the orientation of the molecule. This method, combined with unit-cell measurements, was used to study steroid dimensions and packing long before any complete structure determination was initiated.

Since a crystal is built up of an extremely large number of regularly stacked unit cells, each of which has identical contents, the problem of determining the structure of a crystal is reduced to that of determining the spatial arrangement of the atoms within a single unit cell, or within the smaller asymmetric unit, if (as is usual) the unit cell has some internal symmetry. If there is some static disorder in the structure, the arrangement of the matter in different unit cells may not be identical, varying in an apparently random fashion. Thus what one finds for the arrangement of the matter in a single cell is a space-averaged structure. There may also be dynamic disorder in a structure; since the frequencies of atomic vibrations are of the order of 10^{13} per second, and since sets of X-ray intensity data are collected over periods ranging from seconds to weeks, time-averaging of the atomic distribution is always involved.

The unit cells of most crystals are, of course, extremely small, because they contain comparatively few molecules or ions, and because normal interatomic distances are of the order of a few Å. For example, a diamond is built up of a three-dimensional network of tetrahedrally linked carbon atoms, 1.54 Å apart. This arrangement may be described by a cubic unit cell, 3.56 Å on an edge. A one-carat diamond, which has approximately the volume of a cube a little less than 4 mm on a side, thus contains about 10^{21} unit cells of the diamond structure. A typical crystal suitable for X-ray structure analysis, a few tenths of a millimeter in average dimension, contains 10^{15} to 10^{18} unit cells, each with identical contents that can diffract X rays in unison. It is this fact—that there are so many identically constituted unit cells in a small crystal—that leads to the effectiveness of a crystal as a diffraction grating, which in turn permits a determination of the arrangement of the atoms within each unit cell.

SUMMARY

A crystal is, by definition, a solid that has a regularly repeating internal structure (arrangement of atoms). This internal periodicity was surmised in the seventeenth century from the regularities of the shapes of crystals, and was proved in 1912 when it was shown that a crystal could act as a three-dimensional diffraction grating for X rays, since X rays have wavelengths comparable to the distances between atoms in crystals.

The basic "building-block" in a crystal is called the *unit cell.* The *crystal lattice* is a regular three-dimensional array of points upon which the contents of the unit cell ("the motif") may be considered to be arranged by infinite repetition to build up the crystal structure. The concept of the lattice is important when diffraction by a crystal is considered. Only 14 distinct kinds of lattices—the Bravais lattices—may be constructed consistent with the condition that the view from each lattice point is the same in a given direction as the view in that direction from any other lattice point. These lattices have seven different unit cell shapes with different symmetry properties, corresponding to the seven crystal systems (see Appendix 1). The arrangement of structures on these lattices must be consistent with one of the 230 different combinations of symmetry elements (the 230 space groups) that are possible for arranging objects in a regularly repeating manner in three dimensions.

3

Diffraction

The usual approach to crystal structure analysis by X-ray diffraction presented in texts written for nonspecialists involves the Bragg equation and a discussion in terms of "reflection" of X rays from lattice planes. While the Bragg equation has proved extremely useful, it does not really help in understanding the process, and we will proceed instead by way of an elementary consideration of *diffraction* phenomena generally, and then diffraction from periodic structures (such as crystals), making use of optical analogies.

Some General Remarks The eyes of most animals, including man, comprise efficient optical systems for forming images of objects by the recombination of visible radiation scattered by these objects. Many things are, of course, too small to be detected by the unaided human eye, but an enlarged image of some of them can be formed with a microscope—using visible light for objects with dimensions comparable to or larger than the wavelength of this light (about 6×10^{-7} m), or using electrons of high energy (and thus short wavelength) in an electron microscope. In order to "see" the fine details of molecular structure (with dimensions 10^{-8} to 10^{-10} m), it is necessary to use radiation of a wavelength comparable to, or smaller than, the dimensions of atoms. Such radiation is readily available in the X rays produced by bombarding a target composed of an element of intermediate atomic number (for example, between Cr and Mo in the Periodic Table) with fast electrons, and also in "thermal" neutrons from a nuclear reactor, or in electrons with energies of 10–50 keV. More recently synchrotron* radiation has provided another intense and versatile source of X rays. Each of these kinds of radiation is scattered by the atoms of the sample, just as is ordinary light, and if we could recombine this scattered radiation, as a microscope can, we could form an image of the scattering matter. (X rays are scattered by the electrons in the atom†; neutrons are scattered by the nuclei and also, by virtue of their spin, by any unpaired electrons in the atom; and electrons are scattered by the electric field of the atom, which is of course a consequence of the combined effects of both its nuclear charge and its extranuclear electrons.)

However, neither X rays nor neutrons can be focused by any known lens system, and electrons of such high energy cannot (at least at present) be focused sufficiently well to give, in general, resolution of individual atoms. Thus, the formation of an image of the object under scrutiny, which is the self-evident aim of any method of structure determination—and is a process that we take for granted when we use our eyes or any kind of microscope—is not directly possible when X rays, neutrons, or high-energy electrons are used as a probe to provide atomic resolution.

Diffraction by Slits The pattern of radiation scattered by any object is called the *diffraction pattern* of that object. We are accustomed to think of light as traveling in straight lines and thus casting sharply defined shadows, but that is only because the dimensions of the objects normally illuminated in our experience are much larger than the wavelength of visible light. When light from a point source passes through a narrow slit or a very fine pinhole, the light is found to spread into the region that normally would be expected to be in shadow. To explain this effect each point on the wave front within the slit or pinhole is considered to act as a secondary source, radiating in all directions. The secondary wavelets so generated interfere with each other, either reinforcing or partially destroying each other. It is assumed that any

*Synchrotron radiation is emitted by high-energy electrons, such as those in an electron storage ring, when their path is bent by a magnetic field. The radiation is characterized by a continuous spectral distribution (which can, however, be "tuned" by appropriate selection), a very high intensity (many times that of conventional X-ray generators), a pulsed time-structure and a high degree of polarization.

†When X rays hit an atom, its electrons are set into oscillation about their nuclei as a result of perturbation by the rapidly oscillating electric field of the X rays. The frequency of this oscillation is equal to that of the incident X rays. The oscillating dipole so formed acts, in accord with electromagnetic theory, as a source of radiation with the same frequency as that of the incident beam. This is referred to as "coherent scattering" and is the type of scattering discussed in this book. When there is energy loss, resulting in a wavelength change on scattering, the phenomenon is described as "incoherent scattering." This effect is generally ignored by crystallographers interested in structure and will not be discussed in this book.

phase change on scattering* is the same for each atom and so may be ignored. However there are, under certain conditions (see Chapter 10), exceptions to this assumption.

There is a reciprocal relation between the angular spread of the scattering or diffraction pattern in a particular direction and the corresponding dimension of the object causing the scattering. The smaller the object, the larger the angular spread of the diffraction pattern. Actually, what is involved is the ratio of the wavelength of the radiation used, λ, to the minimum dimension, a, of the scattering object (for example, the width of the slit): the larger the value of λ/a, the greater the spread of the pattern. Figure 3.1a shows the diffraction pattern of a narrow slit, illuminated with radiation of a single wavelength. Figure 3.1b shows the narrower pattern of a slit that is about 2.2 times wider than that used in Figure 3.1a illuminated with the same radia-

*It was shown by J. J. Thomson (in "Conduction of Electricity through Gases") that when radiation is scattered by an electron, there is a phase change of 180° in the sense that the electric field in the scattered wave at a given point is opposed to that of the direct (incident) wave at that same point. This is described in detail by James, ref. 18, Chapter 2.

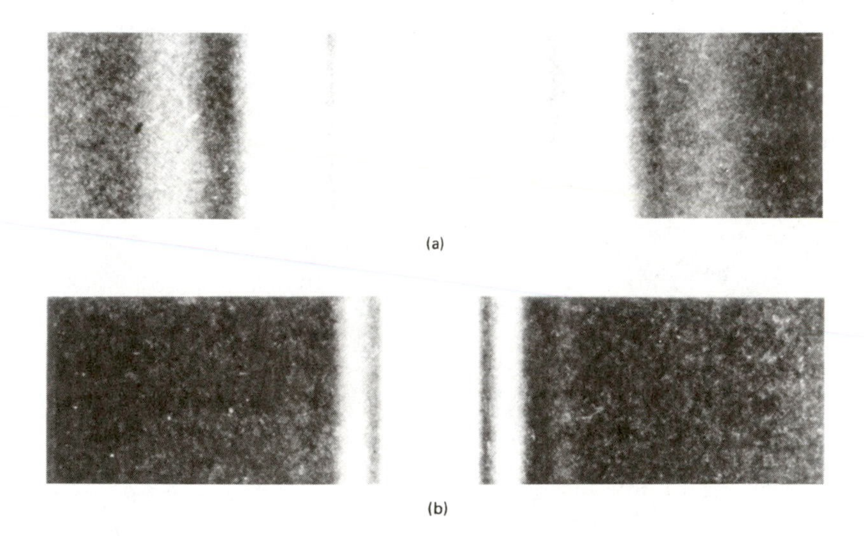

FIGURE 3.1 Diffraction Patterns of Single Narrow Slits.

The diffraction patterns of single slits of different width, illuminated with radiation of a single wavelength. Note that the wider slit gives the narrower diffraction pattern.

(a) The diffraction pattern of a narrow slit.

(b) The diffraction pattern of a slit 2.2 times wider than that used in (a). The diffraction pattern is now narrower by a factor of 2.2.

From *Fundamentals of Optics* by Francis A. Jenkins and Harvey E. White, 3rd edition (1957), (Fig. 16A).

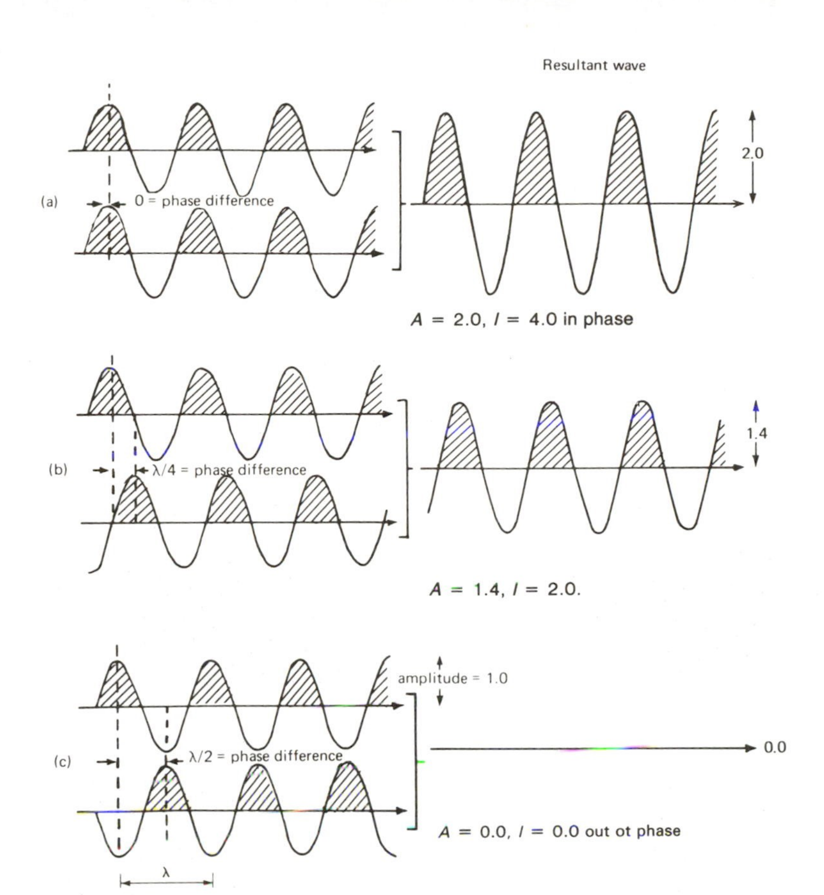

FIGURE 3.2 Interference of Two Waves.

Three examples are shown of what occurs when two waves of the same wavelength and equal amplitude add. In each example, the two separate waves are shown on the left and their sum or resultant wave on the right. The different examples are characterized by varying phase differences. The *phase* of a wave is the position of a crest relative to some arbitrary point. [This position is usually expressed as a fraction of the wavelength, and often this fraction is multiplied by 360° or 2π radians, so that the phase is given as an angle. Thus a phase difference of $\lambda/4$ may be given as $\frac{1}{4}$, 90°, or $\pi/2$ radians.] Although the phase of a given wave varies with time as the wave travels, the difference in phase of two waves of the same wavelength with the same velocity is independent of time. Such waves can interfere with one another. The resultant wave has the same wavelength, λ. The intensity, I, of the resultant wave is proportional to the square of its amplitude, A.

(a) Phase difference zero. In this case there is total reinforcement and the waves are said to be "in phase" or to show constructive interference. If the original waves are of unit amplitude, the resultant wave has amplitude 2, intensity 4.

(b) Phase difference $\lambda/4$. Partial reinforcement occurs in this case to give a resultant wave of amplitude 1.4, intensity 2.

(c) Phase difference $\lambda/2$. The waves are now "out of phase" and there is destructive interference to give no resultant wave (or a wave with amplitude 0, intensity 0).

tion. The *wider slit gives the narrower pattern;* only the scale of the pattern has been changed. Figure 3.1b might equally well be a view of the same slit as in Figure 3.1a, illuminated with radiation of wavelength about 2.2 times shorter than that used in Figure 3.1a. In fact, it is possible to produce this change of scale by any change in a and in λ whose combined effect is to decrease the value of λ/a by a factor of 2.2. The variations in intensity seen in Figure 3.1 arise from the interference of the secondary wavelets generated within the slit. Since this radiation is generated by the same incident wave, the secondary wavelets act as synchronous sources with a constant phase difference dependent on the angle of scattering. The phenomenon of interference is illustrated in Figure 3.2; the amplitude of the wave resulting from the interaction of two separate waves, of the same wavelength and with a constant phase difference, depends markedly on the size of this phase difference. Figure 3.2 shows how such waves may be summed* for three cases of different phase differences. The intensity of the resulting beam is proportional to the square of the amplitude of the resultant beam. In the direction of the direct beam the waves scattered by the slit are totally in phase and reinforce one another to give maximum intensity. However, at other scattering angles, as illustrated in Figure 3.3a, the relative phases of the waves are changed and therefore interference occurs and the intensity falls off as a function of scattering angle. At certain angles each scattered wave may be completely out of phase with another scattered wave and cancel it out, so that the resultant wave for all the scattered waves has zero amplitude, and no illumination occurs at these angles. At most scattering angles, where the different scattered waves are neither completely in phase nor completely out of phase, there is partial reinforcement and thus intermediate intensity. The result is illustrated in the single-slit diffraction pattern of Figures 3.1 and 3.4b, and, more clearly, as the "envelope" in Figure 3.4a.

Diffraction by Regular Arrays Figures 3.3b, 3.4a, and 3.4b show how the diffraction pattern of a single slit is modified by interference effects when increasing numbers of slits are placed side by side in a regular manner to form a one-dimensional grating. The important point to note is that *the diffraction pattern from a grating of slits represents a sampling of the single-slit pattern in narrow regions.* With even as few as 20 slits in the "grating" the small subsidiary maxima vanish almost completely and a sharp diffraction pattern is obtained. The overall diffraction pattern of a series of slits is thus composed of an "envelope" and a series of "sampling regions." The

*The displacements from the mean (zero), parallel to the vertical axis (ordinates), are directly summed at many points along the horizontal axis (abscissae) to give the resultant wave.

"envelope" results from the interference of waves scattered by any individual slit (as described above) and is equivalent to the diffraction pattern of a single slit. The "sampling regions" result from interference of waves scattered from equivalent points in different slits; the spacing of these sampling regions (marked in Figure 3.4a) is inversely related to the spacing of the slits. If the intensity of the single-slit pattern has a certain magnitude (weak or strong) in a region where it is being sampled, then the resulting multiple-slit pattern will have a similar relative magnitude in this region.

The phenomenon of diffraction by a regular two-dimensional pattern may be illustrated by holding a linen handkerchief taut between the eye and a

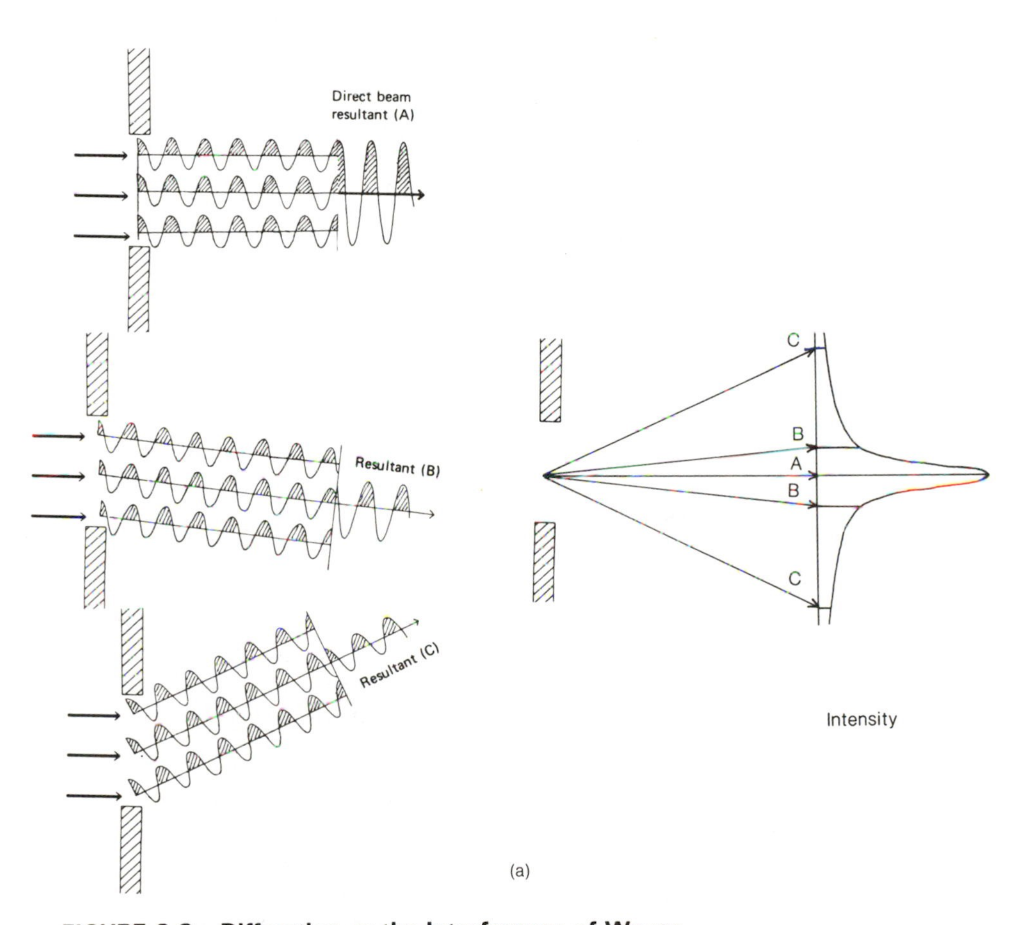

FIGURE 3.3 Diffraction as the Interference of Waves.

(a) Diffraction from a single slit is illustrated by the superposition of selected waves generated within the area of the slit. The variation in intensity with increasing angle is shown by the different amplitudes of resultant waves at different angles.

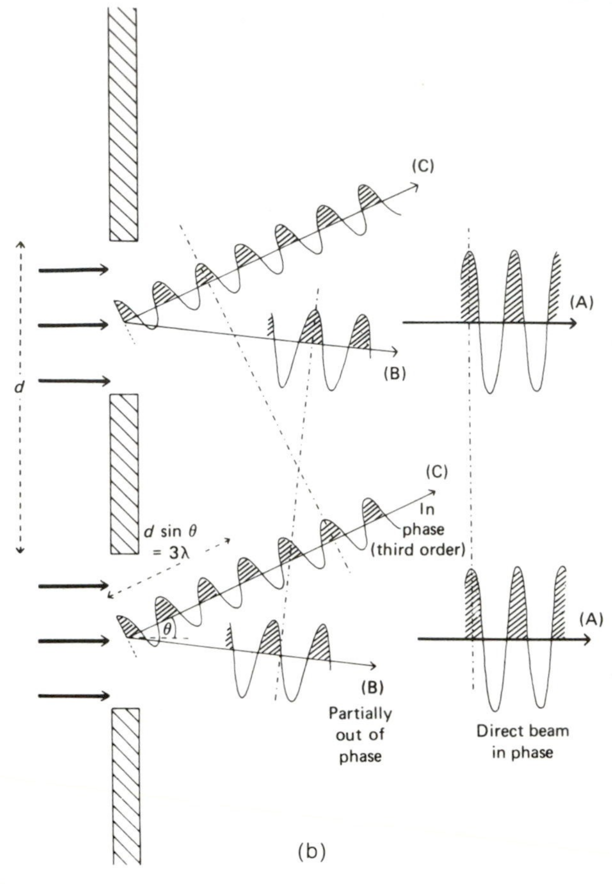

(b) When diffraction occurs from two slits there are two effects to consider.

1. The variation in intensity with angle as a result of interference of the waves generated within each slit separately, as shown in (a) above. This gives the "envelope" illustrated in Figure 3.4a.
2. The interference of scattered waves at a given angle with those at the same angle from the adjacent slit. At angles of constructive interference when the two resultant waves are in phase, "sampling" of the "envelope" occurs. At certain other angles no diffraction is observed.

This figure should be compared with Figure 3.4a. Only the resultant waves, illustrated in (a) above, are shown in this diagram. Those labelled A are undeviated, and produce a bright central image. Those labelled C are, as shown, in phase with each other because the path length difference is exactly three wavelengths, and therefore a diffraction maximum is produced at the corresponding angle. Those labelled B are about 90° out of phase, as indicated, so that the path length difference is not an integral multiple of the wavelength, and consequently the waves interfere partially with each other. If there are many slits equidistantly spaced, the intensity in direction B will

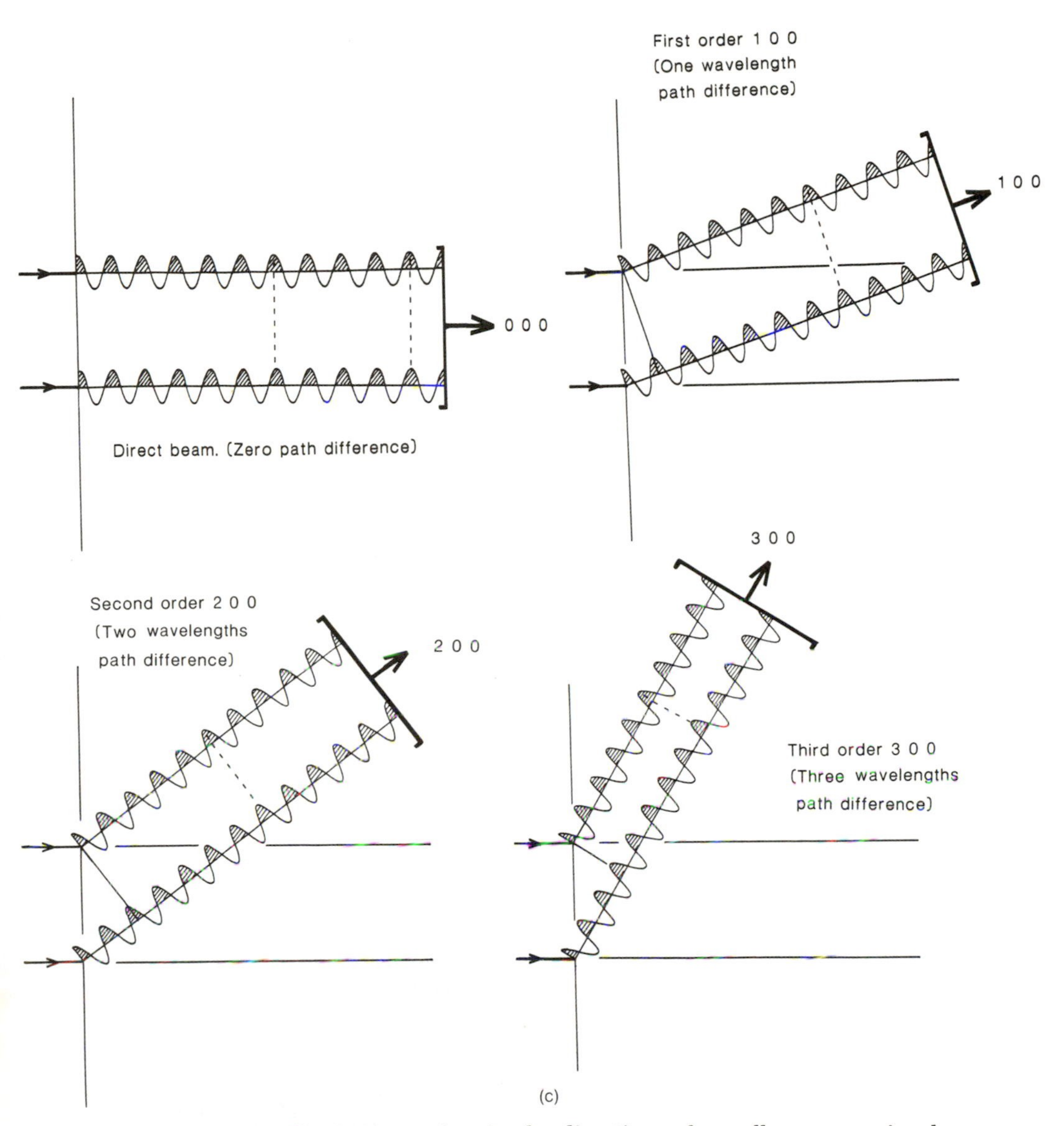

(c)

be so small relative to that in the directions where all waves are in phase ("diffraction maxima") that only the latter will be discernible (cf. Figure 3.4b). The reason the intensity in direction B is then so small is that the phase differences in this direction range over all possibilities if there are many slits.

(c) First, second, third, and higher orders of diffraction are obtained as scattered waves differ by one, two, three, and more wavelengths. Readers should satisfy themselves that with a smaller spacing, a, between scattering objects, the angle at which a given order of diffraction occurs is increased.

Diffraction pattern of two slits

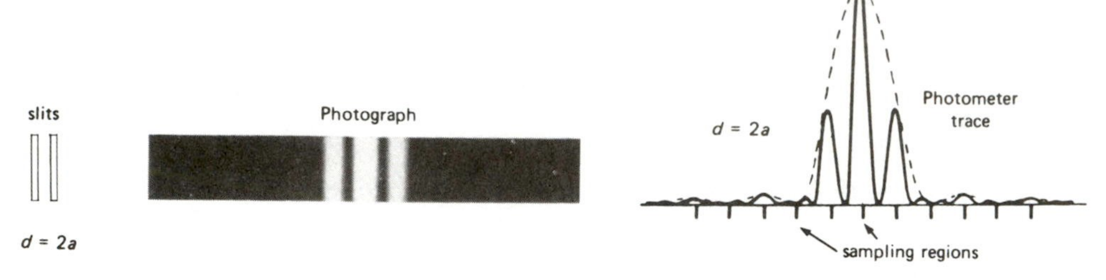

Narrow spacing between slits. Wide spacing between sampling regions

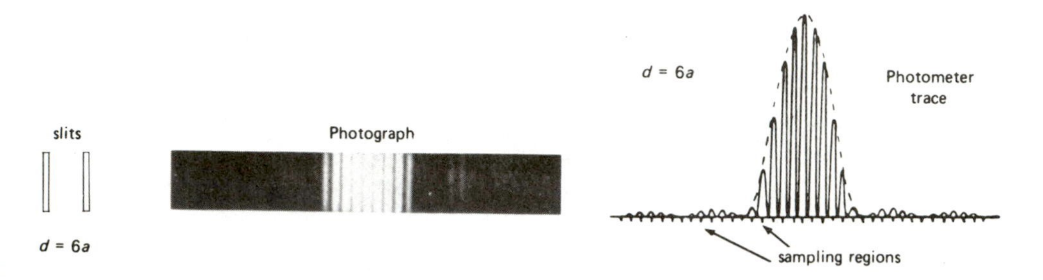

Wide spacing between slits. Narrow spacing between sampling regions.

(a)

Varying numbers of slits

1 slit

5 slits

2 slits

20 slits

(b)

distant point source of light, such as a street light. Instead of just one spot of light for the source, a cluster of lights will be seen. The same phenomenon can also be demonstrated with fine sieves. The larger the spacing between the wires of the sieve the closer the spots around the central pattern. The scattering of light by the areas between the narrowly and regularly spaced threads of the fabric or wires of the sieve may be considered to produce the diffraction effect.

Figure 3.5 shows schematically how a two-dimensional regular arrangment of simple scattering objects, in this case holes in an opaque sheet, produces a two-dimensional diffraction pattern. Each of the one-dimensional gratings in Figures 3.5a and 3.5b produces (in two dimensions) a pattern of scattered light (the diffraction pattern) consisting of lines. These lines are perpendic-

---

FIGURE 3.4 Diffraction Patterns from Equidistant Parallel Slits.

(a) The effect of varying the distance, d, between two slits of constant width, a, is shown. On the left is a diagram of the slits with spacings of $2a$ and $6a$, respectively. In the center is shown a photograph of the diffraction pattern. On the right a photometer tracing of the diffraction pattern for the combination of the two slits is drawn as a solid line, and the diffraction pattern for a single slit, referred to in the text as the "envelope," is drawn as a dashed line. The regions of the "envelope" that are sampled are indicated by short vertical lines at the lower edge of the drawings on the right. When there is a relatively narrow spacing between slits ($d = 2a$), the distance between sampling regions is relatively large, as shown in the upper diagram. When there is a relatively wide spacing between slits ($d = 6a$), the distance between sampling regions has decreased, that is, there is an inverse relationship of the spacing of the sampling regions to the spacing of the slits. The "envelope" (or diffraction pattern of a single slit) that is being sampled remains constant.

(b) Diffraction patterns are shown for gratings containing 1, 2, 5, and 20 equidistant slits, illuminated by parallel radiation. As increasing numbers of equidistant slits are used, the radiation is concentrated in increasingly narrower regions, at first with subsidiary maxima between them (visible in the 5-slit pattern). The diffraction pattern for a grating composed of 20 (or more) slits consists only of sharp images, the intervening minor maxima having disappeared; similarly, the diffraction pattern for a crystal composed of many unit cells contains sharp diffraction maxima.

Summary of key points:

1. The size and shape of the "envelope" are determined by the diffraction pattern of a single slit.
2. The positions of the regions in which the "envelope" is sampled are determined by the spacings between the slits.
3. The diffraction pattern is increasingly sharp the greater the number of slits.

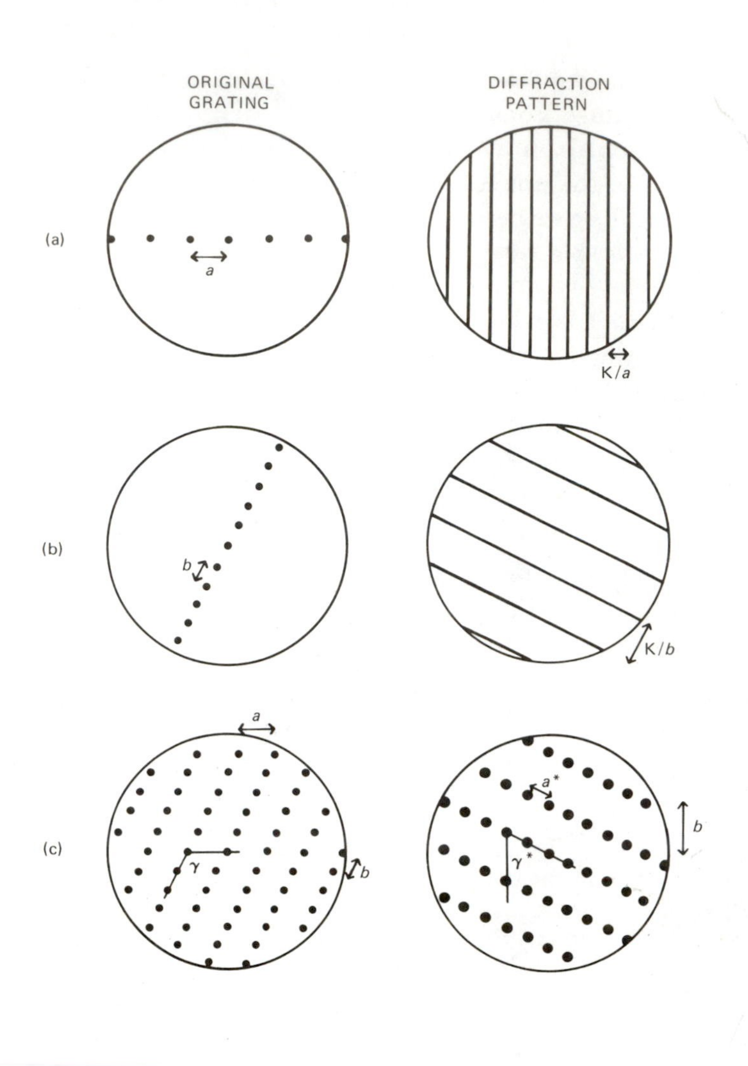
ORIGINAL GRATING
DIFFRACTION PATTERN
(a)
a
K/a
(b)
b
K/b
(c)
a
γ
b
a*
γ*
b

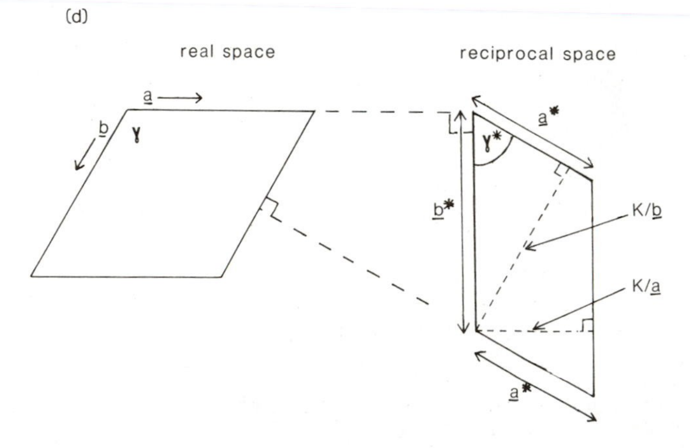
(d)
real space
reciprocal space
a
b
γ
a*
γ*
b*
K/b
K/a
a*

ular to the direction of the original grating because interference effects between light scattered from adjacent holes reduce the scattered light intensity effectively to zero in all directions except that perpendicular to the repeat direction of the original grating. Hence lines of diffracted light are formed. In Figure 3.5c we have both kinds of one-dimensional gratings present at once. This gives a regular two-dimensional grating. Each of the linear gratings in Figures 3.5a and 3.5b may be considered to act independently, and the only regions in which there is not destructive interference are the intersections of the two sets of lines produced in Figures 3.5a and 3.5b, shown in the diffraction pattern in Figure 3.5c. The lattice of the diffraction pattern is necessarily then the "reciprocal" of the lattice of the original scattering objects (see Figure 3.5d).

Thus, in addition to the lattice of the crystal structure in *real* or *crystal* space (discussed earlier), there is a second lattice, related to the first, that is

---

FIGURE 3.5 Schematic Examples of Diffraction Patterns from One- and Two-Dimensional Arrays. Relation of the Crystal Lattice and Reciprocal Lattice.

(a, b, c) Negatives of the original gratings are shown on the left and of the corresponding diffraction pattern (such as might be obtained by holding the original grating in front of a point source of light) on the right. **a** and **b** are direct lattice vectors in the crystal or grating and **a*** and **b*** are vectors in the diffraction pattern (a and b are the spacings of the original gratings, and a^* and b^* are spacings in the diffraction pattern). The reciprocal relationships of a and b to the spacings of certain rows in the diffraction pattern are shown. The fact that the intensity is the same at all reciprocal lattice points in the diffraction pattern in Figure 3.5c should not be thought to be general; it happens here because the scattering objects in the original "crystal lattice" are all particularly simple (just a single isotropic hole in each unit cell) and are much smaller than the wavelength of the radiation used in this hypothetical experiment. Consequently, the intensities of the diffraction maxima show no variation in different directions and do not vary significantly with angle of scattering (which increases with increasing distance from the center of the pattern).

(d) The relationships of **a** and **b** to **a*** and **b*** are shown. The *reciprocal lattice,* which is, in this two-dimensional example, the lattice of spacing a^* and b^*, is of great importance in diffraction experiments. For a particular diffraction pattern, the scale factor K depends upon the wavelength of the radiation used and upon the geometry of the experimental arrangement. (See Figures 3.6–3.8, Figures 4.3c and 4.3d, Figure 4.4b, Figure 4.5b, and Appendix 4, for experimental details and further examples.)

Note: In these diagrams, the black dots on the left-hand side represent holes that cause diffraction, giving the pattern on the right-hand side, in which black lines or spots represent appreciable intensity for diffracted light.

Adapted from Lipson and Cochran, ref. 14.

of importance in diffraction experiments and in many other aspects of solid state physics. This is the *reciprocal lattice;* its definition in terms of the crystal lattice vectors is given in Appendix 2. The relationship between these two important lattices is a particularly simple one when the fundamental translations of the crystal lattice are all perpendicular to one another: then the fundamental translations of the reciprocal lattice are parallel to those of the crystal lattice, and the lengths of these translations are inversely proportional to the lengths of the corresponding translations of the crystal lattice. With nonorthogonal axes, the relations are not hard to visualize geometrically; a two-dimensional example is given in Figure 3.5. As we shall see shortly, the fundamental importance of the reciprocal lattice in crystal diffraction arises from the fact that if a structure is arranged on a given lattice, then its diffraction pattern is necessarily arranged on the lattice that is reciprocal to the first.*

Diffraction by Atoms in Crystals It is a principle of optics that the diffraction pattern of a mask with very small holes in it is approximately equivalent to the diffraction pattern of the "negative" of the mask—that is, an array of small dots at the positions of the holes, each dot surrounded by empty space. This equivalence is discussed lucidly by Feynman (ref. 65). In a crystal the *electrons in the atoms act, by scattering, as sources of X rays, just as the wave front in the slits in a grating may be regarded as sources of visible light.* There is thus an analogy between atoms in a crystal, arranged in a regular array, and slits in a grating, arranged in a regular array. In diffraction by crystals, as by slits, the intensities of the diffraction maxima show a variation in different directions and also vary significantly with angle of scattering (unlike the simplification of Figure 3.5c). Most unit cells contain a complex assembly of atoms, and each atom is comparable in linear dimensions to the wavelength of the radiation used. Figure 3.6 shows a typical X-ray diffraction photograph, taken by the "precession method," which records the diffraction pattern without distortion. Considerable variation in intensity of the individual diffracted beams is evident. The analogy with Figures 3.1, 3.4, and 3.5 holds; that is, *the X-ray photograph is merely a scaled-up sampling of the diffraction pattern of the contents of a single unit cell.* The "envelope" is the diffraction pattern of the scattering matter

*This may alternatively be stated as follows. The diffraction pattern of a molecular crystal is the product of the diffraction pattern of the molecule (also called the molecular transform) with the diffraction pattern of the crystal lattice (which is also a lattice, the reciprocal lattice described above). The result is a sampling of the molecular transform at each of the reciprocal lattice points. The diffraction pattern of a single molecule is too weak to be observable. However, when it is reinforced in a crystal (containing many billions of molecules in a regular array) it can be readily obesrved, but only at the reciprocal lattice points.

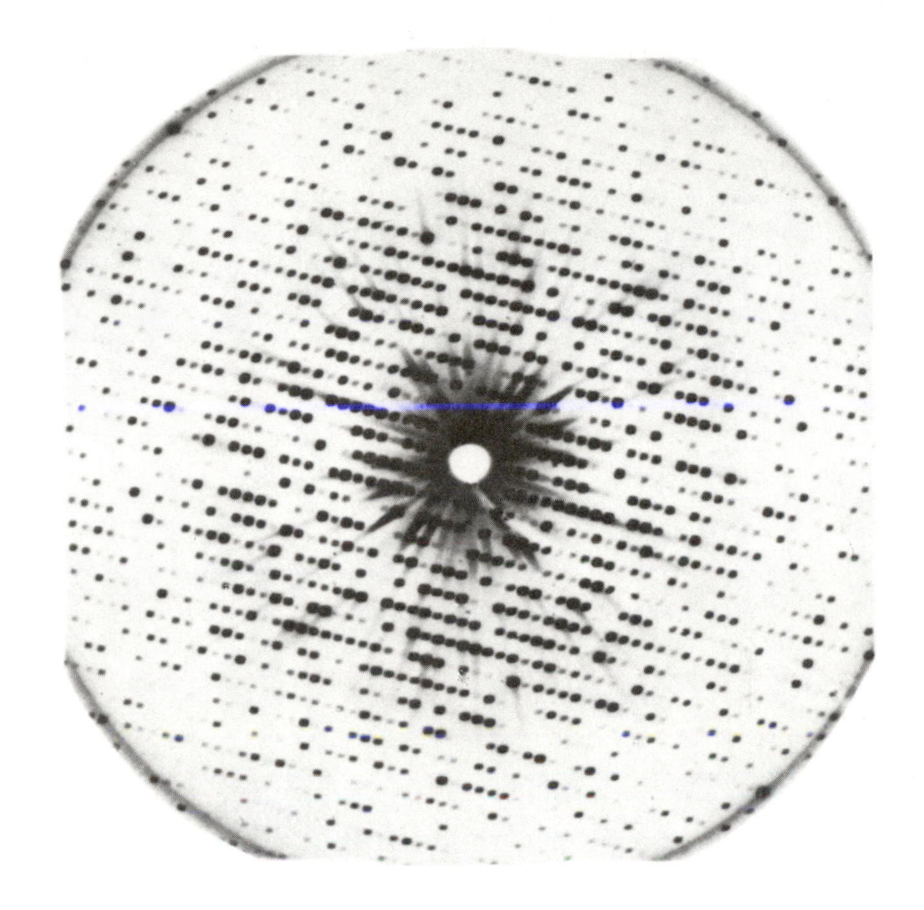

FIGURE 3.6 X-Ray Diffraction Photograph Taken by the Precession Method. This method gives an undistorted representation of one layer of the reciprocal lattice. An X-ray photograph of a crystal of myoglobin is shown here. The direct X-ray beam, which might otherwise "fog" the film, has been intercepted, hence the white hole in the middle of the photograph. The radial streaks, which occur for very intense reflections, occur because the X rays are not truly monochromatic (one wavelength) but contain background radiation of varying wavelength but lower intensity. As a result the spot appears somewhat smeared out (that is, for each reflection, $\sin\theta/\lambda$ is constant but since λ varies for the background "white radiation," $\sin\theta$ must also vary, giving rise to a streak rather than a spot on the film). Note the regularity of the positions of spots in this photograph but the wide variation in intensity (from a very black spot to one that is apparently absent). The positions of the spots (diffracted beams) give information on unit cell dimensions and the intensities of the spots give information on the arrangement of atoms in that unit cell.

(Photograph courtesy of Dr. J. C. Kendrew.)

(the electrons of the atoms) in a single unit cell. The "sampling regions" are arranged on a lattice that is "reciprocal" to the crystal lattice in the sense described in Figure 3.5d.

This sampling is illustrated in Figures 3.7 and 3.8, which have been prepared using a special optical device that permits photographs to be made of the diffraction patterns of arrays of holes cut in an opaque sheet. By an appropriate choice of the optical components, the effective ratio of the wavelength of the light used to the sizes of these holes can be made similar to the ratio of X-ray wavelengths to the sizes of atoms. One can, then, simulate X-ray diffraction photographs of crystals by making patterns of holes in opaque

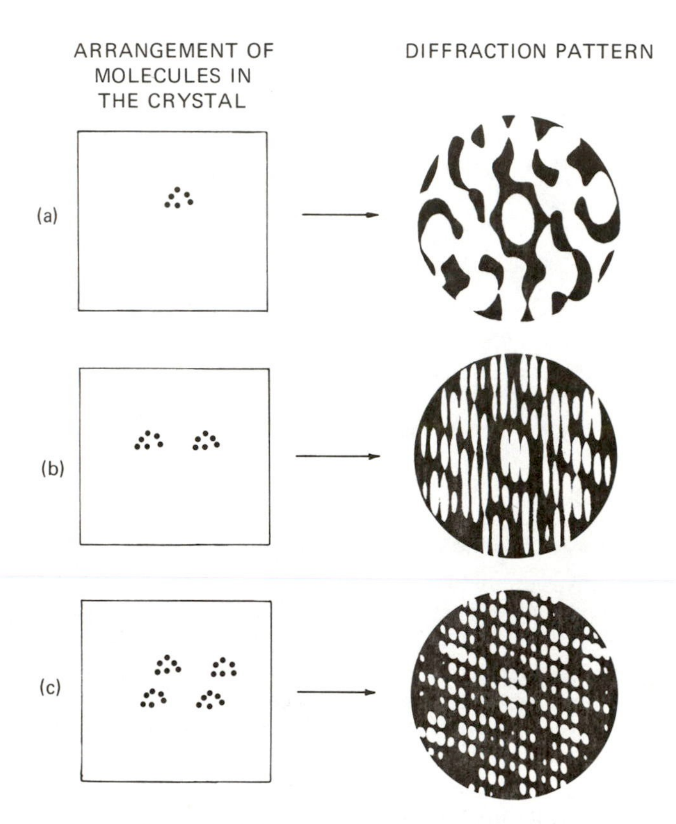

FIGURE 3.7 The Effect of Different Lattice Samplings on the Diffraction Pattern.

This shows the relationship between the diffraction pattern of a "molecule" and various regular arrangements of such molecules.

(a) A single molecule.

(b) Two molecules horizontally side by side.
In comparing (b) with (a), note the analogy with the 2-slit and 1-slit patterns of Figure 3.4b.

(c) Four molecules arranged in a parallelogram.

sheets that are similar, except for scale, to the patterns of arrangement of the atoms in the crystals.

Figure 3.7 shows the relationship between the diffraction pattern of a single "molecule" (3.7a) and various regular arrangements of such molecules. The left-hand side of each part of the figure shows different arrangements of molecules and the right-hand side shows the corresponding diffraction patterns. Figure 3.7b is the diffraction pattern of two "molecules" side by side (horizontally in the orientation shown here) and illustrates the interference arising from the interaction of the scattering by the two molecules,

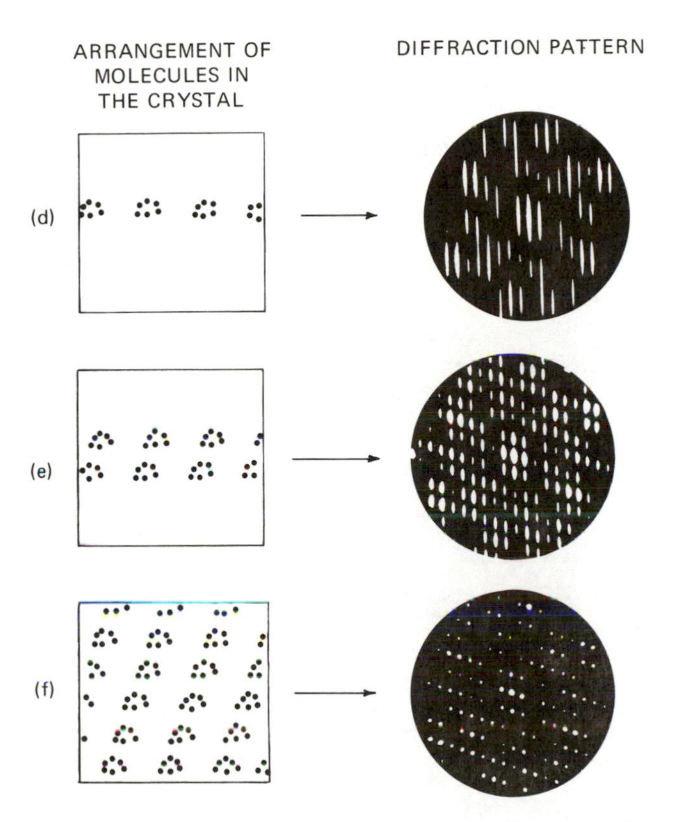

(d) Many molecules horizontally side by side (a one-dimensional crystal). Only part of the row is shown.

(e) Two rows of molecules arranged on an oblique lattice. Only parts of the rows are shown.
In comparing (e) with (d), note again the analogy with the relation of the 2-slit and 1-slit patterns of Figure 3.4b.

(f) Two-dimensional crystal of molecules. Only part of the crystal and part of the diffraction pattern are shown.

From C. A. Taylor and H. Lipson, *Optical Transforms. Their Preparation and Application to X-ray Diffraction Problems,* Plate 26. G. Bell and Sons., London (1964).

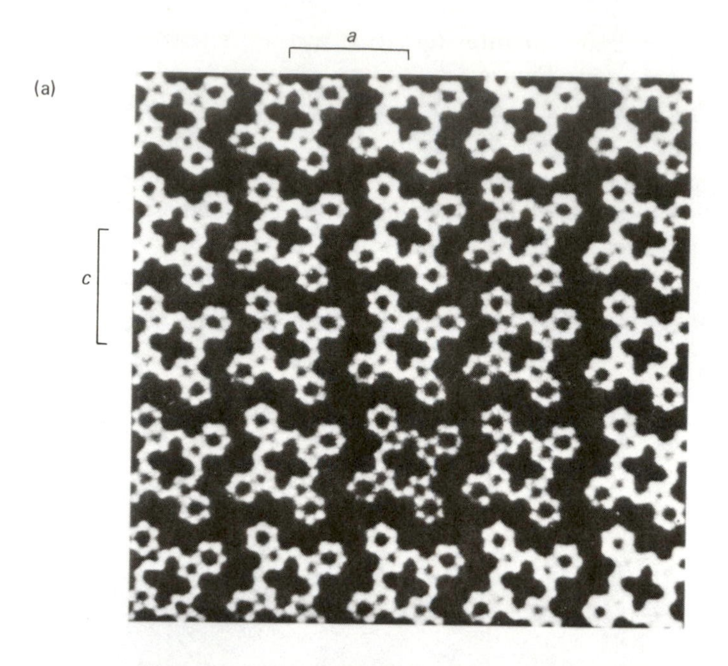
(a)
a
c

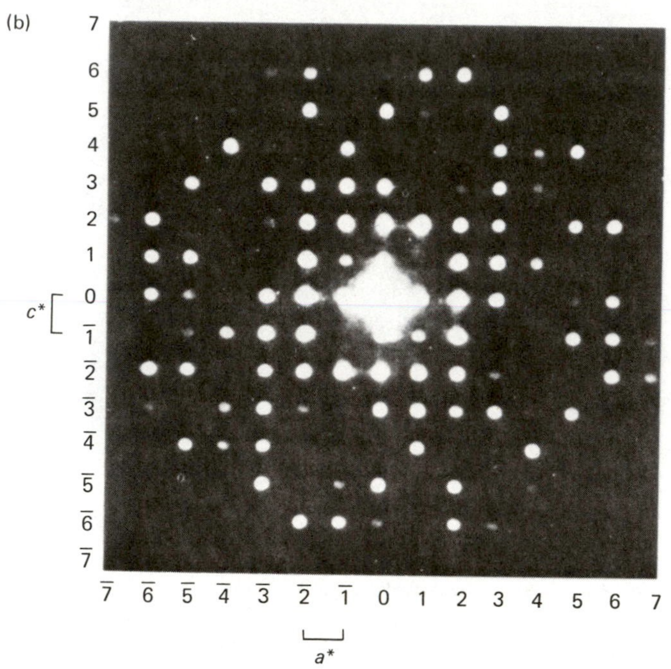
(b)
7 6 5 4 3 2 1 0 $\bar{1}$ $\bar{2}$ $\bar{3}$ $\bar{4}$ $\bar{5}$ $\bar{6}$ $\bar{7}$
c^*
$\bar{7}$ $\bar{6}$ $\bar{5}$ $\bar{4}$ $\bar{3}$ $\bar{2}$ $\bar{1}$ 0 1 2 3 4 5 6 7
a^*

exactly analogous to the interference caused by the presence of two adjacent slits that give rise to Figure 3.4a. Figure 3.7c is the pattern arising from four "molecules" arranged in a parallelogram; now there is interference parallel to each of the two axes of the incipient crystal lattice. Figure 3.7d is the diffraction pattern of an extended regularly spaced row of the molecules—that is, from a one-dimensional crystal; there is sharpening of the diffraction effects parallel to the direction of ordering, but no interference at all in other directions. Figure 3.7e is the pattern obtained by placing two lengthy rows side by side, and finally Figure 3.7f is the pattern obtained from a two-dimensional crystal of these "molecules." The resemblance to the precession photograph in Figure 3.6 is marked. The lattice of the diffraction pattern is reciprocal to that of the "crystal." Figure 3.7f could be derived from Figure 3.7a simply by placing over Figure 3.7a a mask with holes in it at the positions of the reciprocal lattice points; that is, Figure 3.7a is being sampled at

---

FIGURE 3.8 The Optical Diffraction Pattern of an Array of Templates Resembling the Skeleton of a Phthalocyanine Molecule.

(a) The array used to obtain the optical diffraction pattern.
(b) The optical diffraction pattern obtained from (a).
(c) Relative intensities measured from the X-ray diffraction pattern of a phthalocyanine crystal. Qualitative comparison of these values with the intensities of the corresponding spots in the optical diffraction pattern shown in (b) indicates that the model used is a surprisingly good one. *Note:* Intensities for $h0l$ and $\bar{h}0\bar{l}$ are equal. $\bar{h}$ means $-h$ and $\bar{l}$ means $-l$.

Relative intensities for the phthalocyanine crystal

$l\uparrow$								
7	6	0	2	7	0	6	0	0
6	25	52	45	11	4	0	3	0
5	36	1	0	58	0	1	2	0
4	3	17	0	14	0	38	0	9
3	15	1	2	14	4	4	2	1
2	72	85	21	16	0	8	27	1
1	61	0	64	30	2	2	1	3
0		94	72	10	0	2	17	1
$\bar{1}$	61	29	55	0	2	7	10	5
$\bar{2}$	72	46	23	3	0	0	18	14
$\bar{3}$	15	37	14	10	2	21	2	0
$\bar{4}$	3	13	0	10	18	2	1	0
$\bar{5}$	36	0	18	3	19	0	0	0
$\bar{6}$	25	5	35	5	2	0	1	0
$\bar{7}$	6	0	2	0	14	2	0	0
$h\rightarrow$	0	1	2	3	4	5	6	7

From C. W. Bunn, *Chemical Crystallography. An Introduction to Optical and X-ray Methods*, 2nd edition, Plate XIV, Oxford at the Clarendon Press (1961).

reciprocal lattice points to give Figure 3.7f. In Figure 3.8a arrays of holes, each of which has the shape of the skeleton of a phthalocyanine molecule, are shown, together with the optical diffraction pattern (visible light) obtained from these arrays (Figure 3.8b). Note that the intensity variation in the optical diffraction pattern (shown in Figure 3.8b) parallels that found in the corresponding pattern obtained by the diffraction of X rays (listed in Figure 3.8c) because the orientation and spacing of the arrays of holes in Figure 3.8a were chosen to parallel those of the arrangement of phthalocyanine molecules viewed in a particular direction in the crystal.

Diffraction and the Bragg Equation Von Laue, who, with Friedrich and Knipping, discovered the diffraction of X rays by crystals in 1912, interpreted the observed diffraction patterns in terms of a theory analogous to that used to treat diffraction by gratings, extended to three dimensions. On the other hand, W. L. Bragg, who worked out the first crystal structures with his father during the summer of 1913, showed that the angular distribution of scattered radiation could be understood by considering that the diffracted beams behaved *as if they were reflected* from planes passing through points of the crystal lattice. This "reflection" is analogous to that from a mirror, for which the angle of incidence of radiation is equal to the angle of reflection. Waves scattered from adjacent lattice planes will be just in phase (i.e., the difference in the paths travelled by these waves will be an integral multiple of the wavelength, $n\lambda$) only for certain angles of scattering (Figure 3.9). From such considerations Bragg derived the famous equation that now bears his name:

$$n\lambda = 2d \sin \theta \tag{3.1}$$

with λ = the wavelength of the radiation used, n an integer (analogous to the order of diffraction from a grating so that $n\lambda$ is the path difference between waves scattered from adjacent lattice planes with equivalent indices), d, the perpendicular spacing between the lattice planes in the crystal, and θ, the complement of the angle of incidence of the X-ray beam (and thus also the complement of the angle of scattering or "reflection"). Since it appears as if reflection has occurred from these lattice planes, with the direct beam deviated by the angle 2θ from its original direction, diffracted beams are commonly referred to as "reflections." Because the Bragg equation is easily visualized, it is commonly presented in elementary discussions in diagrams such as those in Figures 3.9a and 3.9b; we have attempted in Appendix 3 to show how it can be related to diffraction theory. The Bragg equation can be derived by considering that the path difference between waves scattered from adjacent parallel lattice planes must be an integral number of wavelengths. The equation is satisfied, and thus diffraction maxima occur,

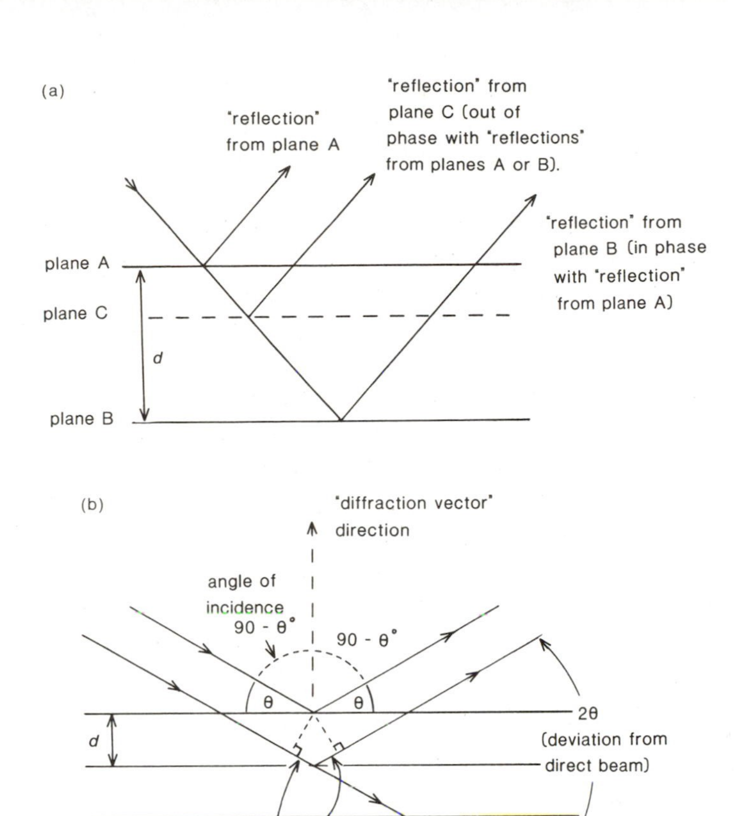

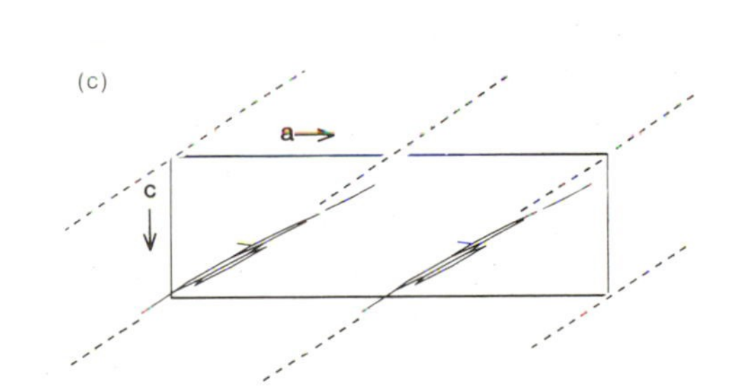

FIGURE 3.9 Diagram of "Reflection" of X Rays by Imaginary Planes Through Points in the Crystal Lattice.

(a) Constructive and destructive interference as waves are "reflected" from imaginary planes, spacing d, in a crystal.

(b) This leads to the Bragg equation $n\lambda = 2d \sin \theta$. Since the path difference (causing phase differences) of waves scattered by two adjacent planes is $2d \sin \theta$, this must equal $n\lambda$ for reinforcement to occur to give a diffracted beam (as illustrated in Figure 3.2).

(c) Planes 2 0 1 in a crystal (see Figure 9.3b). The planes lie perpendicular to the plane of the paper. Note that the planes intersect the unit cell edges once in the **c** direction and twice in the **a** direction. A portion of the crystal structure, which diffracts to give an intense 2 0 1 reflection, is shown.

only when the relation of wavelength, interplanar spacing, and angle of incidence is appropriate. If a nearly monochromatic beam of X rays is used with a single crystal specimen (and thus only particular sets of lattice spacings are present), diffraction maxima will be observed only for special values of the angle of incidence of the beam of X rays, and not for any arbitrary angle. If the crystal is rotated in the beam, additional diffracted beams will be formed at certain rotation angles. These "reflections" will only occur when the angle of incidence of the X-ray beam is such as to satisfy Eq. (3.1) for some set of lattice spacings present in the crystal, that is, λ, d, and θ must all be such that the Bragg equation holds. The chance of this happening for a perfect crystal is low. However, real crystals have a *mosaic spread* (as if composed of minute blocks of unit cells, each block being misaligned by a few tenths of a degree with respect to its neighbors) and the X rays used are never truly monochromatic, so that, in practice, a reflection can be observed over a small range of θ and therefore some reflections are observed in almost any orientation of a single crystal. With a powdered crystalline specimen many different orientations of tiny crystallites are present simultaneously, and for any set of crystal planes, Eq. (3.1) will be satisfied in some of the crystallites so that the complete diffraction pattern will be observed for any orientation of the specimen with respect to the X-ray beam. It is also possible to get a diffraction pattern from a stationary single crystal by the use of a wide range of wavelengths simultaneously. This was, in fact, the way in which von Laue, Friedrich, and Knipping did their original experiment; the technique is known as the Laue method. It is not now generally used for structure analysis, although the method shows promise for use in the study of short-lived intermediates with synchrotron radiation.

The Bragg equation says nothing about the intensities of the diffraction maxima that will be observed when it is satisfied. However, if a particular set of lattice planes happens to coincide, in orientation and position, with some densely populated planar or nearly planar arrays of atoms in a crystal, and if there are no intervening densely populated planes, the corresponding diffraction maximum will be an intense one because the scattering from all atoms is just in phase. The chance that such densely populated planes can be found is far greater in structures with isolated atoms, monatomic ions, or planar molecules than in structures with complex polyatomic species. In an example cited in Chapter 9 (Figure 9.3b), involving a planar organic molecule, the "reflection" with indices $h = 2$, $k = 0$, $l = 1$ (written 2 0 1, i.e., second order in h, direct for k, and first order for l) is very intense because the molecules lie nearly parallel to the lattice plane with indices (2 0 1) and are separated by a spacing very nearly the same as the interplanar spacing of this lattice plane. This is shown in Figure 3.9c.

SUMMARY

To explain what happens when a crystal diffracts X rays, we examined first optical analogies with slits and then with templates resembling two-dimensional crystals. The pattern of radiation scattered by any object is called the diffraction pattern of the object. For diffraction from a slit, the wider the slit the narrower the diffraction pattern for a given wavelength of radiation. The diffraction pattern of many parallel and equidistant slits represents a sampling of the single-slit pattern in narrow regions. For a given wavelength:

1. The size and shape of the "envelope" of intensity variation is determined by the characteristic diffraction pattern of a single slit.
2. The spacings between the "samples" taken from the "envelope" are inversely related to the spacings between the slits.
3. The diffraction pattern is increasingly sharp the greater the number of slits.

These principles may be extended to three dimensions and to crystals, in which the *electrons* in the atoms act as scatterers for X rays, just as the areas within the slits behave as if they were scatterers for visible light. The diffraction pattern of a crystal is arranged on a lattice that is reciprocal to the lattice of the crystal. The analogy with the optical example holds: the X-ray photograph is merely a scaled-up "sampling" of the diffraction pattern of a single unit cell, with the "envelope" being the diffraction pattern produced by scattering from the electrons in the atoms of the unit cell, and the "sampling regions" arranged on the lattice reciprocal to the crystal lattice.

The phenomenon of X-ray diffraction by crystals can be considered in terms of a theory analogous to that of diffraction by gratings and extended to three dimensions (von Laue) or in terms of reflection from planes through points in the crystal lattice (Bragg). These two treatments are equivalent. We have chosen to emphasize the first approach because it provides more insight into the process of structure analysis by diffraction methods.

4

Experimental Measurements

The analysis of a crystal structure consists of three general stages:

1. Experimental measurement of the unit-cell dimensions and of the intensities of a large fraction of the diffracted beams from the crystal. These intensities depend only on the nature of the atoms present and their relative positions within the unit cell.

2. The deduction by some method of a suggested atomic arrangement (a "trial structure"). The intensities of the diffraction maxima corresponding to this arrangement can then be calculated and compared with those observed.

3. Modification (refinement) of this suggested arrangement of scattering matter until the agreement between calculated and observed intensities is within the limits of error of the observations. The present chapter is concerned only with the first of these stages. The rest of the book is concerned with the other two stages.

Two types of experimental data may be derived from measurements of the diffraction pattern

1. The angles or directions of scattering (2θ,* the angular deviation from the direct undeviated beam), which can be used to measure the *size* and *shape* of the unit cell.

2. The intensities of the diffracted beams, which may be analyzed to give the positions of the atoms within the unit cell and, eventually, other information about the atoms (such as their vibration parameters,

*See Figure 3.9b.

fractional occupancies, and electron distributions). The positions are usually expressed as fractions of the unit-cell edges.

Both of these kinds of experimental measurements must be made on a crystal in order to determine the molecular geometry of its components, since both the dimensions of the unit cell and the relative positions of the atoms are needed. Knowledge of the density and the cell dimensions of the crystal and the empirical formula of the compound makes it possible to determine how many atoms of each kind are present in the unit cell. The maximum number of reflections that can be measured for a crystal is a function of the wavelength of the radiation used and the unit-cell size. This number is approximately $32\pi V/3\lambda^3 n$, where V is the unit cell volume, λ is the wavelength of the radiation used and n is a constant (1 if the unit cell is primitive, 2 if it is body-centered or single-face-centered, and 4 if it is fully-face-centered). The diffraction pattern measured is a unique characteristic of the atomic identities and arrangement in that particular crystal; that is, it is like a "fingerprint."

Selection and Orientation of a Crystal A crystal whose structure is to be determined must be a single crystal, not cracked or a conglomerate. This may usually be checked by examining it under a microscope. If the crystal is too large and will not be fully bathed by the X-ray beam, it may be possible to cut it safely with a razor blade or with a solvent-coated fiber. Ideally one can try to find a crystal that can be shaped, often by grinding, until it is approximately spherical so that corrections for absorption of X rays are simplified. However, some crystals are too soft, fragile, or sensitive even for a delicate cutting and must be used as they are grown. For example, protein crystals contain 30–60% water, sometimes more, and they break very readily because the forces between protein molecules are weak.

The crystal may be attached to a glass fiber with glue or some similar material. If it is unstable, it is instead put into a thin-walled glass capillary (generally by gentle suction or simple capillary action) and the capillary is then sealed. An appropriate atmosphere may be maintained in the capillary to ensure stability; for example, protein crystals require a small amount of mother liquor to prevent drying out and disordering or collapse of the crystalline structure. The fiber or capillary is fixed onto a brass pin by shellac or glue and this pin is then placed on a goniometer head as shown in Figure 4.1. A *goniometer head* is simply a device to allow the crystallographer to center the crystal in the X-ray beam and, perhaps, to orient it. Translation adjustments permit the crystal to be centered and there may be two arcs, 90° apart, which can be adjusted to orient an axis of the crystal. If there are well-defined faces the crystal may be very nearly aligned visually. In any event, X-ray measurements must be made, either photographically or with a dif-

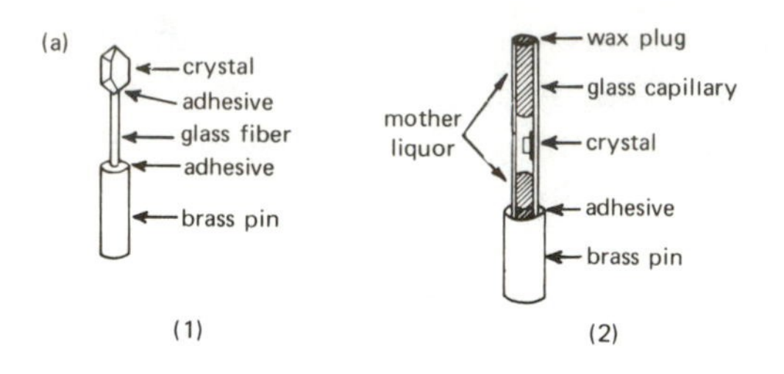
(a)
crystal
adhesive
glass fiber
adhesive
brass pin
(1)
wax plug
glass capillary
mother liquor
crystal
adhesive
brass pin
(2)

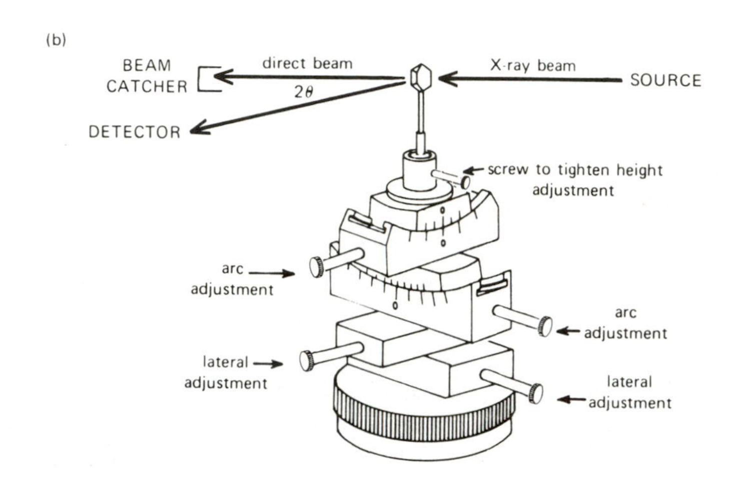
(b)
BEAM CATCHER
direct beam
2θ
X-ray beam
SOURCE
DETECTOR
screw to tighten height adjustment
arc adjustment
arc adjustment
lateral adjustment
lateral adjustment

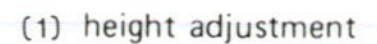
(1) height adjustment

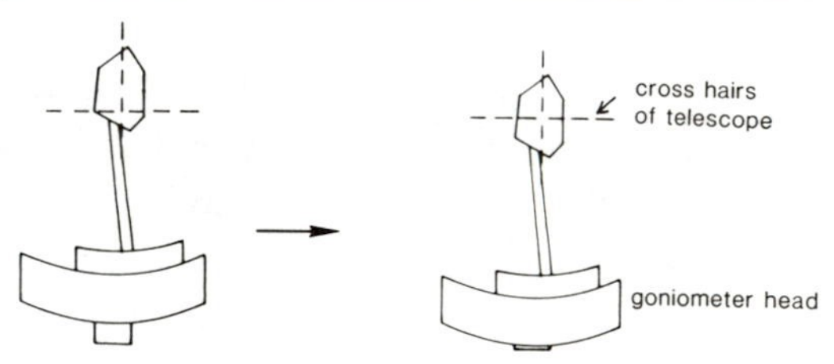
cross hairs of telescope
goniometer head

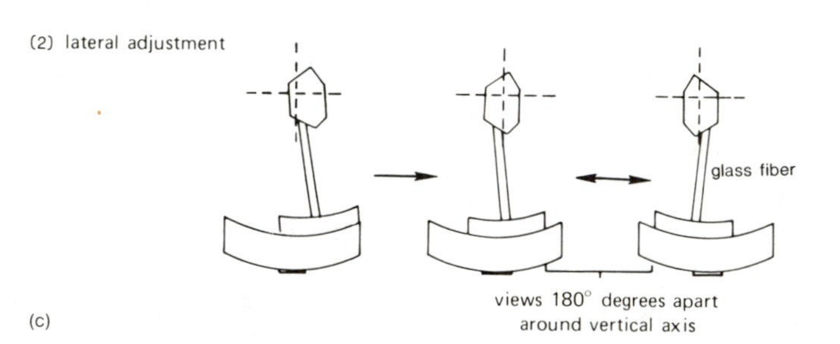
(2) lateral adjustment
glass fiber
views 180° degrees apart around vertical axis

(c)

fractometer, to determine the best goniometer settings for the particular method to be used for diffraction intensity data measurements. Modern diffractometers generally do not require any orientation, only centering of the crystal.

Cell Dimensions and Density The dimensions of the unit cell (a, b, c, α, β, γ) may be found from the angles, 2θ, of the deviations of given diffracted beams from the direction of the incident beam, since each value of 2θ at which a diffraction maximum is observed is a function only of the cell dimensions and of the wavelength of the radiation used. The spatial orientation of these diffracted beams allows for indexing so that the determination of cell dimensions is simplified; however, it is also possible to determine unit-cell dimensions from powder photographs. The situation is illustrated for a single crystal in the example in Appendix 4. One of the easiest ways to measure approximate cell dimensions is from a precession photograph. This gives a direct scaled-up picture of the reciprocal lattice (see Figure 3.6). (The scale factor for the picture is readily obtained from the wavelength, the precisely known geometry of the camera, and the position of the crystal.)

The density of the crystal can usually be measured by flotation. Two miscible liquids in which the crystal is insoluble (one more dense, one less dense than the crystal) are mixed in such proportion that the crystal remains suspended; that is, it neither sinks nor rises to the surface of the resulting mixture. The density of the liquid mixture (with the same density as that of the crystal) is then found by weighing a known volume in a "specific gravity bottle" or "pycnometer." For proteins a "density gradient column" is prepared by layering an organic liquid (in which the protein is insoluble) on another that is miscible with the first. This column can be calibrated by measuring

---

FIGURE 4.1 Goniometer Heads and the Centering of Crystals.

(a) Methods used to mount (1) a stable crystal on a glass fiber; (2) a protein or other macromolecular or unstable crystal, which deteriorates or evaporates on exposure to air, in a glass capillary. (Mother liquor is not needed for some such crystals but is a necessity for many, such as biological macromolecules.)

(b) A goniometer head, showing various possible adjustments that can be made in order to center the crystal in the X-ray beam. The height adjustment is essential in this operation. The arc adjustments are made to *orient* a crystallographic axis in a desired direction, for example, vertically. With most modern diffractometers, goniometer heads without arc adjustments may be used because a part of the instrument itself permits alignment of the crystal in any desired orientation.

(c) The use of the adjustments of the goniometer head, shown in (b), to center the crystal in the crosswires of a telescope and, finally, in the X-ray beam, so that the crystal will be bathed in radiation.

the equilibrium positions along the column of drops of aqueous solution of known density. Some protein crystals are then added to the column and their equilibrium positions read; these positions can be directly converted to densities using the previously prepared chart.

As seen in Appendix 4 the density, combined with the cell dimensions, will give the weight of the contents of the unit cell. If the elemental analysis of the crystal is known, then the number of each type of atom in the cell can be determined. Some crystals, particularly protein crystals, contain a large amount of solvent; the amount of this may be estimated from the loss of weight on drying.

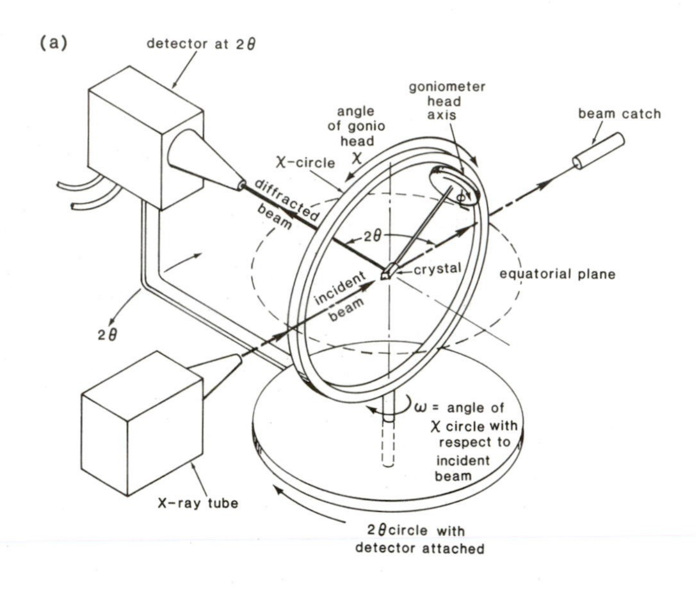

ϕ = spindle axis of goniometer head

2θ = angle between directions of incident and diffracted beams

= angle detector has to be rotated to intercept diffracted beam

ω = angle between diffracted vector and plane of χ-circle

χ = angle between ϕ axis (gonio. head) and diffractometer axis (equatorial plane)

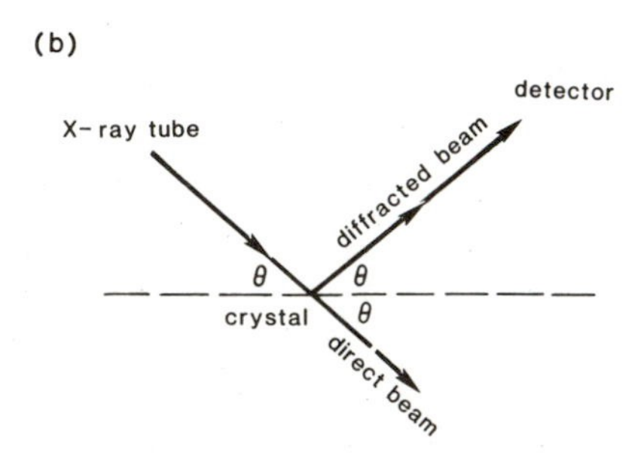

Methods of Intensity Collection The intensities of the diffracted beams are measured by intercepting the beams with something that is sensitive to X rays. The measurements may be done electronically or photographically.

The most common method is to use an automatic diffractometer (Figure 4.2). The diffracted beam is intercepted by a detector, usually a scintillation counter, and the intensity is recorded electronically. The peak may be scanned over a range near the expected scattering angle, or only the intensity at the center of the peak may be measured. Measurements of background counts are also made, or calculated from the profile of the peak, and used to correct the recorded intensities.

The angular settings of the crystal and detector are calculated in advance from precise cell dimensions and from information about the crystal orientation on the diffractometer. Usually the entire instrument is controlled by a computer and the whole data collection is then done automatically.

There are several refinements to these procedures, especially those that ensure that the incident beam is essentially monochromatic (one wavelength). These refinements are used when the highest precision is required. When a crystal monochromator is used, the X-radiation is diffracted first from a standard crystal, and the resulting monochromatized beam (that is, radiation diffracted in a narrow 2θ range) is then used to irradiate the crystal whose structure is being studied.

In the photographic method a camera containing a piece of film is placed behind or around the crystal, with allowance for entry of the direct beam through the camera. The cameras used are so constructed that the indexing

---

FIGURE 4.2 Diffractometry.

In an X-ray diffractometer (a), the crystal is centered in the incident X-ray beam, and the detector can move in a plane parallel to the instrument base, referred to as the equatorial plane, which contains the crystal and the incident and diffracted X-ray beams. The value of 2θ may then be measured directly from the position of the detector at the maximum of a peak due to a diffracted beam (b). In most present diffractometers, the crystal can be rotated around three axes (χ, ϕ, and ω) independently, and the detector can be rotated about a fourth angle (2θ, concentric with ω); this gives rise to the name "four-circle diffractometer." Values of the four corresponding angles may be computed for all possible reflections once a few reflections have been located and identified. The angles for a particular reflection are set automatically under computer control, the intensity of the reflection is measured with the detector and recorded, together with measurements of the background intensity near the reflection, and then a new set of angles is calculated and another intensity measurement made. One normally advances incrementally through the Miller indices, hkl; in this way a systematic scan of all desired reflections is done completely automatically.

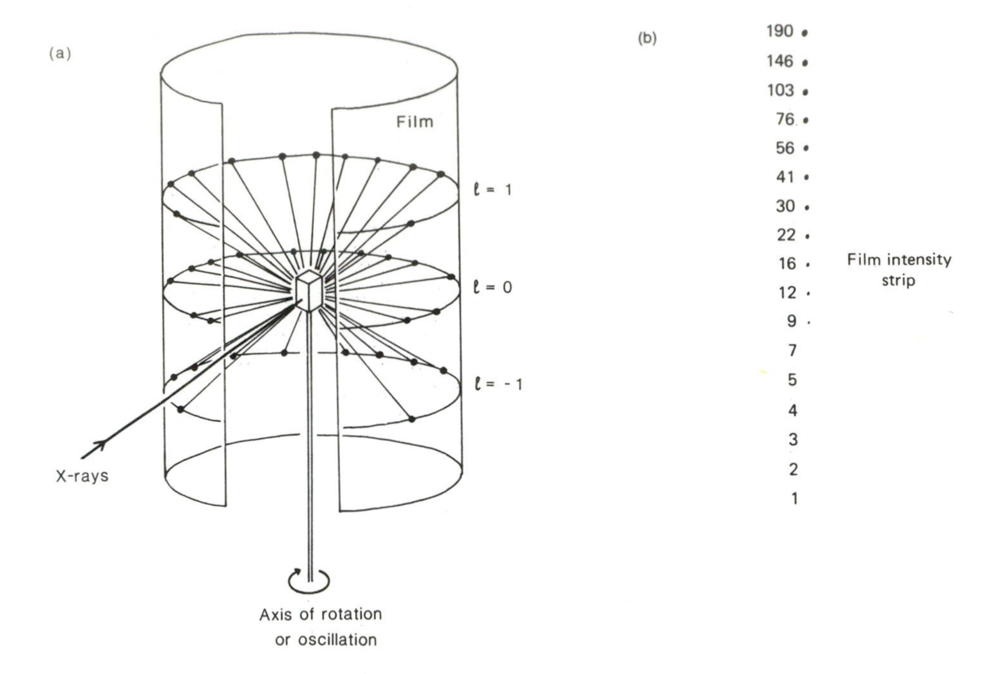

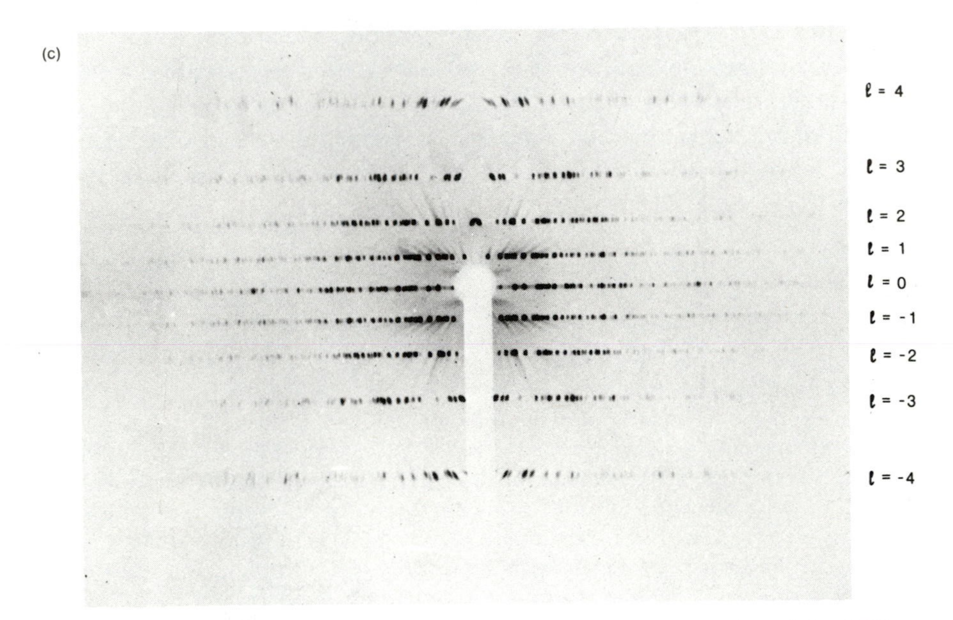

FIGURE 4.3 Rotation and Oscillation Methods for Photography of the Diffraction Pattern.

In these two methods the crystal mounted on a goniometer head is either rotated continuously in one direction (rotation photograph) or is oscillated back and forth through a small angle (oscillation photograph); the resulting diffraction pattern is recorded on a piece of photographic film placed around the crystal. In any one stationary position of the crystal, few or no diffracted beams will normally be obtained because the conditions of the Bragg equation,

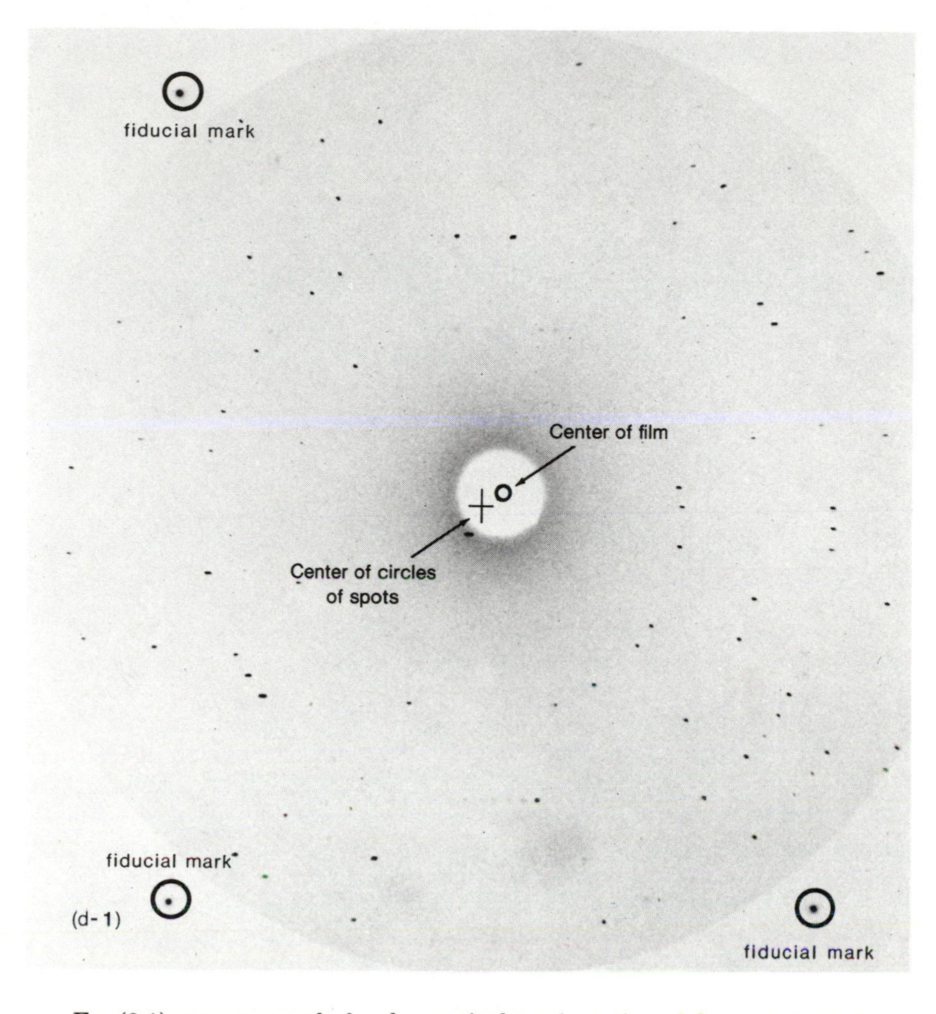

Eq. (3.1), may not apply for the particular orientation of the crystal relative to the incident X-ray beam.

(a) A diagram of the method of obtaining the photograph. The **c**-axis of the crystal is vertical. Each cone of diffracted beams corresponds to a different value of l.

(b) A film intensity strip is shown to illustrate how intensity measurements might be made; the numbers to the left of the spots are relative intensities, obtained by exposing one spot at that position for a length of time proportional to the number shown. Thus each reflection, recorded as a spot on the photograph, may be assigned a number, the relative intensity, from a comparison with the intensity strip.

(c) A typical rotation photograph obtained by unrolling and developing the film illustrated in (a). The indexing of reflections can be very difficult. (Photograph courtesy of Dr. H. L. Carrell.)

(d) Oscillation photographs of a protein crystal.* (Photographs courtesy of Dr. M. J. Adams.)

  (1) A still photograph (no rotation of the protein crystal). In view of the large size of the unit cell there are many reflections possible for any fixed orientation of the crystal. This photograph contains circles of spots that allow

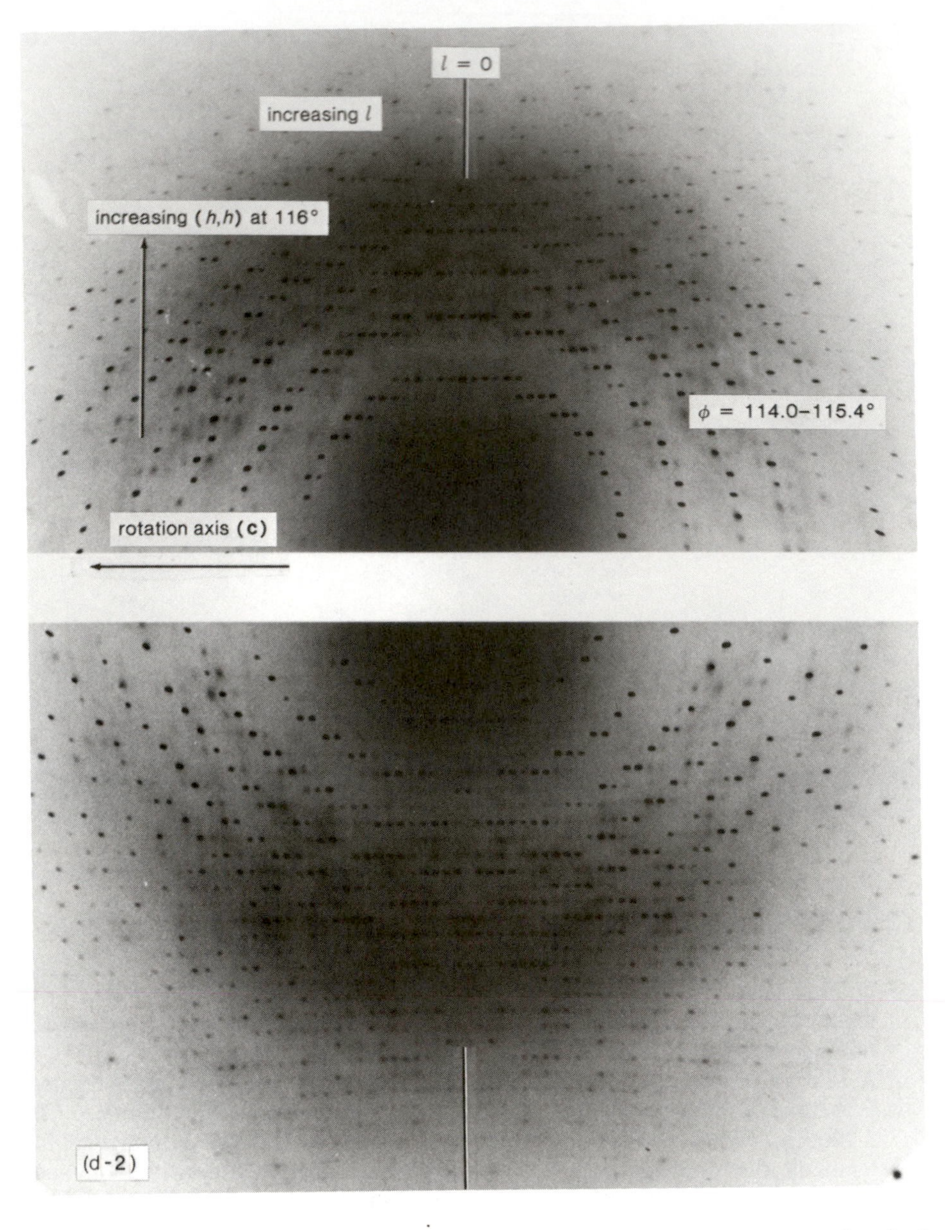

for the reorientation of the crystal so that the center of the circles (marked with a cross) corresponds to the center of the film. A misorientation matrix (to correct this slight misalignment) may be obtained by measuring the positions of some spots relative to the fiducial spots in the corners of the film.

(2 and 3) Oscillation photographs (taken with the synchrotron radiation, wavelength 1.488 Å). These photographs were taken with a Vee (vee-shaped) cassette,† each with an oscillation range of 1.4° about **c**. The strip in the center of the film is a direct-beam stop (at the fold in the cassette). Spots

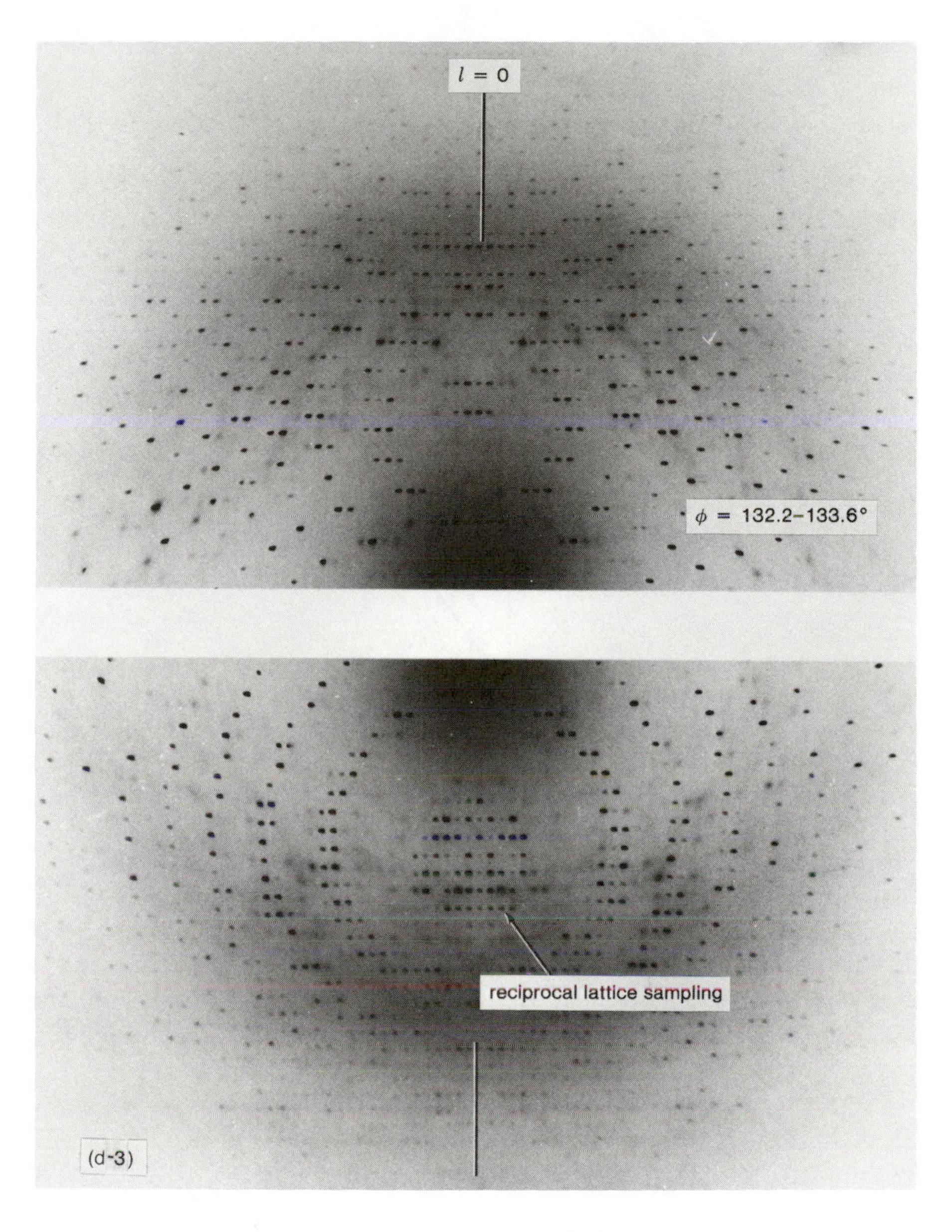

in the center line of the film perpendicular to the strip have $l = 0$ (marked). The spots at the very edge of the film correspond to d values of 1.85 Å. By chance at 116° the hhl planes are perpendicular to the beam so that lines of reflections correspond to h, h, l; $h, (h + 1), l$; etc. l increases horizontally to the left and (h, h) increases vertically up the film. Study of (d) may make it clear that if the V (Vee) were a cylinder the photograph would be similar in appearance to (a3), with layer lines in l. Indexing of such an oscillation photograph is done by computer once the crystal settings with respect to the film are precisely known. Since the unit-cell

(d-4)

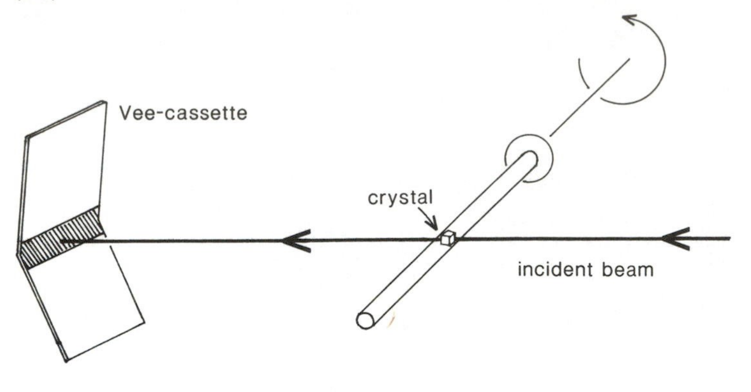

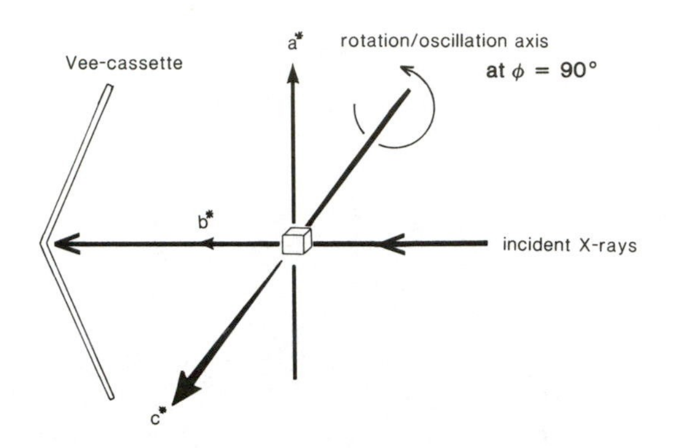

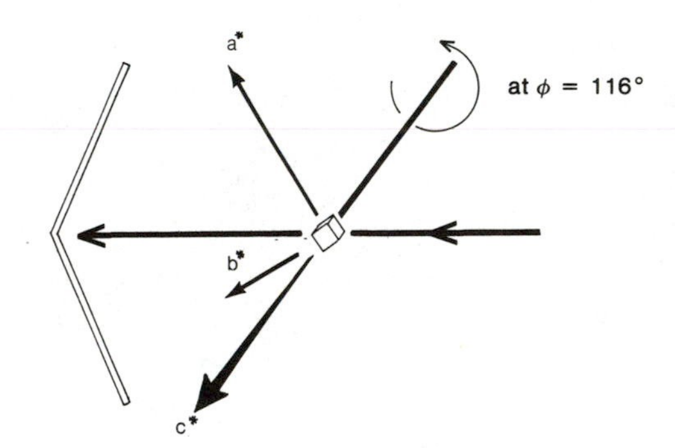

dimensions are so large it is possible to see adjacent points in selected areas of the reciprocal lattice. The oscillation angle range is 114.0–115.4° for (2) and 132.2–133.6° for (3). It can be seen that the zero line (crescent of reciprocal lattice images) is nearly centered in (2) and is lower in (3).

(4) Camera geometry for the above photographs. For further details see ref. 95.

*Sheep liver 6-phosphogluconate dehydrogenase. Space group $C222_1$, a = 72.72, b = 148.15, c = 102.91 Å. M. J. Adams, I. G. Archibald, C. E. Bugg, A. Carne, S. Gover, J. R. Helliwell, R. W. Pickersgill, and S. W. White, *The EMBO Journal* **2**, 1009–1014 (1983).

†See ref. 95 (Arndt and Wonacott, *The Rotation Method in Crystallography*).

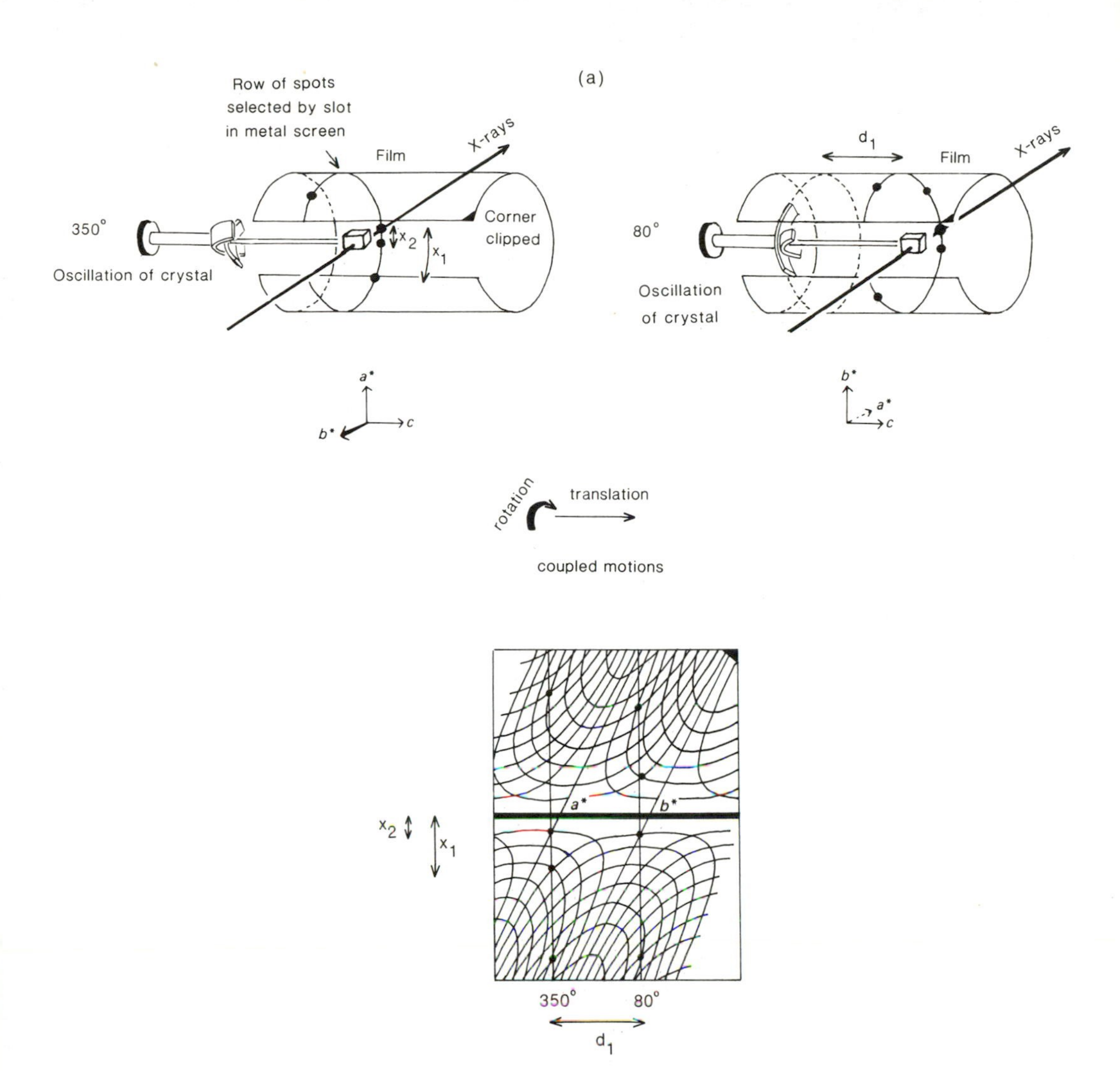

FIGURE 4.4 The Weissenberg Method for Photography of the Diffraction Pattern.

This method is an extension of the oscillation method so that the indexing is simple and unambiguous. Only one layer from the oscillation photograph is selected, by the positioning of a metal screen, with a slit in it, between the film and the X-ray source. The screen may be moved to select different layer lines. The layer lines are orders of diffraction corresponding to the axial distance unit cell repeat that lies parallel to the oscillation axis, which is approximately parallel to the crystal mounting fiber. In this example the crystal is mounted with the **c**-axis parallel to the fiber; hence the layer lines have different values of l and each Weissenberg photograph has a constant value of l. The crystal is oscillated back and forth as in an oscillation photograph. The camera moves parallel to the axis of oscillation and its motion is coupled to the oscillation so that for each angle of the crystal about the rotation axis there is a specific camera position, and hence a particular position parallel to the cylindrical axis of the film, at which spots are recorded.

(b)

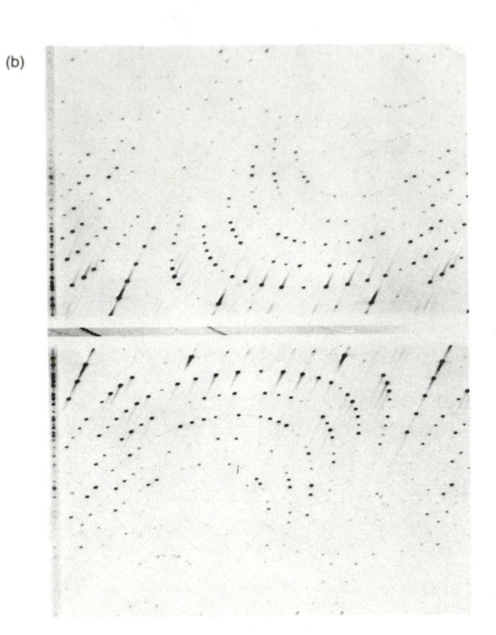

(c)

Oscillation of the crystal over the entire angular range with the camera traverse decoupled.

Line of spots set down for one specific position of the camera in its traverse. The traverse of the camera is coupled to the rotation angle of the crystal

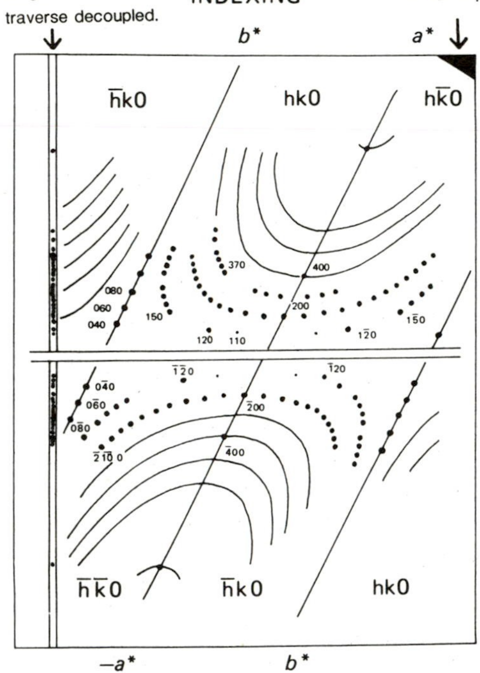

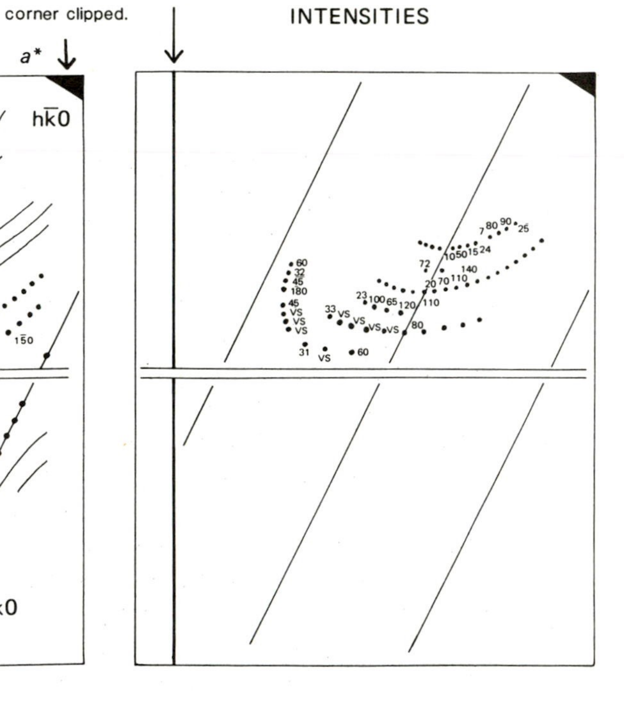

of the reflections is unambiguous and, with the precession camera, simple. The indices of the diffracted beams are simply the orders of diffraction and may be determined by inspection or calculation from the positions of the spots on the film. Some of the photographic methods most commonly used are illustrated in Figures 4.3, 4.4, and 4.5. Precession photographs (see Figure 4.5) are very useful for the indexing of reflections because an undistorted view of the reciprocal lattice is obtained. With most cameras, direct recording of the diffracted beams gives dark spots on the photographic film. In recent years Polaroid film has been used for preliminary indexing and space group determination; in this an intensifier (such as zinc sulfide) faces the X-ray source and the visible light emitted is recorded on the Polaroid film. Such films give positive images rather than negative ones. The degree of blackness (or whiteness) of a spot depends on the exposure time, on the intensity of the diffracted beam, and, to some extent, on the conditions of development of the film.

The intensities of a diffraction pattern can be measured with fair accuracy from the negative of a precession, oscillation, rotation, or Weissenberg photograph. Such photographs may be photometered with a densitometer. (It is

(text continues on p. 58)

(a) The arrangement of the film and layer-line screen in a Weissenberg camera, showing the crystal rotation axis and two selected film positions. Orientations of the reciprocal lattice, selected by convention to be a right-handed system, are shown. It is assumed here that the crystal is being rotated about the **c**-axis, which remains fixed in direction.

(b) A Weissenberg photograph. If the entire photograph were condensed from left to right then the result would resemble a layer of the rotation photograph shown in Figure 4.3c; this condensation is also shown on the left-hand side of this photograph. It was recorded with the camera stationary; it is shown for comparison with the same reflections recorded while the camera was moving.

(c) Indexing of reflections and intensities on a Weissenberg photograph. [*Note:* The different layer lines are separated out (by a metal screen) for Weissenberg photographs. The axis of rotation corresponds to the layer lines of the photograph (i.e., l is constant on a Weissenberg photograph if the rotation axis is **c**).] The diagrams labelled "Indexing" and "Intensities" are schematic and should be used in conjunction with the photograph. "Vs" means very strong (too intense to estimate from this film).

A few reflections have been indexed for clarity; an index with a line above it (e.g., $\bar{5}$) is negative. Thus $1\ \bar{5}\ 0$ means $h = 1, k = -5, l = 0$. Some intensities for reflections, estimated from the intensity strip in Figure 4.3b, are also shown. One corner of the film is clipped so that its orientation in the camera during the experiment is unambiguous. (Photograph courtesy of Dr. H. L. Carrell.)

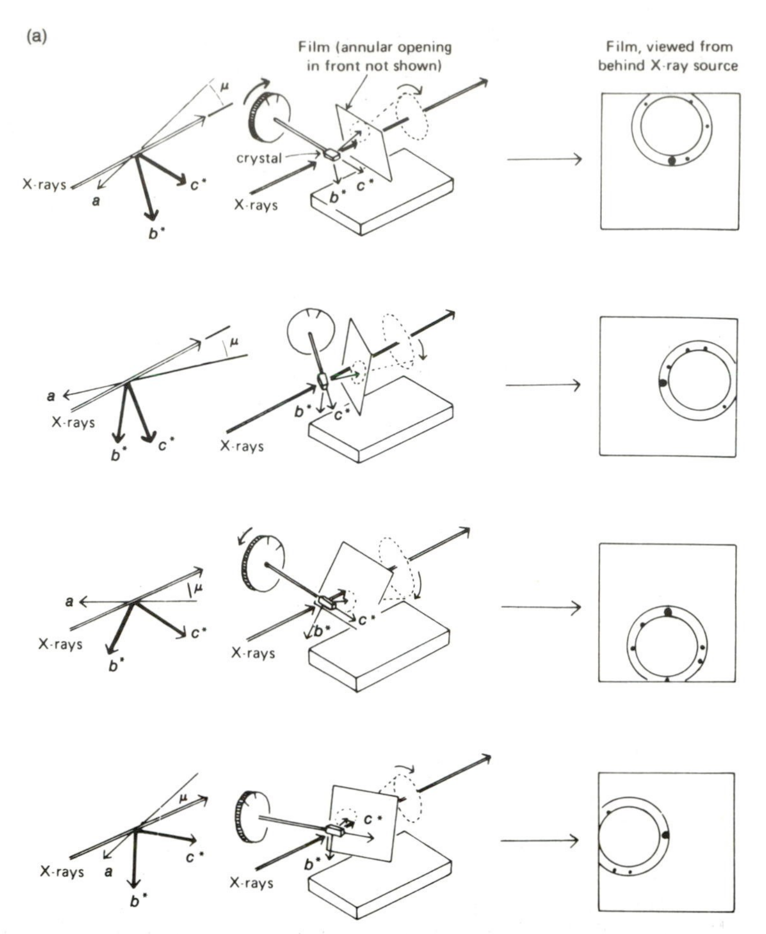

FIGURE 4.5 The Precession Method for the Photography of the Diffraction Pattern.

The camera motion is more complicated in order that the recorded image of the diffraction pattern may be simple. In fact, direct measurement of all reciprocal lattice parameters is possible from a series of precession photographs, with an appropriate scale factor taken into consideration.

(a) A diagram of the motions of a precession camera. The crystal is mounted, for analogy with the Weissenberg example, with the **c***-axis parallel to the axis of the glass fiber on which it is mounted. As the crystal is oscillated about the rotation axis, the flat film holder also undergoes motion so that: (i) The direct beam always hits the center of the film; (ii) The film-to-crystal distance remains constant; (iii) One layer is selected by means of an annular aperture between the film and the X-ray source; (iv) A crystal axis (**a**, in this example) precesses about the X-ray beam; that is, the crystal is moved in such a way that this axis rotates about the X-ray beam making a constant angle, μ, with it. The film follows this precession motion in such a way that *the film is always perpendicular to this axis and hence always parallel to a reciprocal lattice plane* (**b***, **c*** in this example) of the crystal being photographed; (v) Because the film is flat, the camera can be adapted to take Polaroid film, which greatly decreases the time needed to take photographs.

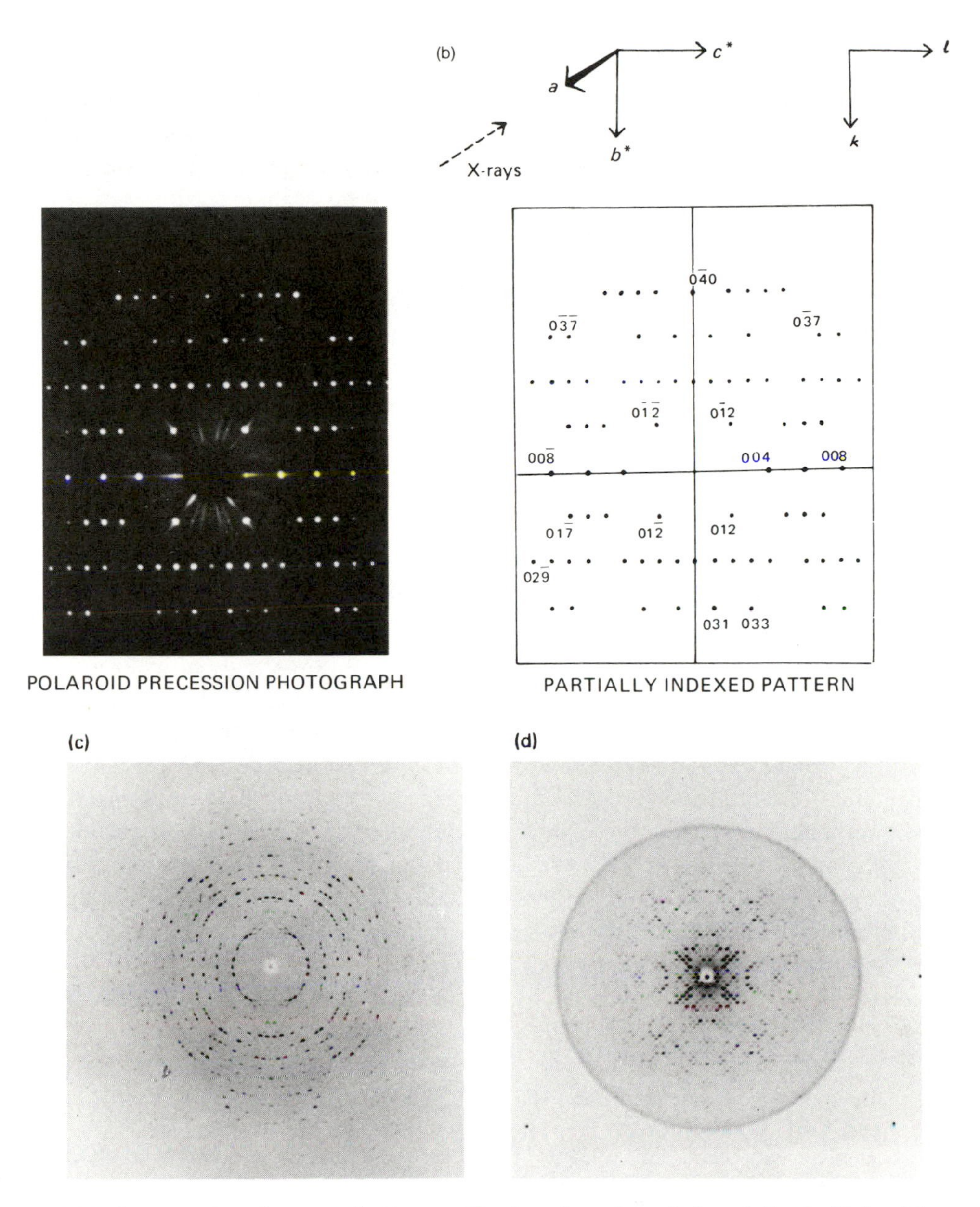

(b) A precession photograph. Some reflections have been indexed. In the Polaroid photograph the intensifier faces the X-ray source. The indexing of reflections for the precession photograph is shown. Note the systematic absences—0 k 0 with k odd, 0 0 l with l odd. By convention the positive direction of **a** is toward the X-ray source so that higher layers of the reciprocal lattice, for example 3 k l, recorded with the film nearer the X-ray source than it is for the zero level, have positive values of h. Note that, when the crystal is mounted with **c** parallel to the fiber mount, oscillation, rotation, and Weissenberg photographs will have layer lines with different values of l and the zero layer will give h k 0 intensities. On the other hand, a precession photograph in the same mounting gives layer

uncommon to measure intensities from Polaroid photographs). The intensity of a source of light is determined photoelectrically and the reduction of intensity of the light by different spots on the photograph is then measured in the same way. Because there is a large intensity variation in the diffraction pattern, from four to six layers of film are usually packed in the camera, so that beams too strong to be measured on the top film will be measurable on one or more of the other films. (The emulsion of each successive film cuts down the intensity of the beam by a constant factor.) Alternatively, the intensities may be estimated visually, the most common procedure several decades ago. If one diffracted beam is exposed in a series of positions on a piece of X-ray film for varying lengths of time, an intensity calibration strip may be prepared in which the intensity is proportional to the exposure time for a given spot. When this is placed over or beside an X-ray photograph on a surface uniformly lighted from behind, the intensities may be estimated visually by comparison with those on the calibration strip; the eye is extremely sensitive to intensity variation. The background intensity is automatically subtracted in this method, which can be surprisingly accurate if appropriate care is taken. However, this method is now chiefly of historical interest.

Generally diffraction data for crystals of small molecules are obtained by diffractometer measurements, and this method can also be used for some proteins. The alternative methods in general use for biological macromolecules (particularly those of high molecular weight) are oscillation photographs and screenless precession photographs with a very high intensity X-ray source (a rotating anode); intensity measurements are then obtained by photodensitometry.

Both the photographic and the diffractometer methods of recording intensities have some disadvantages. The photographic method is limited in that

---

lines in **a** (or **b** but not **c**) and the zero layer photograph is $0\,k\,l$ or $h\,0\,l$ (but not $h\,k\,0$ as it was for the Weissenberg photograph). (*Note:* the axis of rotation corresponds to the horizontal row of spots on the film (**c** in this example).) (Photograph courtesy of Dr. H. L. Carrell.)

(c) Still photograph of a protein crystal.* This photograph shows that the crystal is well-aligned. The **c**-axis is parallel to the incident X-ray beam. (Photograph courtesy C. J. Smart.)

(d) Precession photograph taken with copper $K\alpha$ radiation of the same protein crystal as in (c) and in the same orientation. This is an $hk0$ photograph with **a** vertical and **b** horizontal. The precession angle is 20.25°. The resolution at the outer edge of the photograph is 2.2 Å. (Photograph courtesy C. J. Smart and H. L. Carrell.)

*Xylose isomerase. Space group $I222$, $a = 93.88$, $b = 99.64$, $c = 102.9$ Å. H. L. Carrell, B. H. Rubin, T. J. Hurley, and J. P. Glusker, *Journal of Biological Chemistry* **259,** 3230–3236 (1984).

not all reflections can be measured on one film and the chief irremovable uncertainty comes from the need for scaling the measurements on one film relative to those on others. The inherent precision and accuracy of photographic methods are also somewhat lower than those of the best electronic methods of intensity measurement. However, for proteins, the large unit cells mean that the intensities are weak (partly because there are fewer unit cells in a given crystal volume) and there are large numbers of diffraction maxima. Hence, in order to make the data collection manageable in a finite time, photographic methods and small rotation angles are used for proteins. Any diffractometer measurement of intensities is a *serial* process—the beams are measured one at a time, in sequence—whereas in the photographic method many diffraction maxima are recorded repeatedly and almost simultaneously. Consequently, if there is any deterioration in the crystal (or change in the intensity of the X-ray source) as a function of time, the diffractometer method will not give reliable relative intensities unless repeated checks of standard reflections are made for calibration. Hence with a diffractometer standard reflections are measured at frequent intervals; however, there is no guarantee that they are the ones that would be sensitive to any deterioration that might occur. If damage to the crystal occurs as a function of time (and not X-ray exposure), the rapidity of an automatic diffractometer will offer an advantage relative to the measurement with a manual diffractometer. The overall X-ray exposure time is essentially the same. Radiation damage usually occurs as a result of free radical formation and heating effects; it will continue after X-ray exposure has stopped. It is generally believed that such radiation damage can be reduced by the use of incident monochromatic radiation such as that obtained by reflection from a graphite crystal, by very rapid data collection techniques with the use of, for example, tuned synchrotron radiation, or by lowering the temperature with appropriate attention to the solvent.

The rapid advances in techniques for solving macromolecular structures during the past two decades have led to the need for new methods for collecting the many tens of thousands (or even hundreds of thousands for molecules with molecular weights approaching or exceeding 10^6) of diffraction maxima that are needed. Any method of collecting data in series, as with a conventional diffractometer, is far too slow; thus film methods (such as that illustrated in Figure 4.3) that permit recording many diffraction maxima simultaneously had a renaissance as protein crystallography developed. However, the many problems associated with film techniques have led in the last decade to the development of electronic methods of recording simultaneously many diffraction maxima. Detectors that have an appreciable area and offer the possibility of resolving and individually measuring the intensities of diffraction maxima at different points across this area have been developed and applied. The most successful of these to date is probably the multiwire proportional counter, a carefully engineered device consisting typically of a flat beryllium window, 30 cm square and 1 mm thick, and, parallel to this window, two arrays of fine wires. Each array consists of several hundred parallel wires, the direction of the wires in one array being at 90° to that in the other. During the data collection, the crystal is moved through a very small rotation angle (e.g., 0.08°) while counts are accumulated, then rotated again by a small amount, while counts are again accumulated. Hundreds of these "electronic films" can be recorded in a short time. This detector, and others under current development, are described in ref. 94.

SUMMARY

Two types of experimental diffraction data may be obtained:

1. *The angle of scattering* (2θ, the angular deviation from the direct undeviated beam), which is used to measure the spacings of the reciprocal lattice and hence the spacings of the crystal lattice. These spacings can be used to derive *the size and shape of the unit cell.*
2. *The intensities of the diffracted beams,* which may be analyzed to give the *positions of the atoms* within the unit cell. These positions are usually expressed as fractions of the unit-cell edges.

The diffraction pattern may be recorded either by a radiation detector such as a scintillation counter, or on photographic film. Many different designs of diffractometers and cameras are used to facilitate the indexing of "reflections" and the measurement of intensities.

II

Diffraction Patterns and Trial Structures

5

The Diffraction Pattern Obtained

The result of the collection of X-ray diffraction data is a relative intensity, I, for each "reflection" with indices, h, k, l, together with the corresponding value of the scattering angle, 2θ, for that reflection. All the values of I are on the same relative scale. As already indicated from a study of the diffraction patterns from slits and from various arrangements of molecules (Figure 3.5 and 3.7 especially), the *angular positions* at which scattered radiation is observed (related to the scattering angle 2θ) depend only on the dimensions of the *crystal lattice,* while the *intensities* of the different diffracted beams depend (other than for easily calculated geometric factors, and possible corrections for absorption and other minor effects) only on the *nature and arrangement of the atoms within each unit cell.* Since it is this atomic arrangement that comprises what we mean by the structure of the crystal, it is the primary object of our analysis. How can we deduce it from the pattern of intensity variation? Before we answer this question, we must digress briefly on ways of representing the combination of waves scattered from different points in a given direction—for *the intensity scattered at any angle can be calculated from the sum of the waves scattered from different atoms,* and indeed from different points within each atom, in that direction. In other words, each diffracted beam contains information on the entire atomic structure of the crystal. Since a structure determination involves a matching of the observed intensity pattern to that calculated from a postulated model, it is imperative that we understand how this intensity pattern can be calculated for any desired model.

If we know the atomic arrangement in a crystal, we can, by the methods to be described, calculate the intensities of reflections in the diffraction pattern; relative phases of these reflections are computed at the same time.

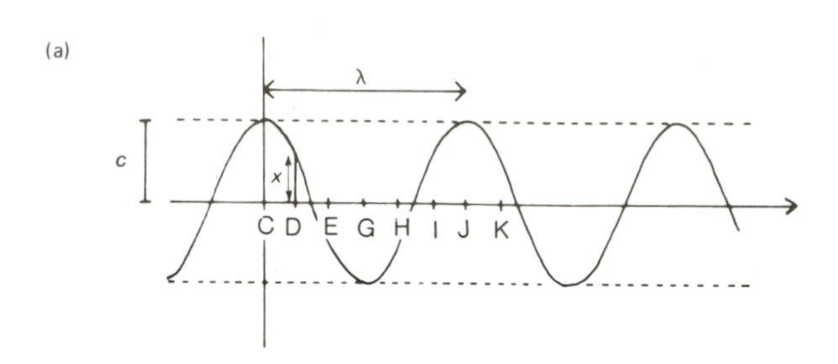

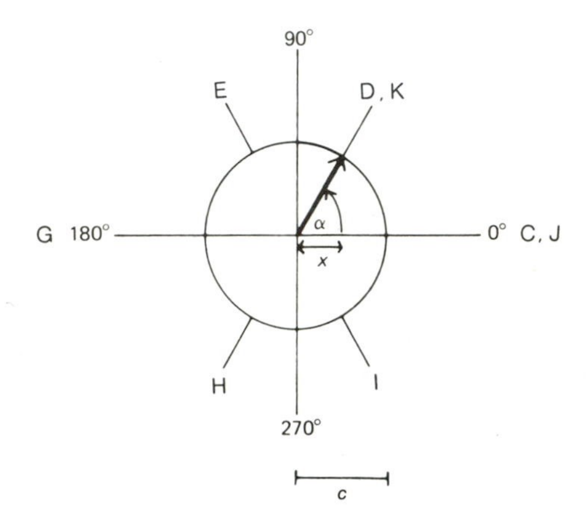

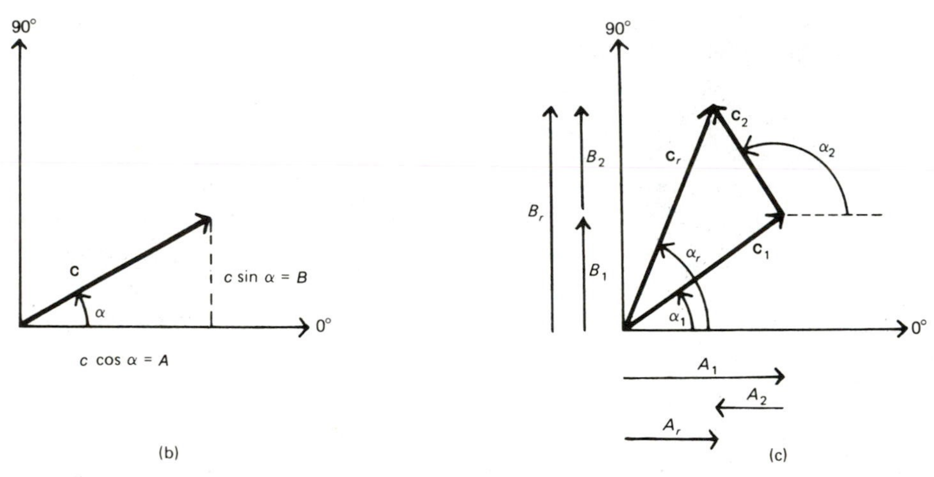

FIGURE 5.1 The Representation of Waves and of Their Superposition as Two-Dimensional Vectors.

(a) The meaning of the angle α in terms of phase, measured relative to the (arbitrary) origin at position C (or its equivalent, J). Recall that phase is expressed as a fraction of the wavelength, sometimes multiplied by 360° (Figure 3.2). The letters, C, D. E, ... here are positioned at intervals of one-sixth wavelength along the wave, and thus, when expressed in angular measure, at 360°/6 = 60° intervals. Positions D and K are at a phase angle, α, of 60° relative to C and J.

However, when we measure the diffraction pattern, phase information is not obtainable. To compute an electron density map and hence determine the crystal structure, we have to use phases calculated from a "trial structure" together with the measured intensities.

It is helpful at this point to refer back to Figure 1.1 where the analogy to the microscope is given. When a lens is used to magnify the image of an object, the light waves passing through the lens are bent and refocused with no discontinuity in the phase relationships of these waves. On the other hand, when an X-ray diffraction pattern is intercepted by photographic film or some other detecting device, these relative phase relationships are lost; only the amplitudes of the diffracted beams are known. Crystallographers have, however, derived ways (to be described later) of recombining these waves mathematically (rather than physically) with approximately correct phases to give an image of the structure that scattered them.

Superposition of Waves: (1) Graphical Representation Electromagnetic radiation, such as X rays, may be regarded as composed of many individual waves. When this radiation is scattered with preservation of the phase relationships among the scattered waves, the amplitude of the resulting beam in any direction may be obtained by summing the individual waves scattered in that direction, taking into account their relative phases (see Figure 3.2). The phase may be calculated by seeing where the maximum in the resultant wave occurs, relative to some arbitrary origin (Figure 5.1a). However, graphical superposition of waves is not convenient with a digital computer, and so, for speed and convenience in computing, an analytical representation is needed.

(2) Algebraic Representation Since any two waves can be represented as trigonometric functions, we can express the displacement, x_1 or x_2, of each wave at any time algebraically as

$$x_1 = c_1 \cos (\phi + \alpha_1) \tag{5.1}$$

$$x_2 = c_2 \cos (\phi + \alpha_2) \tag{5.2}$$

The value of x has fallen to $0.5c$ at those positions, as seen from the wave, or from the lower figure. It also follows from Eq. (5.1) with $(\phi + \alpha) = 0°$ at C, hence $x = c$, and with $(\phi + \alpha) = 60°$ at D, hence $x = c \cos 60° = 0.5c$.

(b) A wave of amplitude c and phase α may be represented as a vector with components $c \cos \alpha$ and $c \sin \alpha$ in an orthogonal system [see Eq. (5.10)].

(c) The result of adding two waves with amplitudes c_1 and c_2 and phases α_1 and α_2 is the same as the result of adding two vectors; that is, the result is a vector $\mathbf{c}_r$ with phase α_r.

Here c_1 and c_2 are the amplitudes of the waves (the maximum displacements), ϕ is proportional to the time for the travelling wave and is the same for all waves under consideration, and α_1 and α_2 are the phases. We will assume that the wavelengths of the scattered waves are identical, inasmuch as the X rays used in any practical structure analysis are essentially monochromatic. The phases α_1 and α_2 are expressed relative to an arbitrary origin. Because the wavelengths are the same, the phase difference between the two scattered waves, $(\alpha_1 - \alpha_2)$, remains constant (assuming coherence—that is, the preservation of phase relationships).

When the waves are superimposed, the resulting displacement, x_r, is, at any time, simply the sum of the individual displacements, as indicated in Figure 3.2,

$$x_r = x_1 + x_2 = c_1 \cos(\phi + \alpha_1) + c_2 \cos(\phi + \alpha_2) \tag{5.3}$$

which may be rewritten as

$$x_r = c_1 \cos \phi \cos \alpha_1 - c_1 \sin \phi \sin \alpha_1 + c_2 \cos \phi \cos \alpha_2 - c_2 \sin \phi \sin \alpha_2 \tag{5.4}$$

or

$$x_r = (c_1 \cos \alpha_1 + c_2 \cos \alpha_2) \cos \phi - (c_1 \sin \alpha_1 + c_2 \sin \alpha_2) \sin \phi \tag{5.5}$$

If we define the amplitude, c_r, and phase, α_r, of the resulting wave such that

$$c_r \cos \alpha_r = c_1 \cos \alpha_1 + c_2 \cos \alpha_2 = \sum_j c_j \cos \alpha_j \tag{5.6}$$

and

$$c_r \sin \alpha_r = c_1 \sin \alpha_1 + c_2 \sin \alpha_2 = \sum_j c_j \sin \alpha_j \tag{5.7}$$

then we can rewrite eq. (5.5) as

$$x_r = c_r \cos \alpha_r \cos \phi - c_r \sin \alpha_r \sin \phi = c_r \cos(\phi + \alpha_r) \tag{5.8}$$

Thus the resultant of adding two waves is a wave of the same frequency, with a phase, α_r (relative to the same origin), given by Eqs. (5.6) and (5.7), or more compactly by

$$\tan \alpha_r = \frac{c_r \sin \alpha_r}{c_r \cos \alpha_r} = \frac{\sum_j c_j \sin \alpha_j}{\sum_j c_j \cos \alpha_j} \tag{5.9}$$

The amplitude of the resultant wave, c_r, is given by

$$c_r = [(c_r \cos \alpha_r)^2 + (c_r \sin \alpha_r)^2]^{1/2} = \left[\left(\sum_j c_j \cos \alpha_j\right)^2 + \left(\sum_j c_j \sin \alpha_j\right)^2\right]^{1/2} \tag{5.10}$$

(3) Vectorial Representation These relationships can all be expressed alternatively in terms of two-dimensional vectors, as illustrated in Figures 5.1b and 5.1c. The length of the jth vector is its amplitude, c_j, and the angle that it makes with the arbitrary zero of angle (usually taken as the direction of the horizontal axis pointing to the right, with positive angles measured counterclockwise) is the phase angle α_j. The components of the vectors along orthogonal axes are just $c_j \cos \alpha_j$ and $c_j \sin \alpha_j$ and the components of the vector resulting from addition of two (or more) vectors are just the sums of the components of the individual vectors making up the sum, a result expressed in Eqs. (5.6) and (5.7).

(4) Exponential Representation (Complex Numbers) For computational convenience, vector algebra is an improvement on graphical representation, but an even simpler notation is that involving so-called "complex" numbers, often represented as exponentials. The *exponential representation* is particularly simple because multiplication of exponentials involves merely addition of the exponents. Consider Eqs. (5.6) and (5.7), expressing the components of the resulting wave, and Eq. (5.10) in which we expressed the amplitude of the resulting wave as the square root of the sum of the squares of its components, which we will now abbreviate as A and B. These relations may be rewritten as

$$A = c_r \cos \alpha_r = \sum_j c_j \cos \alpha_j \tag{5.11}$$

$$B = c_r \sin \alpha_r = \sum_j c_j \sin \alpha_j \tag{5.12}$$

and

$$c_r = (A^2 + B^2)^{1/2} \tag{5.13}$$

We will, as is conventional, let i represent $\sqrt{-1}$, an "imaginary" number. A complex number, C, is defined as the sum of a "real" number, x, and an "imaginary" number, iy (where y is real),

$$C = x + iy \tag{5.14}$$

The magnitude of C, $|C|$, is defined as the square root of the product of C with its complex conjugate C^*, defined as $x - iy$, so that

$$|C| \equiv [CC^*]^{1/2} = [(x + iy)(x - iy)]^{1/2}$$
$$= [x^2 - i^2y^2]^{1/2} = [x^2 + y^2]^{1/2} \tag{5.15}$$

Comparison of Eqs. (5.14) and (5.15) with Eqs. (5.10) through (5.13) shows that the *vector representations of a wave and the complex number representations are parallel,* provided that we identify the vector itself as $A + iB$, that is, c_r of Eq. (5.13) is identified with $|C|$ of Eq. (5.15), and hence the vector components A and B are identified with x and y, respectively. A and

B [as given by Eqs. (5.11) and (5.12)] represent components along two mutually orthogonal axes (called, with enormous semantic confusion, the "real" and "imaginary" axes, although both are perfectly real). The magnitude of the vector is given, as is usual, by the square root of the sum of the squares of its components along orthogonal axes, $(A^2 + B^2)^{1/2}$, as in Eqs. (5.13) and (5.15).

One advantage of the complex representation follows from the identity

$$e^{i\alpha} \equiv \cos \alpha + i \sin \alpha \qquad (5.16)$$

(which can easily be proved using the power-series expansions for these functions). We then have our expression for the total scattering as

$$A + iB = c_r \cos \alpha_r + ic_r \sin \alpha_r \equiv c_r e^{i\alpha_r} \qquad (5.17)$$

The amplitude of the scattered wave is c_r and the phase angle is α_r, as before, with $\alpha_r = \tan^{-1}(B/A)$, as in Eq. (5.9). It is often said, when this *representation* of the result of the superposition of scattered waves is used, that A is the "real" component and B the "imaginary" component, a terminology that causes considerable uneasiness among those who prefer their science firmly founded and not flirting with the unreal or imaginary. It cannot be stressed too firmly that the *complex representation is merely a convenient way of representing two orthogonal vector components in one equation,* with a notation designed to keep algebraic manipulation of the components in different directions separate from one another. Each component is entirely real, as is evident from Figure 5.1.

Scattering by Atoms Since the electrons are the only components of the atom that scatter X rays significantly and since they are distributed over atomic volumes with dimensions comparable to the wavelengths of X rays used in structure analysis, X rays scattered from one part of an atom interfere with those scattered from another at all angles of scattering greater than 0°. At $2\theta = 0$, all electrons in the atom scatter in phase, and the scattering power of an atom at this angle, expressed relative to the scattering power of a free electron, is just equal to the number of electrons present (the atomic number for neutral atoms). The amplitude of scattering for an atom is known as the *atomic scattering factor,* or *atomic form factor,* and is symbolized as f. The way in which the scattering falls off with increasing scattering angle or, more precisely, increasing values of $(\sin \theta/\lambda)$, is indicated in Figure 5.2. The dependence of the spread of the diffraction pattern on the wavelength is exactly analogous to that mentioned earlier for the diffraction pattern of a slit (Figure 3.1). For most purposes in structure analysis it is adequate to assume that atoms are spherically symmetrical, but with some of the best data now available the small departures from spherical symmetry attributable to covalent bonding are detectable. However, in our discussions we will assume spherical symmetry of atoms. This means that the scattering by an assemblage of atoms—that is, by the structure—can be very closely approximated by summing the contributions to each scattered wave from each atom independently, taking appropriate account of differences in phase. Since the diffraction pattern is the sum of the scattering from all unit cells, and thus represents the average contents of a single one of these unit cells, vibrations or disorder may be considered the equivalent of the smearing out of the electron density, so that there is a greater fall-off at a higher $\sin \theta/\lambda$ (cf. the optical analogy in Figure 3.1: the wider the slit, the narrower the diffraction pattern). Neutrons are scattered by atomic nuclei, rather than by electrons around a nucleus, and, hence, since the nucleus is so small (equivalent to a "point atom"), the scattering for a non-vibrating nucleus is almost independent of scattering angle.

Scattering by a Group of Atoms (a Structure) The X-radiation scattered by one unit cell of a structure in any direction in which there is a diffraction maximum has a particular combination of amplitude and phase, known as the *structure factor* and symbolized by F or $F(hkl)$. It is measured relative to the scattering by a single electron and is the Fourier transform of the scattering density (electrons in the molecule) sampled at the reciprocal lattice point hkl. The intensity of the scattered radiation is proportional to the square of the amplitude, $|F|^2$. In the manner just discussed [see Eq. (5.17)], the structure factor can be represented either exponentially or as an ordinary complex number.

$$F(hkl) = |F(hkl)| e^{i\alpha(hkl)} = A(hkl) + iB(hkl) \tag{5.18}$$

with $|F|$ or $|F(hkl)|$ representing the amplitude of the scattered wave, and α or $\alpha(hkl)$ its phase relative to the origin of the unit cell.* As before (Figure 5.1), $\alpha = \tan^{-1}(B/A)$ and $\mathbf{c}_r = |F| = (A^2 + B^2)^{1/2}$. *The quantities A and B*, representing the components of the wave in its vector representation (see Figure 5.3), *can be calculated, if one knows the structure*, merely by summing the corresponding components of the scattering from each atom separately, which are [by Eqs. (5.6) and (5.7)] the products of the individual atomic scattering-factor amplitudes, f_i, and the cosines and sines of the phase angles of the waves scattered from the individual atoms:

$$A(hkl) = \sum_j f_j \cos \alpha_j \tag{5.19}$$

and

$$B(hkl) = \sum_j f_j \sin \alpha_j. \tag{5.20}$$

*The structure factor **F** may be represented as a vector but it is not conventionally written in boldface so we, as is common, will usually use F for the vector and $|F|$ for its amplitude.

(a)

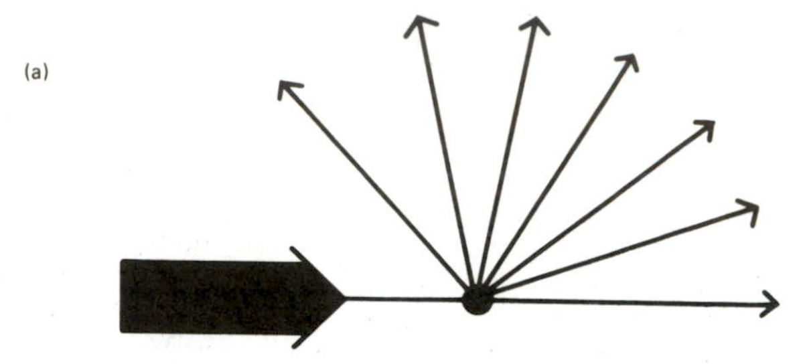

Scattering of light by a particle small relative to the wavelength

Scattering of light by a particle large relative to the wavelength (after Debye)

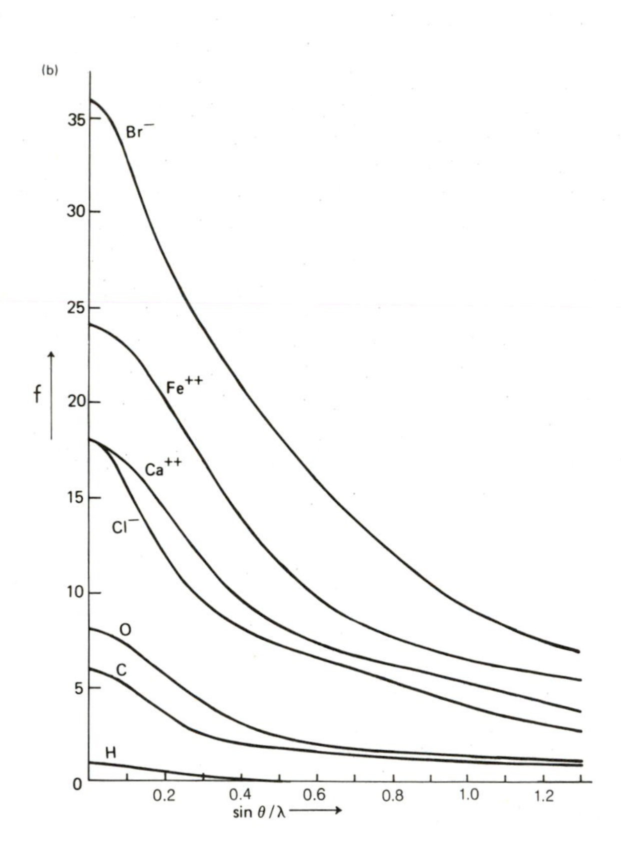

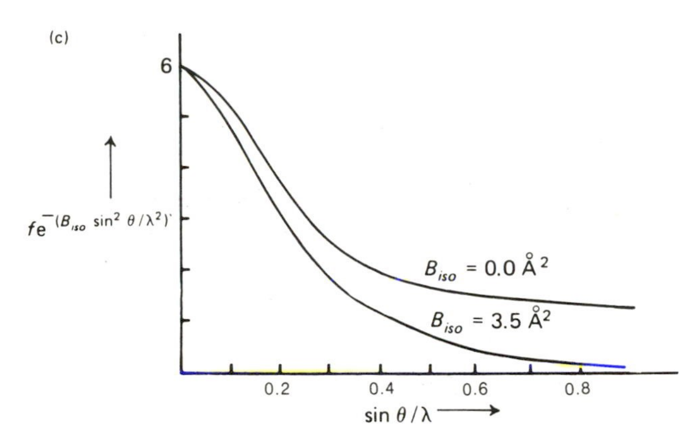

FIGURE 5.2 Atomic Scattering Factors.

(a) When light is scattered by particles that are very small relative to the wavelength of the light, the scattered radiation has approximately the same intensity in all directions. When it is scattered by larger particles the radiation scattered from different regions of the particle will still be in phase in the forward direction, but at higher scattering angles there is interference between radiation scattered from various parts of the particle. The intensity of radiation scattered at higher angles is thus less than for that scattered in the forward direction. This effect is greater the larger the particle relative to the wavelength.

(b) Some atomic scattering factor curves for atoms, given as a function of $\sin\theta/\lambda$ so that they will be independent of wavelength. (Remember that 2θ is the deviation of the diffracted beam from the direct X-ray beam.) The scattering factor for an atom is the ratio of the amplitude of the wave scattered by the atom to that of the wave scattered by a single electron. At $\sin\theta/\lambda = 0$ the value of the scattering factor of a neutral atom is equal to its atomic number, since all electrons then scatter in phase. Note that calcium (Ca^{++}) and chloride (Cl^-) are isoelectronic; that is, they have the same number of extranuclear electrons. The positively charged calcium ion pulls electrons closer to the nucleus than does the chloride ion, which is negatively charged and has a lower atomic number. The resulting "narrower atom" for Ca^{++} will, as illustrated earlier in Figure 3.1, give a broader diffraction pattern. This is shown at high values of $\sin\theta/\lambda$ by higher values of f for Ca^{++} than for Cl^-.

(c) The effects of isotropic vibration on the scattering by a carbon atom. Values are shown for a stationary carbon atom and for one with a typical room temperature isotropic temperature factor, B_{iso}, of 3.5 Å^2, which corresponds to a root-mean-square amplitude of vibration of 0.21 Å. Vibration results in an apparently relatively greater size for the atoms, and consequently a decrease in scattering intensity with increasing scattering angle (for reasons shown in (a) above). If B_{iso} is large, no reflections may be detectable in the outer regions of the diffraction pattern, at high values of 2θ; that is, a narrower diffraction pattern is obtained from the "smeared-out" electron cloud of a vibrating atom (cf. Figure 3.1).

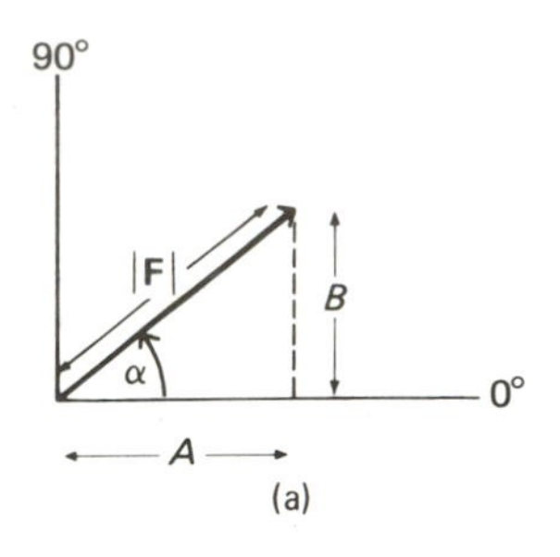

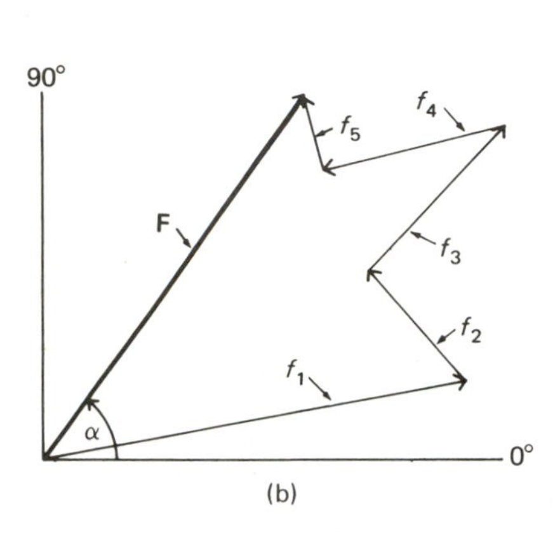

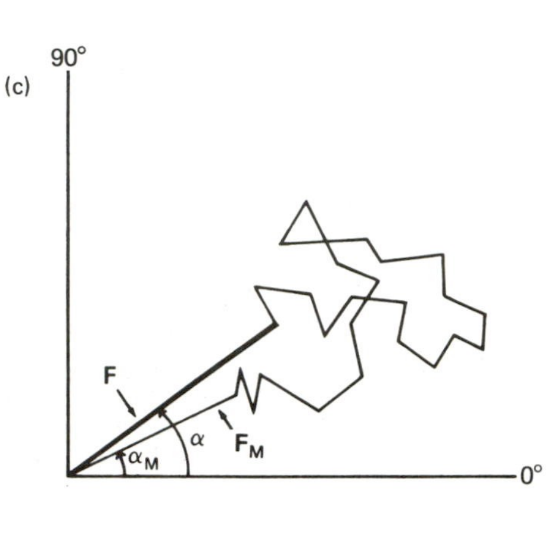

FIGURE 5.3 Vector Representation of Structure Factors.

(a) The relation of $|F|$, A, B, and α is illustrated; $|F|$ is the magnitude (or length) of the vector **F**.

(b) The vector addition of the contribution of each atom to give a resultant **F**. The lengths of the **f**-vectors here are proportional to the f-values for the individual atoms at the value of $\sin\theta/\lambda$ appropriate for the "reflection" in question, including any effect due to thermal motion.

(c) If a "heavy" atom (M) has a much higher atomic number, and hence a much longer vector in a diagram like (b), than any of the other atoms present, then the effect on the vector diagram for **F** is normally as if a short-stepped random walk had been made from the end of $\mathbf{F}_M$. Since the steps or f-values for the lighter atoms are relatively small, there is a reasonable probability that the angle between **F** and $\mathbf{F}_M$ will be small and an even higher probability that α will lie in the same quadrant as α_M. Thus the heavy atom phase, α_M, may be used as a first approximation to the true phase, α.

The phase of the wave scattered in the direction of a reciprocal lattice point (hkl) by an atom situated at the position x,y,z in the unit cell (where x, y, and z are expressed as fractions of the unit cell lengths a, b, and c, respectively) can be shown to be just $2\pi(hx + ky + lz)$ radians, relative to the phase of the wave scattered in the same direction by an atom at the origin (see Appendix 5). Thus Eqs. (5.19) and (5.20) can be rewritten as

$$A(hkl) = \sum_j f_j \cos 2\pi(hx_j + ky_j + lz_j) \tag{5.21}$$

$$B(hkl) = \sum_j f_j \sin 2\pi(hx_j + ky_j + lz_j) \tag{5.22}$$

where the sum is over all atoms in the unit cell, and the value of f_j chosen is that corresponding to the value of $\sin \theta/\lambda$ for the reflection in question, modified to take into account any thermal vibration of the atom. The magnitude of $|F|$ depends only on the *relative* positions of the atoms in the unit cell, except to the extent that f is a function of the scattering angle. The size and shape of the unit cell do not appear as such in the expressions for A and B. In Figure 5.3, F is represented as a vector. A shift in the chosen origin of the unit cell will add a constant to the phase angle of each atom [see Eqs. (5.21) and (5.22)]; that is, it will rotate the diagram in Figure 5.3 relative to the coordinate axes but will leave the length of $|F|$, and hence the values of $|F|^2$ and the intensity, unchanged.

The fall-off in intensity with higher scattering angle (Figure 5.2c) increases as the vibrations of atoms become greater, and these vibrations in turn increase with rising temperature. The reduction in intensity from such vibrational motion can be approximated by an exponential function that has a large effect at high 2θ values (illustrated in Figure 5.2). If the vibration amplitude is sufficiently high, essentially no diffracted intensity will be observed beyond some limiting value of the scattering angle, that is, the "slit" is effectively widened by the vibration and so the "envelope" is narrow (Figure 3.1). If the vibrations are nearly isotropic—that is, do not differ greatly in different directions—the exponential factor can be written as $\exp(-2B_{iso} \sin^2 \theta/\lambda^2)$, with B_{iso} called the *temperature factor.** It is equal to $8\pi^2 \langle u^2 \rangle$, where $\langle u^2 \rangle$ is the mean square amplitude of displacement of the atom from its equilibrium position. It is usually assumed that this displacement is due to vibrations of the atoms, but static displacements that vary randomly from one unit cell to another may simulate vibration.

Since the intensity of any radiation propagated as a wave is proportional to the square of its amplitude,† the intensity of the diffracted beam corresponding to the diffraction maximum *hkl* is proportional to $|F|^2$, except for geometric factors, absorption corrections (if needed), and other minor

*Many crystallographers omit the subscript "*iso*," relying on the context to avoid confusion with the quantity B defined in Eqs. (5.20) and (5.22).

†The type of diffraction discussed in this book is referred to as "kinematical diffraction" and assumes that the incident beam is diffracted and leaves the crystal. In "dynamical diffraction," which is particularly evident in electron diffraction, the diffracted beams interact with the crystal and each other in a way that makes analysis of the diffraction pattern much more complicated.

effects. The geometric correction (a combination of Lorentz and polarization corrections) is a function of 2θ and respectively takes into account both the time the crystal was in a position to diffract that particular beam and the effects of polarization of the X-radiation when it is scattered by electrons. Corrections for the effect of absorption of X-radiation by the crystal are also made in the most careful work; the path lengths through the crystal of many component waves of each diffracted beam must be computed, and the diminution in intensity resulting from absorption is then calculated. If the crystal is strongly absorbing for the radiation used, it is sometimes ground or otherwise shaped until it is spherical because dimensions are then known more precisely and corrections are more uniform and may be calculated more simply.

A general expression for the intensity, taking into account these factors, is

$$I = \mathrm{K}|F|^2(\mathrm{Lp})(\mathrm{Abs}) \tag{5.23}$$

where (Lp) is an abbreviation for the aforementioned geometric factors, (Abs) is the absorption factor, and K is a scale factor. Thus, values of $\mathrm{K}|F|^2$ (where K is some initially unknown scale factor), and hence of $\mathrm{K}^{1/2}|F|$, are immediately available once intensity measurements have been made. Values of (Lp) and (Abs) contain only known quantities and therefore may be computed for each reflection, and it turns out to be easy to get good estimates of K. If the value of I, corrected for (Lp) and (Abs), is called I_{corr}, we can say

$$I_{\mathrm{corr}} = I/(\mathrm{Abs})(\mathrm{Lp}) = \mathrm{K}|F|^2 = \mathrm{K}|F_{\mathrm{novib}}|^2 \exp(-2B_{iso} \sin^2 \theta/\lambda^2) \tag{5.24}$$

where F_{novib} is the value of F for a structure composed of nonvibrating atoms. The application of the (Lp) correction involves *no knowledge of the structure;* an estimation of (Abs) can be made from a knowledge of the shape, orientation and composition of the crystal. The value of F so derived contains information on the temperature factor in a manner illustrated in Figure 5.2c. Thus we could consider that $F = |F_{\mathrm{novib}}|\exp(-B_{iso} \sin^2 \theta/\lambda^2)$. It is possible to derive B_{iso} and K from the experimental data—by assuming that, to a first approximation, the average intensity of diffraction at a certain value of 2θ depends only on the atoms present in the cell, not on their positions—that is, that the structure is random. By comparison of the averages of the observed intensities in ranges of $\sin^2 \theta/\lambda^2$ with the theoretical values for a unit cell with the same contents, approximate values for K and B_{iso} can be found (a Wilson plot). The resulting scale factor K can then be used for preparation of a full list of values of $|F|$ on an approximately absolute scale (relative to the scattering by one electron) for all reflections measured.

The reader should note that the intensity, I, is a simple function of the structure amplitude $|F|$; however, an inspection of Eqs. (5.21) and (5.22) shows that *each value of $|F|$, and hence of the intensity, I, of the dif-*

fracted beams contains, with few exceptions, *a contribution from every atom in the unit cell.* The unraveling of these contributions makes the structure solution complicated.

SUMMARY

When X rays are diffracted by a crystal, the intensity of scattering at any angle is the result of the combination of the waves scattered from different atoms to give various degrees of constructive and destructive interference. A structure determination involves a matching of the observed intensity pattern to that calculated from a postulated model, and it is thus imperative to understand how this intensity pattern can be calculated for any desired model. The combination of the scattered waves can be represented in various ways:

1. The waves may be drawn graphically and the displacements (ordinates) at a given position (abscissae) summed.
2. The waves may be represented algebraically as

$$x_j = c_j \cos(\phi + \alpha_j) \qquad \text{(for the } j\text{th wave)}$$

and the displacements, x_j, summed to give a resultant wave

$$x_r = c_r \cos(\phi + \alpha_r)$$

with

$$c_r \cos \alpha_r = c_1 \cos \alpha_1 + c_2 \cos \alpha_2 + \cdots$$

and

$$c_r \sin \alpha_r = c_1 \sin \alpha_1 + c_2 \sin \alpha_2 + \cdots$$

3. The waves may be expressed as two-dimensional vectors in an orthogonal coordinate system, amplitude c_j, with the phase angle α_j measured in a counterclockwise direction from the horizontal axis. This is the equivalent of representing one complete wavelength as 360°, so that the periodicity of the wave is expressed. The phase relative to some origin is given as a fraction of a revolution. The vectors may then be summed by summing their components.
4. The waves may be represented in complex notation

$$A_j + iB_j = c_j e^{i\alpha_j}$$

which is merely a convenient way of representing two orthogonal vector components (at 0° and 90°) in one equation. By convention A is the component at 0° and B the component at 90°.

X rays are scattered by electrons. The extent of scattering depends on the atomic number of the atom and the angle of scattering, 2θ, and is represented by an atomic scattering factor f. For a group of atoms the amplitude (relative to the scattering by a single electron) and the phase of the X rays scattered by one unit cell are represented by the structure factor $F(hkl) = A(hkl) + iB(hkl)$ for each reflection. For a known structure with atoms j at positions x_j, y_j, z_j this may be calculated from

$$A(hkl) = \sum_j f_j \cos 2\pi(hx_j + ky_j + lz_j)$$

$$B(hkl) = \sum_j f_j \sin 2\pi(hx_j + ky_j + lz_j)$$

where the summation is over all atoms in the unit cell. The phase angle α is $\tan^{-1}(B/A)$ and the structure factor amplitude $|F|$ is $(A^2 + B^2)^{1/2}$. The value of F may be reduced as a result of thermal vibration so that if F_{novib} is the value for a structure containing stationary atoms, the experimental values will correspond to $F = F_{\mathrm{novib}} \exp(-B_{iso} \sin^2 \theta/\lambda^2)$, where B_{iso}, the temperature factor, is a measure of the amount of vibration ($B_{iso} = 8\pi^2 \langle u^2 \rangle$, where $\langle u^2 \rangle$ is the mean square amplitude of vibration).

6

The Phase Problem

In order to obtain an image of the scattering matter in three dimensions (the electron distribution), which is the aim of these studies, one must perform a three-dimensional Fourier summation ("synthesis"). Fourier series* are used because they can be applied to a description of regularly periodic functions, and crystals contain periodic distributions of scattering matter. The number of electrons per unit volume or electron density at any point x,y,z, represented by $\rho\,(xyz)$, is then given by the following expression:

$$\rho(xyz) = \frac{1}{V_c} \sum\sum\sum_{\text{all } hkl} F(hkl) \exp[-2\pi i(hx + ky + lz)] \qquad (6.1)$$

Here V_c is the volume of the unit cell and $F(hkl)$ is the structure factor for the particular set of indices h, k, and l. The triple summation is over all values of the indices h, k, and l. This summation represents a mathematical analogy to the process effected physically in the microscope, shown in Figure 6.1. As mentioned in Chapter 5, the *amplitude* of F is easily derived [Eq. (5.23)] from the intensity of the diffracted beam. The relative *phase*, however, is not.

If we put

$$\phi = 2\pi(hx + ky + lz) \qquad (6.2)$$

*Fourier's theorem states that a continuous, single-valued, periodic function can be represented by the summation of cosine and sine terms.

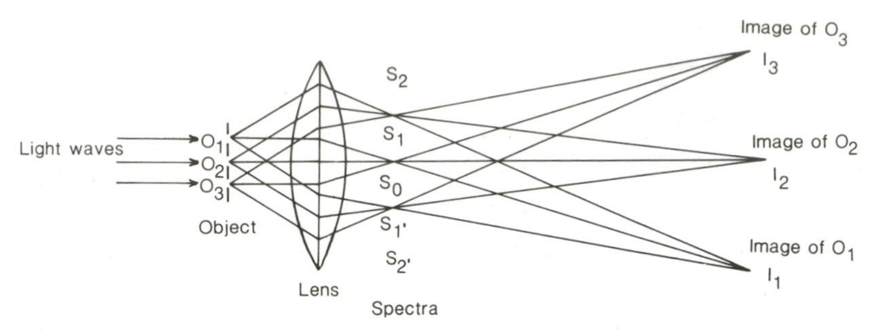

FIGURE 6.1 Formation of an Image by a Lens.

Formation of the image of a diffraction grating by the interference of the orders of diffraction that are brought together by the lens. The orders of diffraction are represented by numbers. They are brought to foci in the focal plane of the lens at points S. (After W. H. Bragg.)

The formation of an image of an aperiodic object is similar. With a grating, as illustrated here, the Fourier transform of the scattering object is imaged in the focal plane (the points S); *note that each point S has rays coverging at it that originate from every point of the object.* With a nonperiodic scattering object, the focal plane also contains the Fourier transform of the scattering object, with rays originating from *each* point of the object contributing to *every* point of the transform. In the image plane, however, all rays emanating from a *given* point of the scattering object are brought back together at a *single* point.

we get, from Eqs. (5.16) and (5.18),*

$$\begin{aligned} Fe^{-i\phi} &= (A + iB)(\cos\phi - i\sin\phi) \\ &= A\cos\phi + B\sin\phi - i(A\sin\phi - B\cos\phi) \end{aligned} \tag{6.3}$$

Because the summation in Eq. (6.1) is over all values of the indices, it includes, for every reflection hkl, the corresponding reflection with all indices having opposite signs, $-h$, $-k$, $-l$ (also denoted $\bar{h}$, $\bar{k}$, $\bar{l}$). The terms for these two reflections, as expanded in Eq. (6.3), can conveniently be paired in the summation. The *magnitude* of each term (A, B, $\cos\phi$, and $\sin\phi$) is normally the same† for a reflection with indices hkl as for that with indices $-h$ $-k$ $-l$. The *sign* of the term will change if the term involves sines [since $\sin(-x) = -\sin x$] but will remain unchanged if it involves cosine functions [since $\cos(-x) = \cos x$]. Both A [the sum of cosines, by Eq. (5.19)] and $\cos\phi$ have the same sign for hkl as for $-h$, $-k$, $-l$, whereas B [the sum of sines,

*It should be noted that the exponential terms in the expressions for F and ρ are opposite in sign. $F = \Sigma f e^{i\phi}$ and $\rho = (1/V)\Sigma F e^{-i\phi}$, because these are Fourier transforms of each other. The intensity at a particular point of the diffraction pattern of an object (a set of relative $|F|^2$ values) is proportional to the square of the Fourier transform of the object at that point (with the distribution of matter in the object described by ρ).

†This implies that "Friedel's Law" is obeyed; deviations from this are considered in Chapter 10.

by Eq. (5.20)] and sin ϕ have opposite signs for this pair of reflections. Therefore, when Eq. (6.3) is substituted in Eq. (6.1) and the summation is made, the $i(A \sin \phi - B \cos \phi)$ terms cancel for each pair of reflections hkl and $\overline{hkl}$ and thus vanish completely. The remaining terms, $A \cos \phi$ and $B \sin \phi$, need be summed over only half the reflections, omitting all those with any one index (for example, h) negative; a factor of 2 is introduced because half the reflections are omitted. Thus we may write, by Eqs. (6.1) and (6.3),

$$\rho(xyz) = \frac{|F(000)|}{V_c} + \frac{2}{V_c} \sum\sum\sum_{\substack{h \geq 0,\ \text{all } k,\ l \\ \text{excluding } F(000)}}^{\infty} (A \cos \phi + B \sin \phi) \quad (6.4)$$

Since $A = |F| \cos \alpha$ and $B = |F| \sin \alpha$ [by Eq. (5.17)], the above expression for the electron density may be written*

$$\rho(xyz) = \frac{|F(000)|}{V_c} + \frac{2}{V_c} \sum\sum\sum_{\substack{h \geq 0,\ \text{all } k,\ l \\ \text{excluding } F(000)}}^{\infty} |F| \cos(\phi - \alpha) \quad (6.5)$$

Therefore, *if we knew* $|F|$ *and* α (for each h,k,l) *we could compute* ρ *for* all values of x, y, and z and plot the resulting values of $\rho(xyz)$ to give a three-dimensional electron density map. Then, assuming atoms to be at the centers of peaks, we would know the structure.

However, *we can normally obtain only the structure factor amplitudes* $|F|$ *and not the phase angles,* α,† directly from the experimental measurements. We *must derive* α, either from values of A and B that are computed from structures we have deduced in various ways ("trial structures"), or by purely analytical methods. The problem of getting estimates of the phase angles so that an image of the scattering matter can be calculated is called the *phase problem* and is the central one in X-ray crystallography. Chapters 8 and 9 are devoted to methods used to solve the phase problem, either by deriving a trial structure and so calculating approximate values of α for each reflection, or by trying to find values of α directly.

Once the approximate positions and identities of all the atoms in the asymmetric unit are known (that is, once the crystal structure is known), the amplitudes and phases of the structure factors can readily be calculated. These calculated amplitudes, $|F_c|$, may be compared with the observed amplitudes, $|F_o|$. If the structural model is a correct one and the experimentally observed data are reasonably precise, the agreement should be

*A schematic example of the calculation of the function described in Eq. (6.5) is shown in Figure 6.2.

†Post has shown (*Acta Crystallographica* **A35** (1979) 17) that under certain conditions when two-beam diffraction occurs some phase information may be found from experimental measurements.

good. Of course, the calculated phases cannot be compared with the observed phases because normally there are *no observed phases.*

At the start of a structure determination one does not know the positions of all the atoms in the structure (for if one did the structure would probably not need to be investigated), but one can often deduce in various ways (see Chapters 8 and 9) an approximation to the correct structure. The amplitudes and phases of the structure factors calculated from this approximation to the structure will be, to some degree, incorrect, the errors increasing with the imperfections of the model as compared with the actual structure. However, these imperfect calculated phases are at least a crude approximation to the correct ones, and the observed amplitudes are presumably correct (except for random or occasionally unknown systematic errors). Hence one can calculate an *approximation to the true electron density* by a three-dimensional Fourier summation of the *observed structure factor amplitudes,* $|F_o|$, with the *calculated phases.* It has been found that the *general features of an electron density map depend much more on the phase angles than on the structure factor amplitudes.* Thus a map calculated with only approximately correct phases will be an imperfect representation of the structure, primarily because of errors in the phases (and to a much lesser extent because of experimental errors in $|F_o|$). However, it is biased toward the correct structure because the observed structure amplitudes, $|F_o|$, were used. By comparison with a similar synthesis using the $|F_c|$, or even more simply by computing the difference of these two (a "difference synthesis"), one can deduce the changes in the model needed to give better agreement with observation. The positions of some hitherto unrecognized atoms may be indicated, and shifts in the positions of some atoms already included will normally be suggested as well.

We have emphasized the analogy between the action of a lens in collecting and refocussing radiation to give an image of the scattering matter, and the process of Fourier summation, a mathematical technique for forming an image by use of information about the amplitudes and relative phases of the scattered waves. Fourier summation techniques can be applied even when the waves cannot be refocussed, as in the X-ray experiment. With a lens the waves are (ideally) brought together with the same relative phases that they had when they left the object; in the X-ray diffraction experiment these relative phases are usually not measurable, although if they can be found in some way, then it is possible to compute an electron density map as shown in Figure 6.2.

The reader may wonder at this stage how the addition of electromagnetic radiation of fixed wavelength (exemplified in Figure 6.1) can be reconciled with the summation of terms with apparently *different* wavelengths, illustrated in Figure 6.2. These waves that are summed in this way to give the electron density map are, for convenience, referred to as "density waves";

they are individual terms in Eq. (6.5) for each reflection hkl [that is, each $|F|\cos(2\pi(hx + ky + lz) - \alpha)$ as a function of x, y and z is a "density wave"]. In effect, Eq. (6.5) could be rewritten to say that the electron density, $\rho\ (x,y,z)$ at a point in space, x,y,z, is equal to the sum of these "density waves." Thus each reflection (diffracted beam) with its relative phase can be considered to correspond to a "density wave" in the crystal, with an amplitude that can be derived from the intensity of the reflection; the superposition of these "density waves" once their relative phases are known produces the electron density map for the crystal.

The determination of the relative phases of these "density waves" is the subject of much of the rest of this book. But what is the wavelength of a "density wave" and how is it related to the order (h,k,l) of the diffracted beam? One solution of this problem comes from a close examination of Eq. (6.5) that shows that $|F|$ is modified by a cosine function $(2\pi(hx + ky + lz) - \alpha)$; thus it becomes a periodic function of h, k, and l. In the simple case (Figures 6.2a and 6.2b) where k and l are both zero, cos $2\pi(hx)$ is at a maximum value when $x = 1/h$—that is, the cosine term has an apparent wavelength of a/h (where a is the unit cell length in the x direction and x is expressed as a fraction of this). Another way of considering this apparent problem comes from the realization (see Chapter 3, especially Figure 3.3, and Appendixes 3 and 5) that the condition for obtaining diffraction maxima is that there be an integral number of wavelengths in the path difference between diffracted beams from adjacent repeat units. The reflection indices h, k, and l give the numbers of these wavelengths when *measured along the corresponding reciprocal axis directions.* The key fact is that when a wave, or any periodic arrangement is *projected* on a line that is at an angle $(90° - \alpha)$ to the direction of propagation of the wave, the wavelength is enlarged by the factor $1/\sin \alpha$ (Figure 6.2b). The fact that this enlargement occurs is familiar to anyone who has ever looked at the image of a periodic object projected onto a surface at a high angle to the object.

An X ray coming in at a Bragg angle θ to a set of lattice planes (hkl) makes an angle $(90° - \theta)$ with the normal to these planes (the reciprocal lattice vector $ha^* + kb^* + lc^*$, whose magnitude is $1/d_{hkl}$; see Figure 6.2b, 3.5, 3.9 and Appendix 2). The exiting diffracted beam also makes an angle $(90° - \theta)$ with this reciprocal lattice vector. When beams diffracted from adjacent planes have a total path difference of λ (and thus are in phase), the path difference in the direction of the direct (incident) beam is only $\lambda/2$. It is the projection of this path difference onto the reciprocal lattice vector that must match the interplanar spacing if the Bragg condition is to be fulfilled and thus the effective wavelenth is $\lambda/(2 \sin \theta) = d_{hkl}$ (Figure 3.9b). This is the wavelength of the "density wave." If it were possible to have $\theta = 90°$, $\sin \theta = 1.0$, then the "density wavelength" would be $\lambda/2$ (although as seen in Figure 4.2, this represents the situation where the detector coincides with

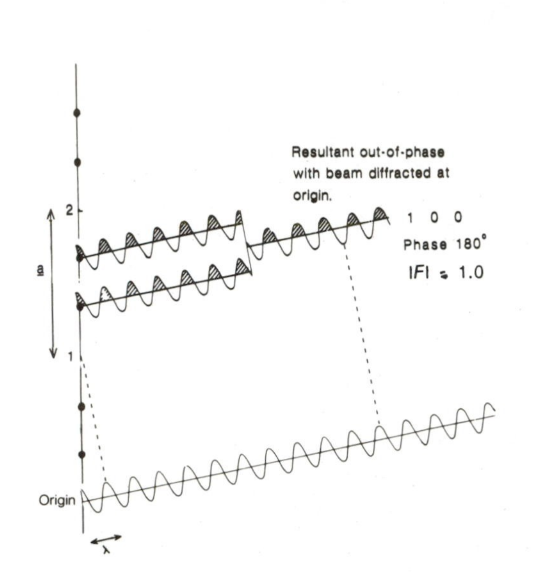
Resultant out-of-phase
with beam diffracted at
origin.
1 0 0
Phase 180°
|F| = 1.0
2
1
a
Origin
λ

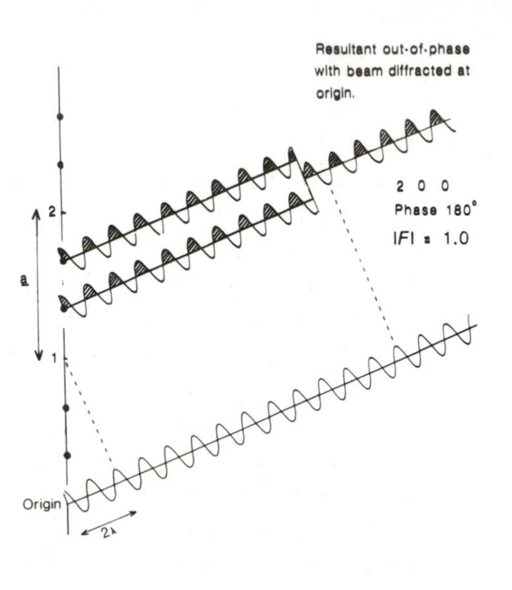
Resultant out-of-phase
with beam diffracted at
origin.
2 0 0
Phase 180°
|F| = 1.0
2
1
a
Origin
2λ

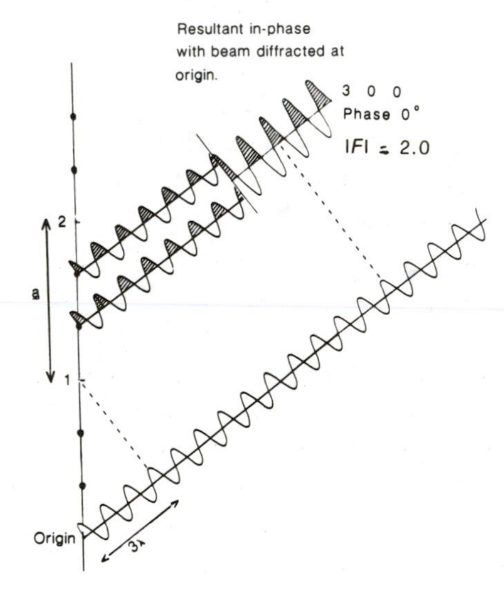
Resultant in-phase
with beam diffracted at
origin.
3 0 0
Phase 0°
|F| = 2.0
2
1
a
Origin
3λ

(a)

(b)

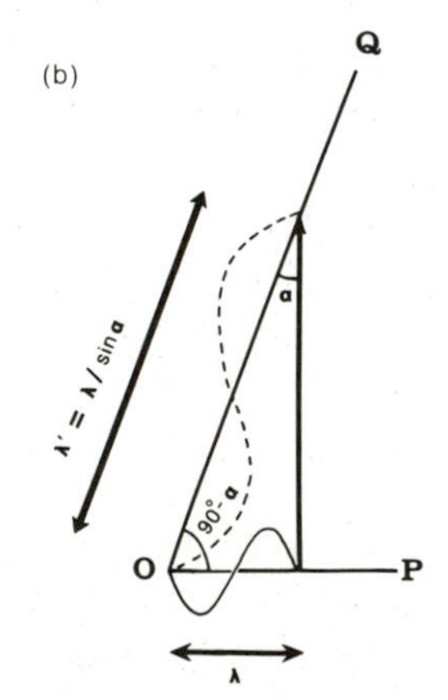
Q
λ' = λ / sin α
α
90°-α
O
P
λ

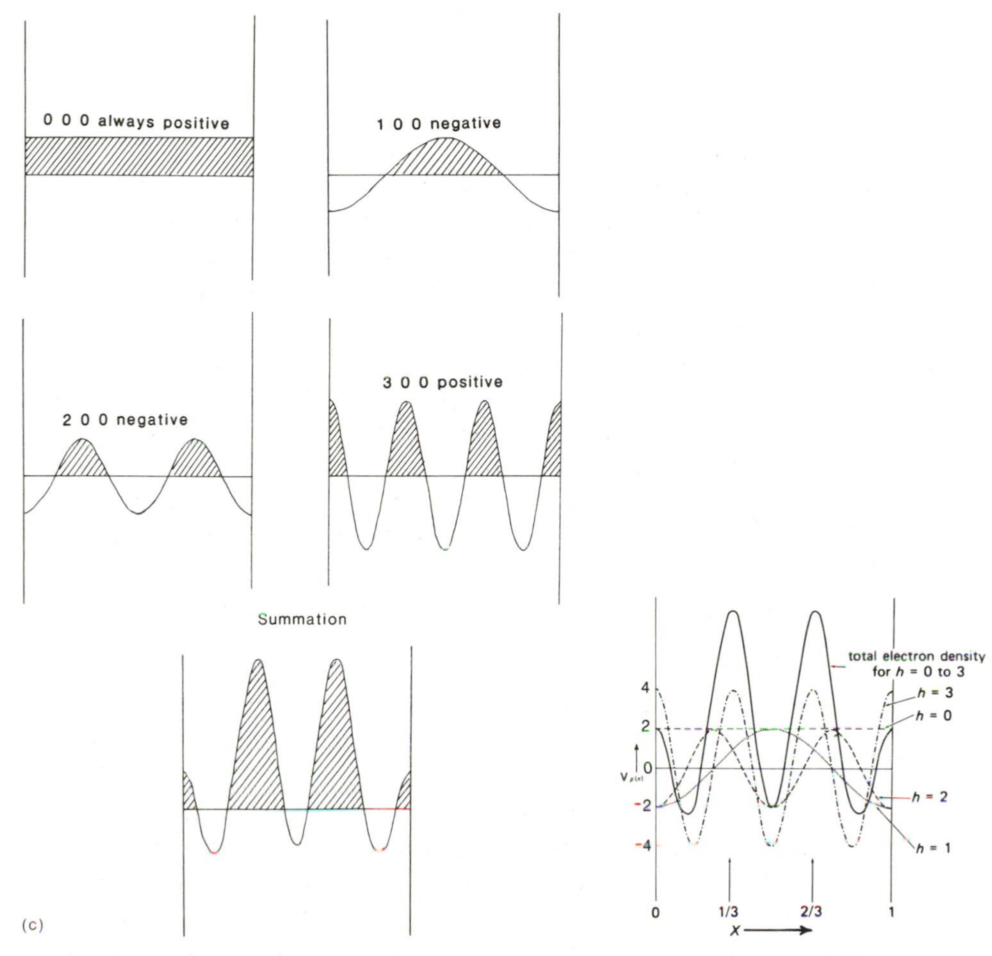

FIGURE 6.2 Formation of an Electron Density Map from Diffraction Information.

(a) The summation of scattered waves to give waves with $V_c\,\rho(xyz)$ the $|F|$ values used in this example.

(b) Apparent wavelengths of density waves. A wave, wavelength λ, moves along a direction OP. Its projection (broken line) on direction OQ, which is at an angle $(90° - \alpha)$ to OP, has a wavelength greater than λ by the factor $1/\sin\alpha$.

(c) The summation of Fourier density waves to give an electron density map with peaks at $x = \pm\frac{1}{3}$. At any point, x, y, z in the unit cell, volume V_c, the electron density, $\rho(xyz)$ may be calculated from Eq. (6.5):

$$V_{c\rho}(xyz) = F(000) + 2 \sum_{\substack{h \geq 0,\ \text{all } k,\ l \\ \text{excluding } F(000)}}\sum\sum |F| \cos\,[2\pi\,(hx + ky + lz) - \alpha]$$

The following data have been used for this one-dimensional example:

h	-3	-2	-1	0	1	2	3
$\|F\|$	2	1	1	2	1	1	2
$\alpha(°)$	0	180	180	0	180	180	0
$\cos\,[2\pi(hx - \alpha)]$	$+\cos 6\pi x$	$-\cos 4\pi x$	$-\cos 2\pi x$	$+1$	$-\cos 2\pi x$	$-\cos 4\pi x$	$+\cos 6\pi x$

$\therefore\ \rho(x) = (4\cos 6\pi x - 2\cos 4\pi x - 2\cos 2\pi x + 2)/V_c$

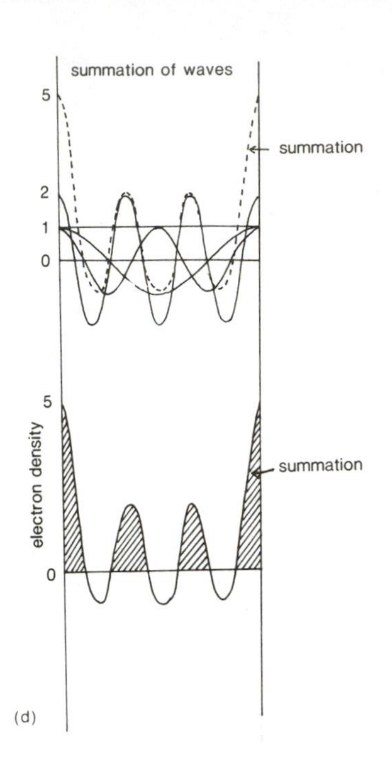
summation of waves
5
summation
2
1
0
5
electron density
summation
0
(d)

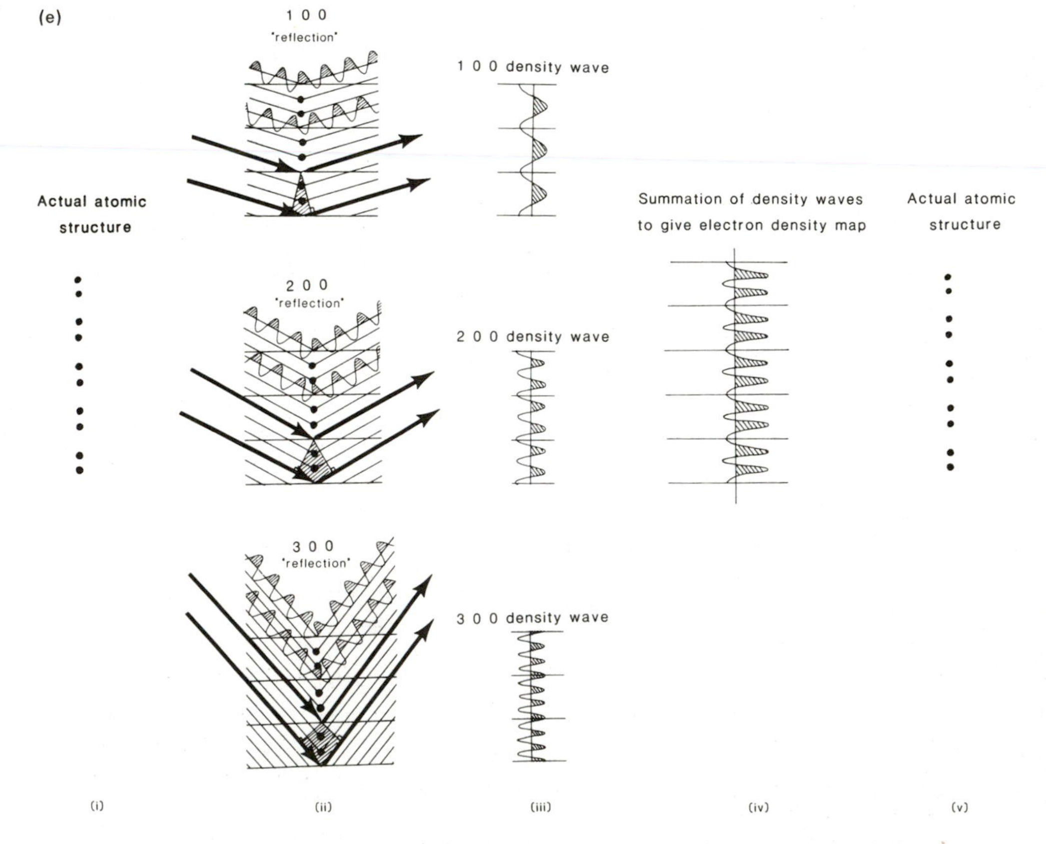
(e)
Actual atomic structure
1 0 0
'reflection'
1 0 0 density wave
2 0 0
'reflection'
2 0 0 density wave
3 0 0
'reflection'
3 0 0 density wave
Summation of density waves to give electron density map
Actual atomic structure
(i)
(ii)
(iii)
(iv)
(v)

the incident beam collimator). At other angles this wavelength may be significantly enlarged (e.g., by a factor of 20 for sin θ = 0.05 for a "low resolution" reflection). "High resolution" implies a high value of sin θ and thus a small value for the effective wavelength of the "density wave;" as we shall see in Chapter 9 (especially Figure 9.10), high-angle reflections, leading to short effective wavelengths of "density waves," are needed to provide high-resolution images of molecules.

Thus, in summary, the wavelengths of the "density waves" are $d_{hkl} = \lambda/(2 \sin \theta)$, the amplitudes are $|F(hkl)|$ and the phase angle is α_{hkl}. For example, the wavelength of the 1 0 0 "density wave" is the repeat distance a (= d_{100}) (see Figure 6.2e), the wavelength of the 2 0 0 "density wave" is $a/2$ because the second order of diffraction occurs at a sin θ value twice that of the first order, and so forth. These are the density waves that are summed to give the electron density map shown in Figure 6.2c.

But we have not yet explained the way in which we derive the relative phases of the "density waves." We attempt to show, in Figure 6.2a, how the X rays scattered from different atoms are summed to give the resultant X ray beams of various amplitudes (and hence intensities). We illustrate the

---

The summation is depicted graphically to illustrate the meaning of the equation above. When h = 0 the function does not depend on x and so is a straight line. The phase angle of this is necessarily 0°. The function for h = 1 is $-\cos 2\pi x$, the negative sign resulting from the phase angle of 180°, and so forth. These functions are summed for each value of x to give the result shown by the heavy solid line. It has peaks at $x = \pm\frac{1}{3}$. Clearly, unless the relative phases were known, it would not be possible to sum the waves. This kind of calculation must be made, with thousands of reflections, at each of many thousands of points to give a complete electron density map in three dimensions. High-speed computers are needed to effect such summations efficiently. For a three-dimensional electron density map it is not possible to plot heights of peaks (because we have no fourth spatial dimension) and therefore contours of equal electron density (or height) are drawn on sections through the three-dimensional map. Such a map is shown schematically in Figure 1.1b; many are given in Chapter 9 and later chapters. The resulting map resembles a topographical map or, in some ways, a weather map. Atomic centers appear at the centers of areas of high electron density, which look like circular mountains on a topographical map. Note that the larger values of F dominate the Fourier summation.

(d) The summation of Fourier density waves, but with wrong relative phases, to give an incorrect structure. In this example, all phases are positive.

(e) Summary of the diffraction experiment showing (i) the atomic structure (one-dimensional in this case), (ii) diffraction of X rays by the crystal structure, (iii) density waves, (iv) summation of the density waves to give the electron density map, and the result is an image of the actual structure, (v).

atoms, one at $x = \frac{1}{3}$ and the other at $x = \frac{2}{3}$. The important fact for the reader to understand is that each resultant wave should be traced back and its phase compared with that of an imaginary wave being scattered at the origin of the repeat unit (with a phase angle of 0°). Recall that, for the third order reflection, the path difference between waves scattered one repeat unit (a) apart (that is, by equivalent atoms in adjacent unit cells) is three wavelengths. Compared to an imaginary atom at the origin, the atom at $x = \frac{1}{3}$ scatters with a path difference of one wavelength and the atom at $x = \frac{2}{3}$ scatters with a path difference of two wavelengths. Thus both scatter in phase with the wave scattered at $x = 0$. However, for the second order, also illustrated in Figure 6.2(a), the atom at $x = \frac{1}{3}$ scatters X rays with a path difference of 0.67 wavelengths from that scattered by an imaginary atom at the origin, and the atom at $x = \frac{2}{3}$ scatters with a path difference of 1.33 wavelengths from the wave scattered at the origin. The resultant wave is then $(0.33 + 0.67)/2 = 0.50$ wavelengths out of phase with the wave scattered by the imaginary atom at the origin. Thus, in summing "density waves" the 3 0 0 wave has a relative phase angle of 0° and the 2 0 0 "density wave" has a relative phase angle of 180°. This is illustrated in Figure 6.2a, 6.2b, and 6.2c.

The "density waves," derived by arguments such as these, are summed, as shown in Figure 6.2 (c), (d) and (e) to give the electron density of the structure and the peaks in such a map correspond to the centers of atoms. The aim of an X-ray diffraction study is to obtain such a map, using measured amplitudes for each of the diffracted waves and phase angles derived by one of a variety of methods to be described in the following chapters. It is more important to have correct phase angles than correct structure amplitudes. The importance of the phases in determining a structure is illustrated in Figure 6.3. An electron density map with incorrect $|F|$ values and correct phases much more nearly approximates the correct structure than does an electron density map with incorrect phases and correct $|F|$ values.

Most of the methods commonly used to determine trial structures (and hence their phase angles) are discussed in Chapters 8 and 9. An essential test of the correctness of the model (trial structure) that has been chosen is the fit of the observed structure factor amplitudes to those calculated for the model [see Eqs. (5.18), (5.21), and (5.22)]. The calculations involved, which must eventually be made for all reflections, are impracticable unless a high-speed computer is available. They generally, however, give a good measure of whether or not the structure is correct. One measure of the correctness of a structure is the so-called discrepancy index or conventional residual, R, defined as

$$R = \frac{\Sigma|(|F_o| - |F_c|)|}{\Sigma(|F_o|)} \tag{6.6}$$

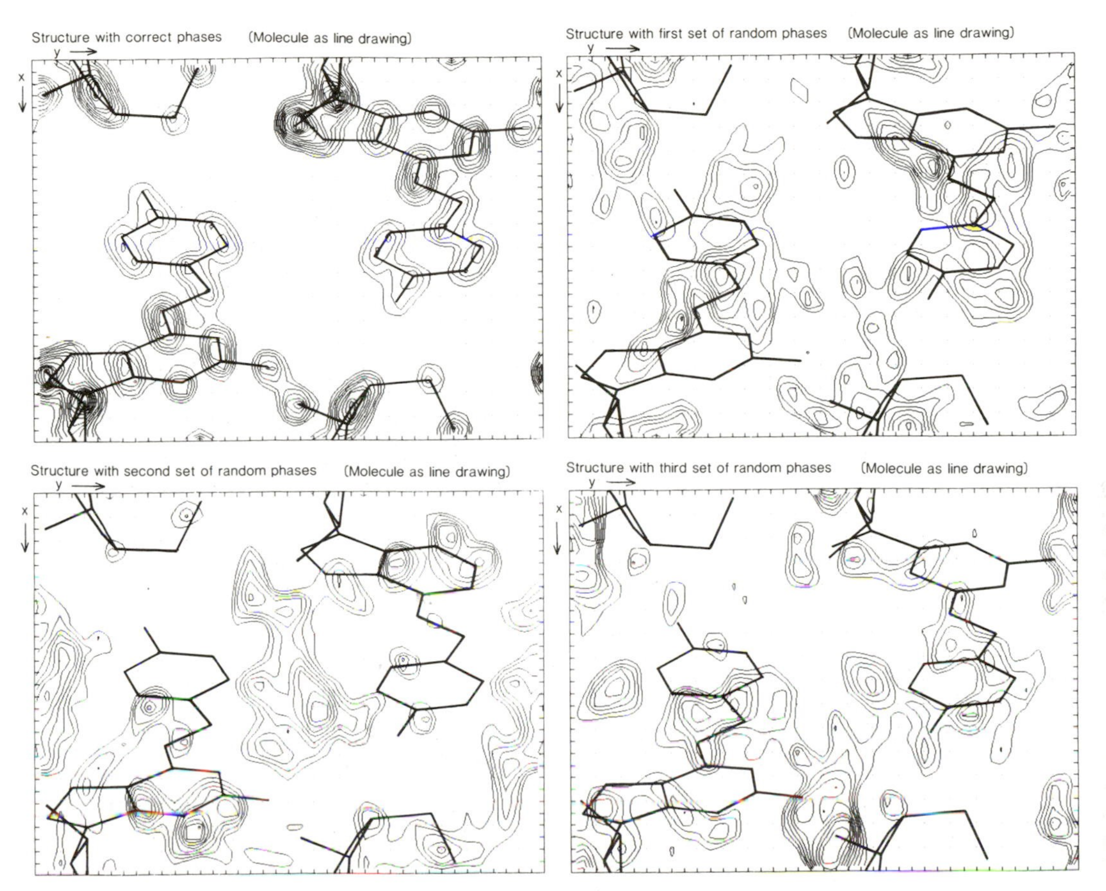

FIGURE 6.3 Comparison of Electron Density Maps When the Phases Are Correct and When They Are Random.

In the computation of all maps shown here the same $|F|$ values but different phases were used. Phases for the three "random phase" maps were found by a computer program for random number generation. Since this is a noncentrosymmetric structure, the phase for each reflection could have any value between 0° and 360°. In each case, the molecular skeleton is shown in the correct position.

It is a measure of how closely the observed structure factor amplitudes are matched by the values calculated for a proposed trial structure. At present, R values in the range of 0.02 to 0.06 (2 percent to 6 percent) are being quoted for the most reliably determined structures. An R value of 0.67 corresponds to a random centrosymmetric structure; that is, with proper scaling a randomly incorrect structure with a center of symmetry would give an R value of about 0.67. A refinable trial structure may have an R value between 0.25 and 0.35, or even somewhat higher. If one atom of high atomic number is present, the initial trial value of R may be much lower because the position of this atom can usually be determined reasonably well even at an early stage, and a heavy atom normally dominates the scattering, as illustrated in Figure 5.3c. If the trial structure is a reasonable approximation to the correct structure, the R value goes down appreciably as refinement proceeds.

The discrepancy index, R, is, however, only one measure of the precision (but not necessarily the accuracy) of the derived structure. It denotes how well the calculated model fits the observed data. Many complications can cause errors in the observed or calculated structure factors or both—for example, absorption of the X-radiation by the crystal, or atomic scattering factors and temperature factors that do not adequately describe the experimental situation. The fit of the calculated structure factors to the observed ones may then be good, but if the observations are systematically in error, the *accuracy* of the derived structure may be low, despite an apparently high *precision*. Hence care must be taken in interpreting R values. In general, the lower the R value the better the structure determination, but if one or more very heavy atoms are present, they may dominate the structure factor calculation to such an extent that the contributions from light atoms may not have noticeable effects on R, especially if the structure has not been refined extensively. The positions of the light atoms may then be significantly in error. Also the resolution of the data (i.e., the maximum value of $\sin\theta/\lambda$) must be taken into account in assessing the meaning of an R value. A few grossly incorrect trial structures have been refined to R values as low as 0.10 and yet have remained seriously in error. Fortunately this situation is not common.

SUMMARY

The electron density at a point x,y,z in a unit cell of volume V_c is

$$\rho(xyz) = \frac{1}{V_c} \sum\sum\sum_{\text{all } h,k,l} |F| \cos[2\pi(hx + ky + lz) - \alpha]$$

Therefore, if we knew $|F|$ and α (for each h,k,l) we could compute ρ for all values of x, y, and z and plot the values obtained to give a three-dimensional electron density map. Then, assuming atomic nuclei to be at the centers of peaks, we would know the entire structure. However, we can usually obtain only the structure factor amplitudes $|F|$ and not the phase angles, α, directly from experimental measurements. This is the phase problem. We must usually derive values of α, either from values of A and B computed from suitable "trial" structures or by the use of purely analytical methods. Approximations to electron density maps can be calculated with experimentally observed values of $|F|$ and calculated values of α. If the trial structure is not too grossly in error, the map will be a reasonable representation of the correct electron density map, and the structure can be refined to give a better fit of observed and calculated $|F|$ values.

The discrepancy index, R, is one measure of the correctness of a structure determination. However, it is at best a measure of the precision of the fit of the model used to the experimental data obtained, not a measure of the accuracy. Some structures with low R values have been shown to be incorrect.

7

Symmetry and Space Groups

A certain degree of symmetry is apparent in much of the natural world as well as in many of man's creations in art, architecture, and technology. Symmetry is perhaps the most fundamental property of the crystalline state. This chapter introduces some of the fundamental concepts of symmetry—symmetry operations, symmetry elements, and the combinations of these characteristics of finite objects (point symmetry) and infinite objects (space symmetry)—as well as the way these concepts are applied in the study of crystals.

An object is said to be symmetrical if after some movement, real or imagined, it is or would be indistinguishable (in appearance and other discernible properties) from the way it was initially. The movement, which might be, for example, a rotation about some fixed axis or a mirrorlike reflection through some plane or a translation of the entire object in a given direction, is called a symmetry operation. The geometrical entity with respect to which the symmetry operation is performed—an axis or a plane in the examples cited—is called a symmetry element. Various possible symmetry operations and symmetry elements of crystals are considered below.

It is possible not only to determine the crystal system characteristic of a given crystalline specimen by analysis of the intensities of the reflections in the diffraction pattern of the crystal, but also to learn much more about its symmetry, including the Bravais lattice and the probable space group. As indicated in Chapter 2, the 230 space groups represent the distinct ways of arranging identical objects on one of the 14 Bravais lattices by the use of certain symmetry operations to be described below. The determination of the space group of a crystal is important because it may reveal some symmetry within the contents of the unit cell being considered, and this symmetry may imply that some particular molecule has several parts that are equivalent to each other by symmetry. Space group determination also vastly simplifies the analysis of the diffraction pattern because different regions of the pattern may then be known to be identical. It also greatly simplifies the required calculations because only the contents of the asymmetric portion of the unit cell (the asymmetric unit) need to be considered in detail.

Scrutiny of photographs of crystal diffraction patterns reveals that there are often systematically related positions where diffraction maxima might occur but where, in fact, the observed intensity is zero. For example, if molecules pack in a crystal so that there is a two-fold screw axis parallel to the a-axis (which means that the molecule is moved a distance $a/2$ and then rotated 180° about the screw axis, so that for every atom at position x there is another at $\frac{1}{2} + x$), then, as far as $h\ 0\ 0$ reflections are concerned, the unit cell size has been halved and the reciprocal lattice spacing has doubled. Reflections will then only be observed for *even* values of h. This situation is made evident by summing, in Eqs. (5.21) and (5.22) for atoms at x and $\frac{1}{2} + x$. When k and l are zero, A and B are zero if h is odd.

Most, but not all, combinations of symmetry elements give rise to systematic relationships among the indices of some of the "absent reflections." For example, the only $h\ k\ 0$ reflections with appreciable intensity may be those for which $(h + k)$ is even. Such systematic relationships imply certain symmetry relations in the packing in the structure. Before continuing with an account of methods of deriving trial structures, we present a short account of symmetry and particularly its relation to the possible ways of packing molecules or ions in a crystal.

SYMMETRY GROUPS

Any isolated object, such as a molecule or a real crystal, can possess *point-symmetry;* that is, any symmetry operation for this object, such as a rotation of, say, 180°, must leave at least *one point* within the object *fixed.* On the other hand, an infinite array of points, such as a lattice (or an ideal unbounded crystal structure), has *translational symmetry* as well, since translation (motion in a straight line, without rotation) along any integral number of lattice vectors moves it into self-coincidence and thus is a symmetry operation. A translation operation leaves no point unchanged since it moves all points equal distances in parallel directions; it is an example of a *space-symmetry* operation. Because most macroscopic crystals consist of 10^{12} or more unit cells, it is a fair approximation to regard the arrangement of atoms throughout most of a real crystal as possessing translational symmetry. Edge effects normally are completely negligible in structure analysis.

POINT SYMMETRY AND POINT GROUPS

The operations of rotation, mirror reflection, and inversion through a point (see below) are point-symmetry operations, since each leaves at least one point of the object in a fixed position. The geometrical requirements of lattices restrict the number of possible rotational-symmetry elements that a crystal can have. These possible rotational symmetries will now be considered:

1. *n-fold rotation axes.* A rotation of $(360/n)^\circ$ leaves the object or structure apparently unchanged (self-coincidence). The order of the axis is said to be n. When $n = 1$—that is, a rotation of 360°—the operation is equivalent to no rotation at all (0°), and is said to be the "identity operation." A four-fold rotation axis is shown in Figure 7.1a, and is denoted 4. It may be proved that only axes of order 1, 2, 3, 4, and 6 are compatible with structures built on three-dimensional (or even two-dimensional) lattices. Isolated molecules sometimes have symmetry axes of other orders, but when crystals are formed from a molecule with, for example, a five-fold axis, the five-fold axis *cannot be a symmetry axis of the crystal.* The molecule may still retain its five-fold symmetry in the crystal, but it can never occur at a position such that this symmetry is a necessary consequence of a five-fold symmetry in the crystalline environment. In other words, the five-fold symmetry is local and not crystallographic—that is, not required by the space group.

2. *n-fold rotatory-inversion axes.* The inversion operation, with the origin of coordinates as the "center of inversion," implies that every point x, y, z becomes $-x$, $-y$, $-z$. An n-fold rotatory-inversion axis implies that a rotation of $(360/n)^\circ$ (where n is 1, 2, 3, 4, or 6) followed by inversion through some point on the axis produces no apparent change in the object or structure. The one-fold case, $\bar{1}$, is the inversion operation itself and is often merely called a center of symmetry. A two-fold rotatory-inversion axis, denoted $\bar{2}$, is shown in Figure 7.1b. In general these axes are symbolized as $\bar{n}$.

The rotatory-inversion operations differ from the pure rotations in an important respect: they convert an object into its mirror image. Thus a pure rotation can convert a left hand only into a left hand; on the other hand, a rotatory-inversion axis will, on successive operations, convert a left hand into a right hand, then that back into a left hand, and so on. Objects that cannot be superimposed on their mirror images cannot possess any element of rotatory-inversion symmetry.

3. *Mirror planes.* These are designated m. They convert a left-handed molecule into a right-handed molecule. As shown in Figure 7.1b, a mirror plane is equivalent to a two-fold rotatory-inversion axis, $\bar{2}$, oriented perpendicular to the plane. The symbol m is more common for this symmetry element.

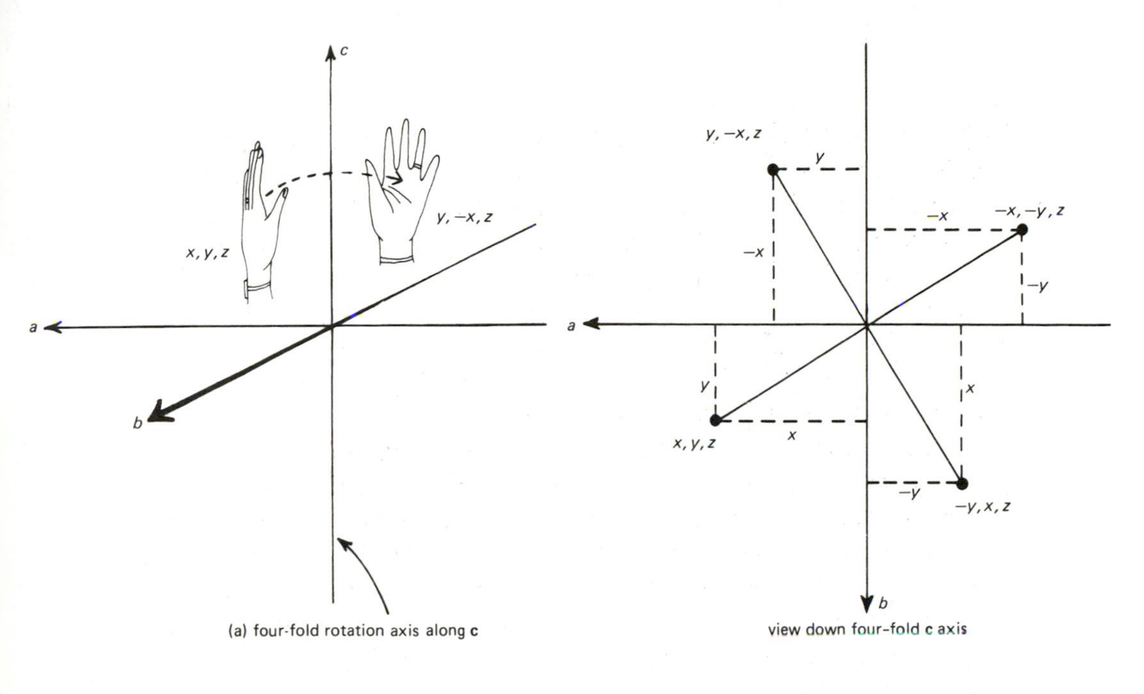

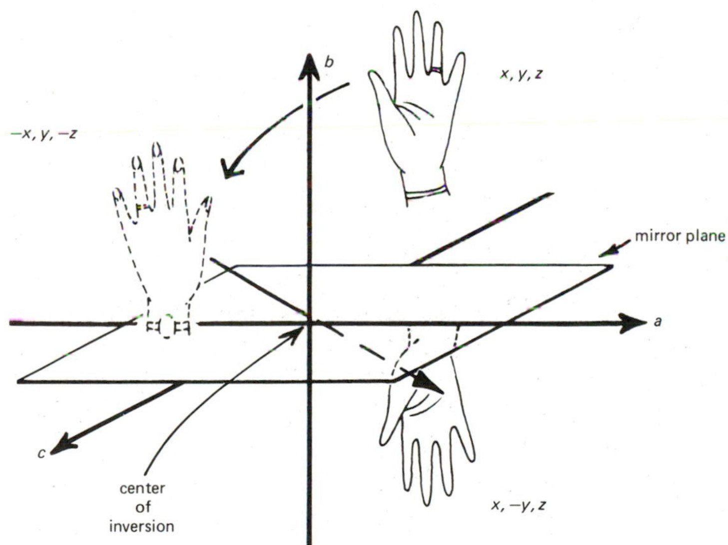

(b) two-fold rotatory-inversion axis *or* mirror plane

FIGURE 7.1 Some Symmetry Operations.

To make the distinction of left and right hands clearer a ring and watch have been indicated on the left hand but not on the right (even after reflection from the left hand).

(a) A four-fold rotation axis, parallel to **c** and through the origin of a tetragonal unit cell ($a = b$), moves a point at x, y, z to a point at $(y, -x, z)$ by a rotation of 90° about the axis. The sketch on the right shows all four equivalent points

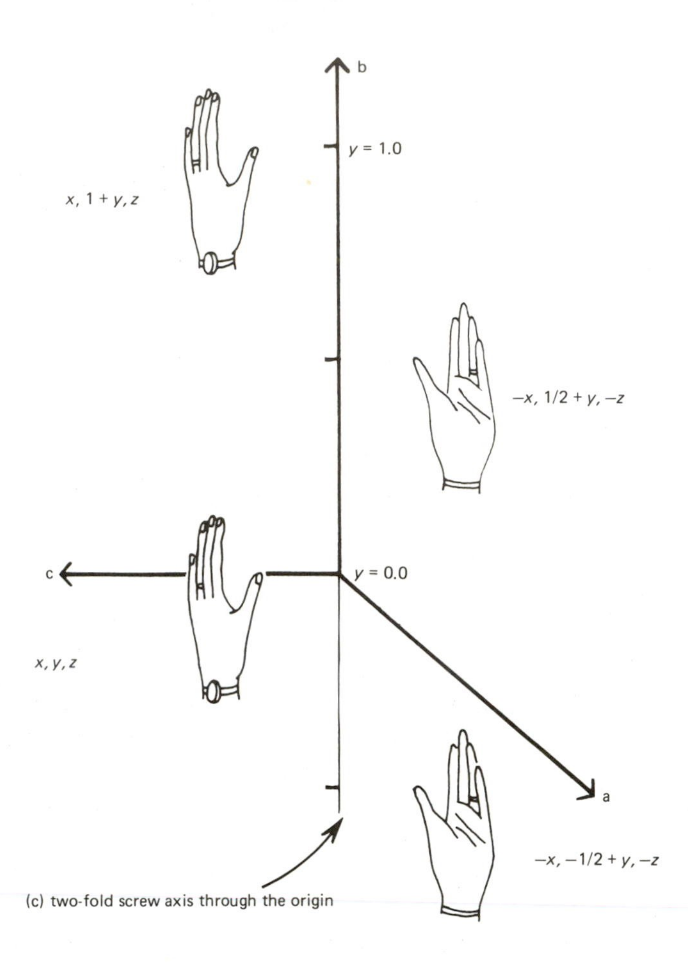

(c) two-fold screw axis through the origin

resulting from successive rotations; only two of these are illustrated in the left hand sketch.

(b) The operation $\bar{2}$, a two-fold rotatory-inversion axis parallel to **b** and through the origin, converts a point at x, y, z to a point at $x, -y, z$. One way of analyzing this change is to consider it as the overall result of, first, a two-fold rotation about an axis through the origin and parallel to **b** (x, y, z to $-x, y -z$) and then an inversion about the origin ($-x, y, -z$ to $x, -y, z$). This is the same as the effect of a mirror plane perpendicular to the **b** axis. Note that a left hand has been converted to a right hand. The hand illustrated by broken lines is an imaginary intermediate for the symmetry operation $\bar{2}$.

(c) A two-fold screw axis, 2_1, parallel to **b** and through the origin, which combines both a two-fold rotation (x, y, z to $-x, y, -z$) and a translation of $b/2$ ($-x, y, -z$ to $-x, \frac{1}{2} + y, -z$). A second screw operation will convert the point $-x, \frac{1}{2} + y, -z$ to $x, 1 + y, z$, which is the equivalent of x, y, z in the next unit cell along **b**. Note that the left hand is *never* converted to a right hand.

(d) Some crystallographic four-fold screw axes showing two identity periods for each. Note that the effect of 4_1 on a left hand is the mirror image of the effect of 4_3 on a right hand. The right hand has been moved slightly to make this relation obvious.

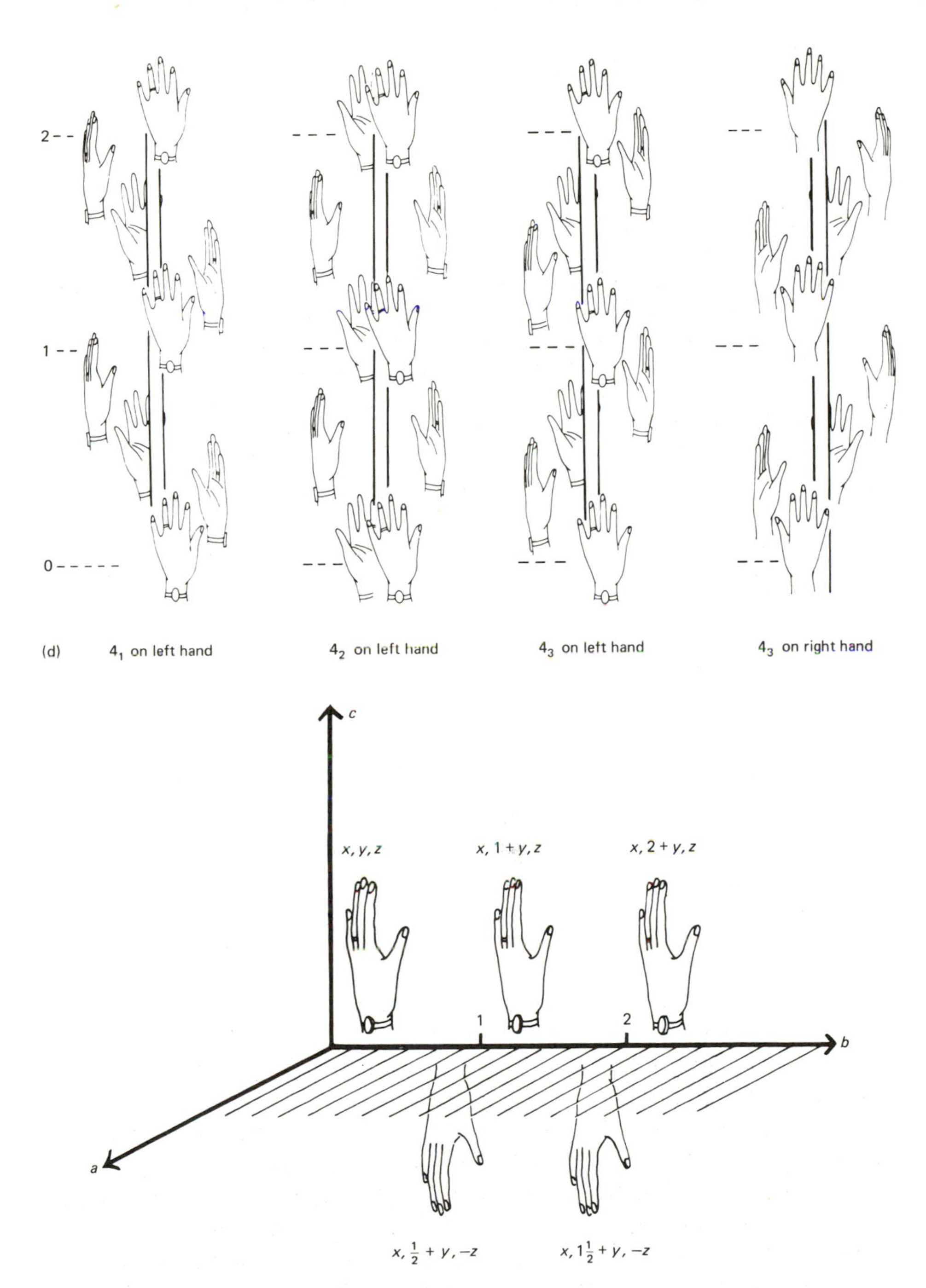

(e) *b*-glide plane through the origin and normal to **c**

(e) A *b*-glide plane normal to **c** and through the origin involves a translation of $b/2$ and a reflection in a plane normal to c. It converts a point at x, y, z to one at x, $\frac{1}{2} + y$, $-z$. Note that left hands are converted to right hands, and *vice versa*.

The point symmetry operations listed above (1, 2, 3, 4, 6, $\bar{1}$, $\bar{2}$ or *m*, $\bar{3}$, $\bar{4}$, and $\bar{6}$) can be combined together in just 32 ways in three dimensions to form the 32 three-dimensional *crystallographic point groups*. There are, of course, other point groups, appropriate to isolated molecules and other figures, containing, for example, five-fold axes. The 32 crystallographic point groups or symmetry classes may be applied to the shapes of crystals or other finite objects; the point group of a crystal may sometimes be deduced by an examination of any symmetry in the development of faces. For example, a study of crystals of beryl shows that each has a six-fold axis perpendicular to a plane of symmetry (6/*m*) with two more symmetry planes parallel to the six-fold axis and at 30° to each other (*mm*). The corresponding point group is designated 6/*mmm*. This external symmetry is a manifestation of the symmetry in the internal structure of the crystal. Frequently, however, the environment of a crystal during growth is sufficiently perturbed that the external form or morphology of the crystal does not reflect, to the extent that it might, the internal symmetry. Diffraction studies then help to establish the point group as well as the space group.

SPACE SYMMETRY

Combination of the point-symmetry operations with translations give rise to various kinds of space-symmetry operations in addition to the pure translations.

1. *n-fold screw axes.* These result from the combination of translation and pure rotation and are symbolized by n_r. They involve a rotation of $(360/n)°$ and a translation parallel to the axis by the fraction r/n of the identity period along that axis. A two-fold screw axis, 2_1, is shown in Figure 7.1c. If $p = n - r$, then the axes n_r and n_p are *enantiomorphous;* that is, they are mirror images of one another, like left and right hands. It is important, however, to note that it is only the *axes* that are enantiomorphous; structures built on them will not be enantiomorphous unless the objects in the structure are themselves enantiomorphous. Thus a *left* hand operated on by a 4_1 will give an arrangement that is the mirror image of that produced by the operation of a 4_3 on a *right* hand, but not, of course, the mirror image of that produced by the operation of a 4_3 on another left hand, as shown in Figure 7.1d.

2. *Glide planes.* These symmetry elements result from the combination of translation with the mirror operation (or its equivalent, $\bar{2}$ normal to the plane), as illustrated in Figure 7.1e. The glide must be parallel to some lattice vector, and, because the mirror operation is two-fold, a point equivalent by a simple translational symmetry operation—a lattice vector—must be reached after two glide translations. Thus these translations may be half of the repeat distance along a unit-cell edge, in which case the glide plane is

referred to as an *a*-glide, *b*-glide, or *c*-glide, depending on the edge parallel to the translation. Alternatively, the glide may be parallel to a face-diagonal. No glide operation involves fractional translational components other than $\frac{1}{2}$ or $\frac{1}{4}$, and the latter occurs only for glide directions parallel to a face diagonal or a body diagonal in certain nonprimitive space groups.

SPACE GROUPS

It is possible to combine the various pure rotations, rotary inversions, screw axes, and glide planes in just 230 ways compatible with the geometrical requirements of three-dimensional lattices. There are thus 230 three-dimensional *space groups,* ranging from that with no symmetry other than the identity operation (symbolized by $P1$, the P implying primitive) to those with the highest symmetry, such as $Fm3m$, a face-centered cubic space group. These 230 space groups represent the 230 distinct ways in which objects (such as molecules) can be packed in three dimensions so that the contents of one unit cell are arranged in the same way as the contents of every other unit cell.

It is interesting to note that these 230 unique three-dimensional combinations of the possible crystallographic symmetry elements were derived independently in the last two decades of the nineteenth century by Fedorov in Russia, Schönflies in Germany, and Barlow in England. It was not until several decades later that anything was known of the actual atomic structure of even the simplest crystalline solid. Since the introduction of diffraction methods for studying the structure of crystals, the space groups of many thousands of crystals have been determined. Although representatives of most, but not all, of the 230 space groups have been found, in fact about 60% of the organic compounds studied crystallize in one of six space groups.*

All 230 space groups and the systematically absent reflections that they imply are listed in *International Tables,* Volume 1 (ref. 20 in the bibliography), which is in constant use by X-ray crystallographers. Part of a specimen page is shown in Figure 7.2. Once the space group is determined from the systematically absent reflections in the X-ray diffraction pattern and by other means if needed, *only the structure of the contents of the asymmetric unit,* not the entire unit cell, *need be determined.* The contents of the rest of the cell (and of the entire structure) are then known by application of the symmetry operations of the space group. An example is shown in Figure 7.3.

Space Group Ambiguities The principal method used to determine the space group of a crystal is that of determining which reflections are system-

*Centrosymmetric space groups $P2_1/c$, $P\bar{1}$, $C2/c$, $Pbca$ and noncentrosymmetric space groups $P2_12_12_1$ and $P2_1$.

Orthorhombic 222 $P\,2_1\,2_1\,2_1$ No. 19 $P2_12_12_1$ D_2^4

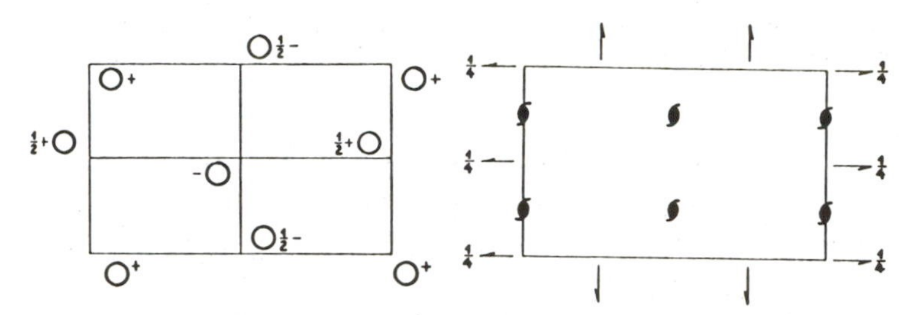

Origin halfway between three pairs of non-intersecting screw axes

Number of positions, Wyckoff notation, and point symmetry			Co-ordinates of equivalent positions	Conditions limiting possible reflections	
4	*a*	1	$x,y,z;\ \tfrac{1}{2}-x,\bar{y},\tfrac{1}{2}+z;\ \tfrac{1}{2}+x,\tfrac{1}{2}-y,\bar{z};\ \bar{x},\tfrac{1}{2}+y,\tfrac{1}{2}-z.$	hkl:	No conditions
				$0kl$:	
				$h0l$:	
				$hk0$:	
				$h00$:	$h=2n$
				$0k0$:	$k=2n$
				$00l$:	$l=2n$

Symmetry of special projections

(001) pgg; $a'=a, b'=b$ (100) pgg; $b'=b, c'=c$ (010) pgg; $c'=c, a'=a$

FIGURE 7.2 Part of a Page from "*International Tables for X-Ray Crystallography,*" *Volume I.*

Information on the space group $P2_12_12_1$. The crystal is orthorhombic and there are three sets of mutually perpendicular nonintersecting screw axes. P denotes a primitive lattice (that is, one lattice point per cell with no face- or body-centering) and 2_1 denotes a two-fold screw axis. The origin of the cell, chosen so that it lies halfway between these three pairs of nonintersecting screw axes, lies in the upper left-hand corner with the x-direction down and the y-direction across to the right; x is parallel to **a** and y is parallel to **b**. The symbol (⸙) refers to a two-fold screw axis perpendicular to the plane of the paper. The symbol (⇀) refers to a two-fold screw axis in a plane parallel to the plane of the paper; the fractional height of this plane above the plane $z = 0$ is shown (unless the screw axes are in the plane $z = 0$). The operations of the space group on the point (x, y, z) give three additional equivalent positions, whose coordinates are listed. Thus the screw axis parallel to **c** at $x = \frac{1}{4}$, $y = 0$, converts an atom at x, y, z to one at $\frac{1}{2} - x$, $-y$, $\frac{1}{2} + z$. Similar transformations are effected by the other two sets of screw axes (parallel to **a** and **b**, respectively). The diffraction patterns of crystals with this space group show systematic absences *only* for $h\,0\,0$ when h is odd, $0\,k\,0$ when k is odd, and $0\,0\,l$ when l is odd. Such crystals contain only molecules of one handedness (chirality).

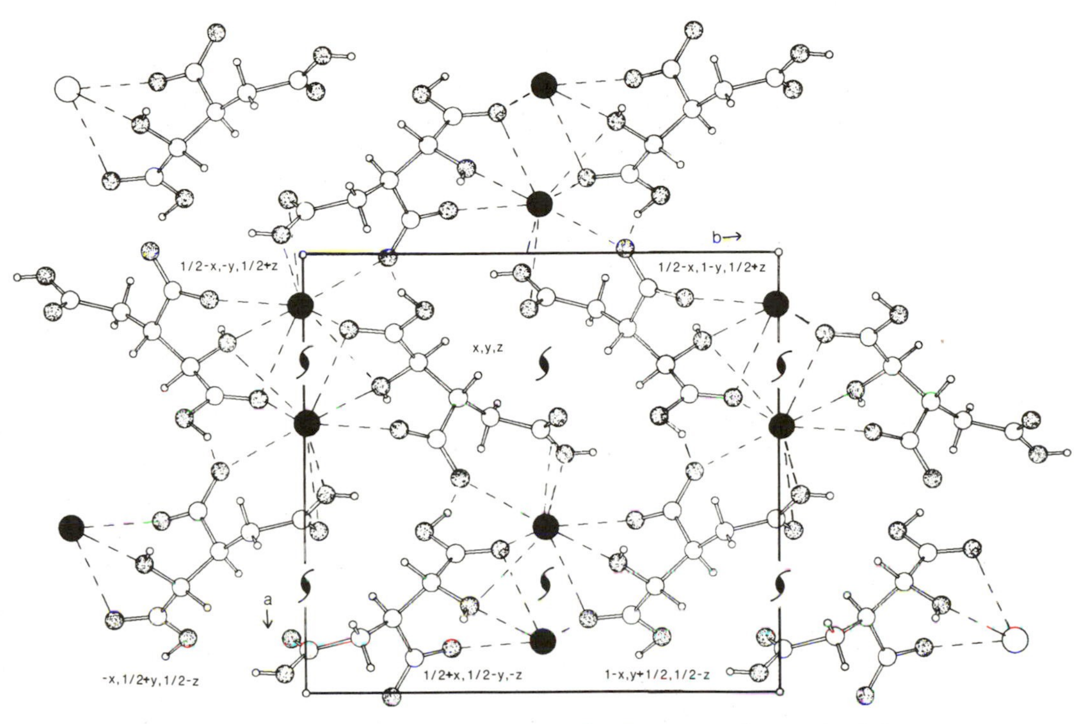

FIGURE 7.3 A Structure that Crystallizes in the Space Group $P2_12_12_1$.

Contents of the unit cell of potassium dihydrogen isocitrate. [See *Acta Crystallographica B24* (1968), 578.] The space group requires that, for each atom at x, y, z there should be equivalent atoms at: $\frac{1}{2} - x$, $-y$, $\frac{1}{2} + z$; $\frac{1}{2} + x$, $\frac{1}{2} - y$, $-z$; and $-x$, $\frac{1}{2} + y$, $\frac{1}{2} - z$. These are indicated on the diagram (oxygen stippled, potassium dark, hydrogen small). Interactions via hydrogen bonding and metal coordination are indicated by broken lines. This figure illustrates how anions cluster around a cation (K^+) and how this clustering, together with hydrogen bonding, is a major determinant of the structure.

atically absent in the space group. These are listed in *International Tables*, Volume 1. As shown in the example on page 91, these systematic absences depend on the translational symmetry of the space group (screw axes, glide planes, face- or body-centering). That is, a two-fold screw axis resulted in systematic absences for $h00$ when h is odd. Therefore, space groups with the same translational symmetry elements (for example, $P2_1$ and $P2_1/m$)* will have the same systematic absences, giving rise to space group ambiguities.

However, there are ways of overcoming this problem. If the crystal contains only one enantiomorph of an asymmetric molecule, then the space group cannot contain a mirror or glide plane or a center of symmetry, since

*Equivalent positions for $P2_1$ are x, y, z and $-x$, $\frac{1}{2} + y$, $-z$. Equivalent positions for $P2_1/m$ are x, y, z; $-x$, $-y$, $-z$; $-x$, $\frac{1}{2} + y$, $-z$; and x, $\frac{1}{2} - y$, z.

these symmetry elements convert one enantiomorph into the other. As a result, if the ambiguity involves a centrosymmetric and a noncentrosymmetric structure (such as $P1$ and $P\bar{1}$ or $P2_1$ and $P2_1/m$), then a distinction can be made if the structure contains molecules of only one chirality. Even if it does not, the distinction can usually be made, as described in more detail in Chapter 8, page 108, by a consideration of the distribution of intensities in the diffraction pattern since centrosymmetric structures have a higher proportion of reflections of very low intensity than do noncentrosymmetric structures. Other diagnostic methods involve tests of physical properties including the piezoelectric and pyroelectric effects (that is, the development of charges on opposite faces when the crystal is subjected to stress or heat respectively) or second harmonic generation (that is, the generation of a second beam with twice the frequency of the incident light on passage through the crystal). These effects are only found for noncentrosymmetric crystals. Still another method of distinguishing between space groups is to analyze the vectors in the Patterson map, described in Chapter 9. Finally, a consideration of the chemical identity of the contents of the unit cell may help resolve any ambiguity.

The following example of such an ambiguity may be of interest. The protein xylose isomerase, consisting of four subunits bound in a tetramer, crystallizes in the space group $I222$ or $I2_12_12_1$ with two molecules in the unit cell. There is ambiguity in the space group because the systematic absences in reflections are the same for both. However, each unit cell for each space group contains eight asymmetric units, and therefore one subunit (one quarter of the molecule) must be the asymmetric unit (together with solvent, not considered here). It follows that the space group cannot be $I2_12_12_1$ because then the protein would need to be an infinite polymer, contrary to physical evidence; all groups related by a 2_1-axis must be equivalently related, so if any two are related in a certain way, all groups along the 2_1-axis are similarly related. Thus it is concluded that the space group is $I222$.

CHIRALITY

Chirality is the handedness of a structure; that is, if a structure cannot be superimposed on its mirror image it is said to be chiral or enantiomorphous. We are most familiar with this in the example of the asymmetric carbon atom—that is, a carbon atom connected to four different chemical groups so that two types of molecules, related to each other by a mirror plane, are found. This chirality, however, can also extend to the crystal structure itself. For example, silica crystallizes in a helical arrangement that has a handedness shown in the external shape of the crystal—small hemihedral* faces

*Called hemihedral because only half the number of faces expected for a centrosymmetric structure are observed.

appear in such a way as to give crystals that are mirror images of each other. The observation of such hemihedry was used by Louis Pasteur in 1848 to separate sodium ammonium tartrate into its left- and right-handed enantiomers. Solutions of these rotate the plane of polarization of light in opposite directions. When such resolution* occurs the space group must contain no mirror planes, glide planes, or centers of inversion (i.e., any symmetry operation that would convert a left-handed structure into a right-handed structure). Such crystals also exhibit pyroelectric and piezoelectric properties as a result of their asymmetry. Pasteur's resolution of sodium ammonium tartrate was possible because the space group was that for a noncentrosymmetric structure. If an asymmetric molecule crystallizes in a centrosymmetric space group, then there are equal numbers of left- and right-handed molecules in the crystal structure. This will be discussed further in Chapter 10.

SUMMARY

Symmetry in the contents of the unit cell is revealed to some extent by the symmetry of the diffraction pattern and by the systematically absent reflections. The probable space group of the crystal can be deduced from this information about the diffraction pattern. Knowledge of the space group may also give information on molecular packing, even before the structure has been determined.

1. There are 14 distinct three-dimensional lattices, corresponding to seven different "crystal systems."
2. Point-symmetry operations, which leave at least one point within an object fixed, characteristic of crystals comprise:
   (a) n-fold rotation axes (1, 2, 3, 4, 6).
   (b) n-fold rotatory inversion axes ($\bar{1}$, $\bar{2}$ or m, $\bar{3}$, $\bar{4}$, $\bar{6}$).
3. These point-symmetry operations can be combined in 32 and only 32 distinct ways to give the three-dimensional crystallographic point groups.
4. Combination of point-symmetry operations with translations gives space-symmetry operations.
   (a) n-fold screw axes, n_r.
   (b) Glide planes.
5. All these operations may act on a given motif in the asymmetric portion of the structure. They can be combined in just 230 distinct ways, giving the space groups.

*The term "resolution" is used in a different sense from that in the caption to Figure 9.10. Here it is used to mean the separation of enantiomers. The term is also used to describe the process of distinguishing individual parts of an object, as when viewing them through a microscope.

8

The Derivation of Trial Structures

I. Analytical Methods for Direct Phase Determination

As indicated at the start of Chapter 4, after the diffraction pattern has been recorded and measured in some appropriate fashion, the next stage in a crystal structure determination is solving the structure—that is, finding a suitable trial structure (approximate positions of most atoms in a unit cell of known dimensions and space group). The term "trial structure" implies that the structure found first is only an approximation to the correct or "true" structure; "suitable" implies that the trial structure is close enough to the true structure that it can be smoothly refined to a good fit to the data set. Methods of finding suitable trial structures form the subject of this chapter and the next; structure refinement is considered in Chapter 11. In the early days of structure determination, trial and error methods were, of necessity, almost the only available way of solving structures. Structure factors for a suggested model were calculated and compared with those observed. This method of determining "trial structures" is only of minor importance now, although it still plays some role, particularly in structures in which a molecule or ion is known, from symmetry considerations, to occupy a particular site.

We begin with a discussion of so-called "direct methods," analytical techniques for deriving an approximate set of phases from which a first approximation to the electron density map can be calculated. Interpretation of this may then give a suitable trial structure. "Direct methods" make use of the fact that the intensities of reflections contain structural information and that the electron density in a real crystal cannot be negative. Centrosymmetric structures, with each atom at x, y, z, matched by an equivalent atom in the structure at $-x$, $-y$, $-z$, are considered first, since the problems presented by noncentrosymmetric structures are more formidable. Other techniques for deriving trial structures and the principles upon which they are based are discussed in the following chapter.

In structures with a center of symmetry at the origin, each structure factor has a phase angle of 0° or 180°, so that $\cos \alpha$ is just $+1$ or -1 and $\sin \alpha = 0$. Therefore, $|F| \cos \alpha = F = +|F|$ or $-|F|$. If N reflections have been observed for the structure, 2^N electron density maps could be calculated, representing all possible combinations of signs for all N independent structure factors. One of these 2^N maps must represent the true electron density, but how could one tell which one it is? For even as few as twenty reflections, more than one million different maps can be calculated ($2^{20} = 1{,}048{,}576$), and most structures of interest have of the order of $10^3 - 10^4$ unique reflections. Since the contributions from reflections with high values for the structure factor amplitude will tend to dominate any map calculated, including that calculated with phases corresponding to the correct structure, only the highest valued terms need be considered initially in trying to get an approximation to the correct map. However, even with as few as ten terms, the number of possible maps is 1024, much too high a number to make any simple trial and error method practicable. With a noncentrosymmetric crystal structure, a phase angle may be anywhere between 0° and 360° and one would have to calculate an impossibly large number of maps to ensure having at least approximately correct relative phase angles for even ten reflections.

It is possible, however, to derive relations among the phases of different reflections. These relations come from the fact that *the electron density can never be negative* and is near zero except for approximately spherical peaks at atomic positions. This, of course, implies that the intensities in the X-ray diffraction pattern contain phase information (because the phases are constrained to give positive electron density). For centrosymmetric structures one often speaks of the *sign* of a structure factor because the phase angle, α, must be 0° or 180°, which means $\cos \alpha = \pm 1$, $\sin \alpha = 0$. The terms used in forming an electron density map involve $|F| \cos \alpha$ and $|F| \sin \alpha$ (Chapter 6), so for this situation only $|F|$ and $-|F|$ occur. Relationships can be found among the signs of the structure factors and these relationships involve the magnitudes of the larger structure factors normalized (that is, modified) in a certain way (to be described in this chapter).

The relationships between the relative phases of certain reflections are illustrated in Figure 8.1a. Suppose that $F(1\,0\,0)$ for a centrosymmetric structure is large. If it has a positive sign (phase angle of 0°), then in the electron density map computed using this reflection a peak appears at $x = 0$ and a hole at $x = \frac{1}{2}$. However if this reflection has a negative sign there is a peak

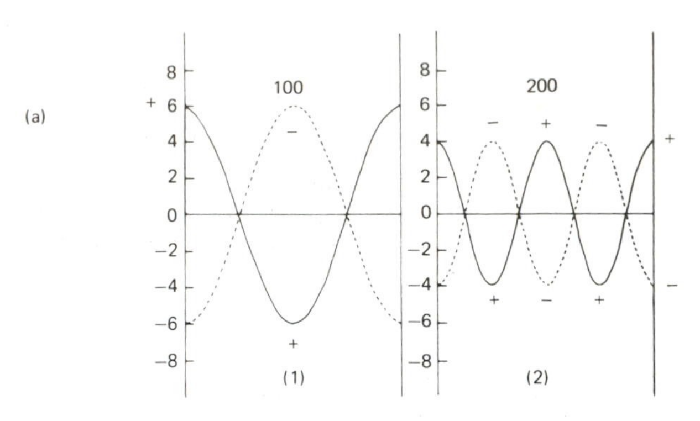

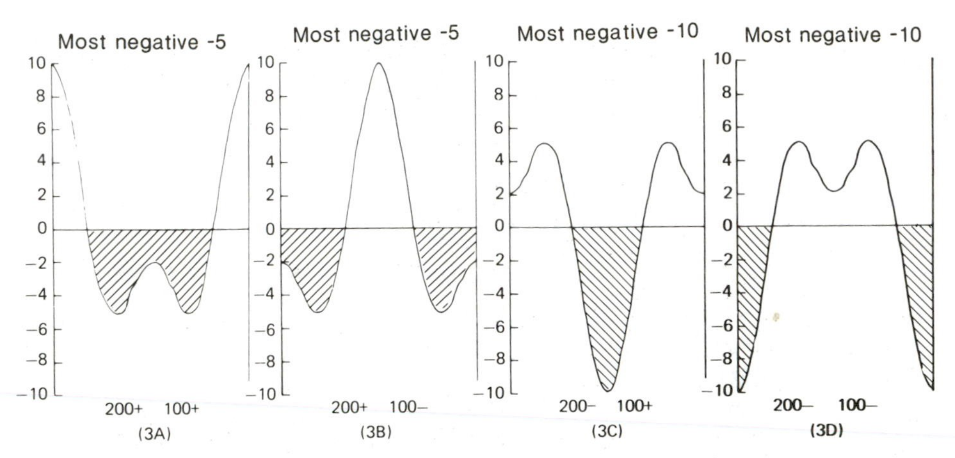

FIGURE 8.1 Direct Sign Determination by Analytical Methods.

(a) In centrosymmetric structures the phase angle of any structure factor, $F(hkl)$, is either 0° or 180°. The "electron density maps" based on one structure factor ("density wave") are shown for $F(1\,0\,0)$ and $F(2\,0\,0)$. In general, for centrosymmetric structures, if $F(hkl)$ is large, whatever its sign, and $F(2h2k2l)$ is also large, then the latter is probably positive (a phase of 0°).

(1) Possible situations for $F(1\,0\,0)$: solid line—F positive, phase 0°; dotted line—F negative, phase 180°.

(2) Possible situation for $F(2\,0\,0)$; solid line—F positive, phase 0°; dotted line—F negative, phase 180°.

(3) Summations for the four combinations of possible situations in (1) and (2), showing the deep negative areas obtained when $F(2\,0\,0)$ is given a phase of 180° (C, D). The $F(0\,0\,0)$ term, which has been omitted, is always positive and therefore when it is included the sum is always more positive at each given point (see Figure 6.2).

Areas of negative electron density are shaded. The inferences on the position of an atom (at x) from these electron density maps are: 3A, $x = 0$; 3B, $x = \frac{1}{2}$; 3C, 3D, $x = \frac{1}{4}, \frac{3}{4}$. The last two have more negative troughs and so are excluded. Therefore we conclude 200 is + (phase angle 0°) and x is 0 and/or $\frac{1}{2}$.

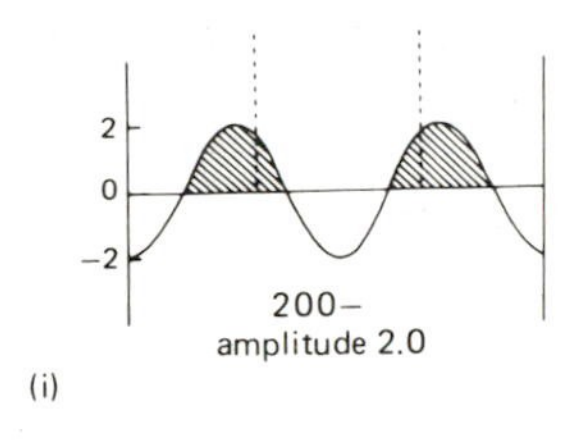

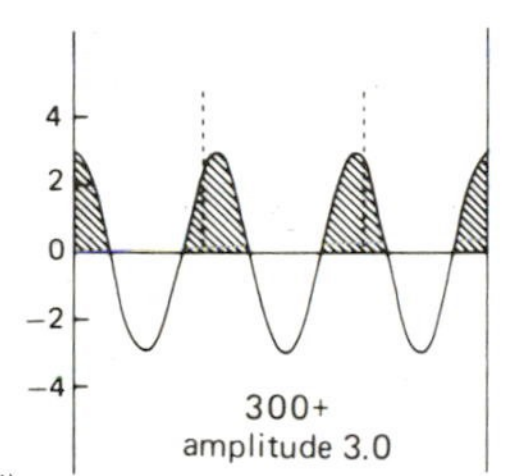

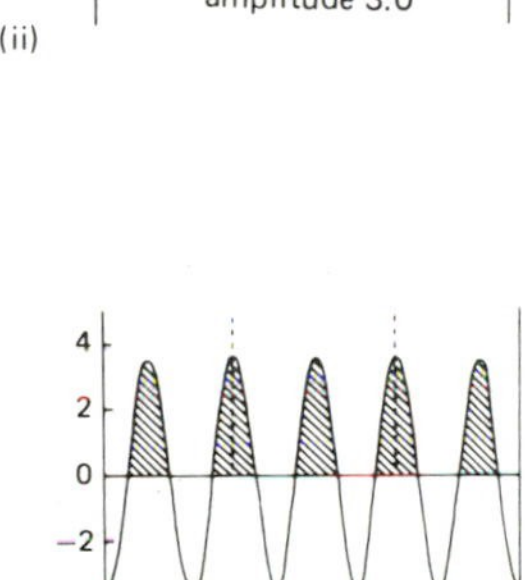

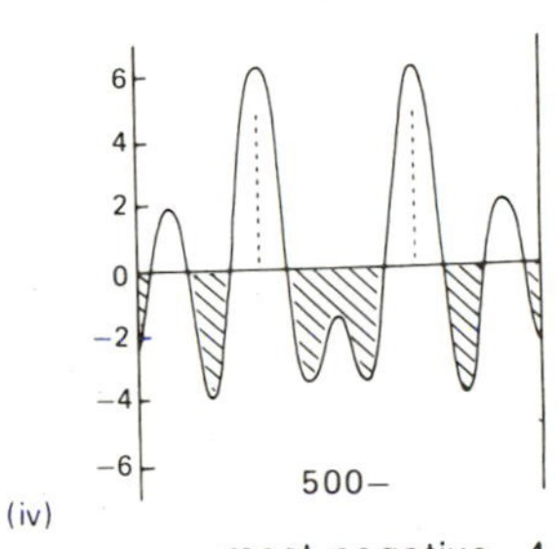

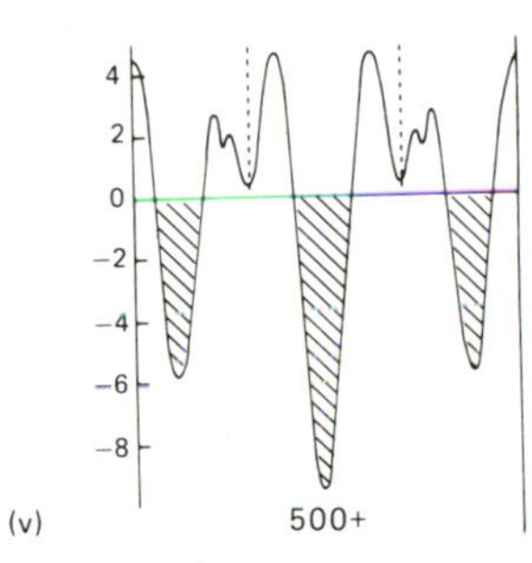

(b)

(b) If $|F(2\,0\,0)|$, $|F(3\,0\,0)|$, and $|F(5\,0\,0)|$ are all large they must contribute significantly to the final electron density map (via "density waves"). Suppose that it is found that $F(2\,0\,0)$ has a negative sign and $F(3\,0\,0)$ has a positive sign; the areas in which each then contributes in a positive manner to the electron density map are dotted in (i) and (ii) on the left. The regions in which these areas overlap, near $x = \pm 0.3$, correspond to regions to which $F(5\,0\,0)$ contributes positively only if the sign of the term $F(5\,0\,0)$ is negative, that is, a phase of 180°, as indicated in (iii). On summation of these terms with the indicated signs the background is reduced, as in (iv); if $F(5\,0\,0)$ has a positive sign, that is, a phase of 0°, the map is far less satisfactory. The relation among these signs may then be written (where s means the sign of)

$$s(5\,0\,0) \approx s(2\,0\,0)\,s(3\,0\,0)$$

which is a special case of Eq. (8.1). This follows from the discussion in the text since deep negative troughs (areas of negative electron density) are not satisfactory or physically meaningful. With proper phasing the background is reduced to a value closer to zero. Thus in (iv) the most negative value of the electron density is -4 e/Å, while for (v), which has a less satisfactory set of phases, the most negative value of the electron density is -9 e/Å.

(c) An illustration of the successive application of Eq. (8.1) (see text) to derive the signs of some of the strongest reflections for a structure. Values of $|E|$ are

FIGURE 8.1 (*continued*)

derived from those for $|F|$, chiefly to eliminate the effects of thermal vibration and to treat each atom as if all its electrons were concentrated at a point. Values of E are used in deriving sign relationships because their magnitudes depend only on the arrangement and relative atomic numbers of the atoms.

Monoclinic example

$$|F(hkl)| \neq |F(\bar{h}kl)|$$

$$k + l \text{ even} \quad s(hkl) = s(h\bar{k}l) = s(\bar{h}\bar{k}\bar{l}) = s(\bar{h}k\bar{l})$$
$$s(\bar{h}kl) = s(\bar{h}\bar{k}l) = s(hk\bar{l}) = s(h\bar{k}\bar{l})$$
$$k + l \text{ odd} \quad s(hkl) = s(\bar{h}\bar{k}\bar{l}) = -s(h\bar{k}l) = -s(\bar{h}k\bar{l})$$
$$s(\bar{h}kl) = s(h\bar{k}\bar{l}) = -s(hk\bar{l}) = -s(\bar{h}\bar{k}l)$$

The compound studied is 2-keto-3-ethoxybutyraldehyde-bis(thiosemicarbazone), space group $P2_1/c$. The above sign relationships for this space group are to be found in *International Tables* Volume 1 (ref. 20).

Relation to be used [Eq. (8.1)]:

$$s(h + h', k + k', l + l') \approx s(h, k, l)\, s(h', k', l')$$

h k l	*E*		
3 3 1	3.74	+	Signs chosen arbitrarily fixing the origin. (If one or all of these signs had been negative another allowable origin would have resulted.)
$\bar{9}$ 6 7	3.25	+	
$\overline{13}$ 1 4	2.92	+	

h k l	*E*	*Relationships used*	*Sign found*	*(Notes)*
12 0 0	4.35	(6 0 0)(6 0 0)	+	(+) (+) = (−) (−) = +
6 0 0	2.80		?	
$\overline{25}$ 1 4	3.49	($\overline{12}$ 0 0)($\overline{13}$ 1 4)	+	++ = +
$\overline{22}$ 4 5	2.22	(3 3 1)($\overline{25}$ 1 4)	+	++ = +
6 4 2	2.86		a	An additional undetermined sign is chosen and is temporarily designated a
18 4 2	2.92	(12 0 0)(6 4 2)	a	+a = a
9 7 3	2.07	(3 3 1)(6 4 2)	a	+a = a
$\overline{22}$ 6 1	2.30	($\overline{13}$ $\bar{1}$ 4)($\bar{9}$ 7 $\bar{3}$)	−a	−(+a) = −a
$\overline{19}$ 3 2	2.84	(3 $\bar{3}$ 1)($\overline{22}$ 6 1)	−a	+(−a) = −a
		($\overline{13}$ $\bar{1}$ 4)($\bar{6}$ 4 $\bar{2}$)		−(+a) = −a
$\bar{7}$ 3 2	2.14	(12 0 0)($\overline{19}$ 3 2)	−a	+(−a) = −a
25 1 0	2.03	(18 4 2)(7 $\bar{3}$ $\bar{2}$)	−	a(−a) = −
		(6 4 2)(19 $\bar{3}$ $\bar{2}$)		a(−a) = −

Eventually the sign of some reflections could be found both in terms of a and independently of it; this established the fact that the sign of a was probably +. If this had not happened it would have been necessary to calculate two maps, one with the sign of a positive and one with the sign of a negative.

at $x = \frac{1}{2}$ and a hole at $x = 0$. The fact that this reflection is strong thus implies that there must be a peak in the electron density map near either $x = 0$ or $x = \frac{1}{2}$, whatever the sign to be associated with $F(1\ 0\ 0)$. If $F(2\ 0\ 0)$ is considered, it can be seen that a peak at *either* 0 or $\frac{1}{2}$ implies a positive sign for $F(2\ 0\ 0)$ (see Figure 8.1a(2). Consequently, *if* $F(1\ 0\ 0)$ *is strong,* $F(2\ 0\ 0)$ *is probably positive* regardless of the sign of $F(1\ 0\ 0)$. Figure 8.1a shows also that when only these two terms are summed, a positive sign for $F(2\ 0\ 0)$ results in an "electron density" that has a shallower negative trough than does the density resulting when a negative sign is assigned to $F(2\ 0\ 0)$, regardless of the sign of $F(1\ 0\ 0)$.

The principle of positivity of electron density may be extended to three dimensions. It can be shown that

$$\mathrm{s}(h_1, k_1, l_1)\mathrm{s}(h_2, k_2, l_2) \approx \mathrm{s}(h_1 + h_2, k_1 + k_2, l_1 + l_2) \qquad (8.1)$$

where s means "the sign of" and (h_1, k_1, l_1), (h_2, k_2, l_2), and $(h_1 + h_2, k_1 + k_2, l_1 + l_2)$ are *all* strong reflections. All of these relations are statistical probabilities rather than exact equations, as implied by the use of $\approx$ rather than $=$. It should be noted that a special case of eq. (8.1) is

$$\mathrm{s}(2h\ 0\ 0) \approx [\mathrm{s}(h\ 0\ 0)]^2 \approx + \qquad (8.2)$$

in agreement with our qualitative argument for 2 0 0 and 1 0 0 in Figure 8.1a. In Figure 8.1b it is shown that if $F(3\ 0\ 0)$ is known to be positive and $F(2\ 0\ 0)$ is known to be negative, then, if all three are strong reflections, $F(5\ 0\ 0)$ is likely to be negative. Again, this is shown to be consistent with the principle of positivity of electron density.

In practice these analytical methods of phase determination are carried out on "normalized structure factors"—that is, values of the structure factor, $|F|$, modified to remove the fall-off in the individual scattering factors, f, with increasing scattering angle, 2θ (see Figure 5.2b).* A normalized structure factor, E, represents the ratio of a structure factor, F, to $(\Sigma_i f_i)^{1/2}$, where the sum is taken over all atoms in the unit-cell at the value of $\sin\theta/\lambda$ appropriate to the (h, k, l) values for the reflection and includes an overall vibration factor. This sum represents the root-mean-square value that all $|F|^2$ measurements at the value of $\sin\theta/\lambda$ would have if the structure were a random one composed of equal atoms. This use of F values is approximately equivalent to considering each atom to be a point atom (an extremely sharp peak occupying almost no volume in the electron density map).

*Contained in the expression for E is a factor, ϵ, that allows for the fact that reflections in certain reciprocal lattice zones or rows (for example, $0\ k\ 0$, $h\ k\ 0$, etc.) have higher average intensities in certain space groups than do general reflections (hkl). This enhancement of intensities is modified when E values are computed.

An analysis of the E values in the diffraction pattern of a compound (called the "distribution of E values") shows that they contain (as, of course, do the F values) information on whether the structure is centrosymmetric or noncentrosymmetric. For example the mean value of E is 0.798 for a centrosymmetric structure and 0.886 for a noncentrosymmetric structure (see Appendix 6 for more details). Information on significant features of the structure is contained in the very intense and very weak reflections; these have different distributions when the structure is centrosymmetric and when it is noncentrosymmetric, the centrosymmetric distribution having a higher proportion of reflections with very low intensities.

Once a table of $|E|$ values has been prepared it is usual to rank these E values in decreasing order of magnitude and work with the strongest 10 per cent or so. Then one chooses groups of three reflections that satisfy the conditions that their indices are related in the manner described in Eq. (8.1); this selection of "triple products" [$E(hkl)$, $E(h', k', l')$, $E(h + h', k + k', l + l')$] is made by computer. Since each of the three reflections in a triple product has a high E value, the product of their signs is probably positive. This listing is called the "Σ_2" listing; Σ is used because, in the probability formula, summations are involved. The "Σ_1" relations are simpler because they involve only pairs of reflections related by $E(hkl)$ and $E(2h, 2k, 2l)$ and contain the implication that the sign of $E(2h, 2k, 2l)$ is probably positive in a centrosymmetric structure [when $E(hkl)$ is also in the list of high values; see Figure 8.1a].

The probability aspects of these sign relationships are very important. The probability* that a triple product is positive is

$$P_+ = \tfrac{1}{2} + \tfrac{1}{2}\tanh(|E_{h,k,l}E_{h',k',l'}E_{h-h',k-k',l-l'}|)/N^{1/2} \tag{8.3}$$

$$P_+ = \tfrac{1}{2} + \tfrac{1}{2}\tanh(|E_H E_K E_{H-K}|)/N^{1/2} \tag{8.4}$$

(where N is the number of atoms in the unit cell, $H \equiv h, k, l$, and $K \equiv h', k', l'$). The probability that $E(hkl)$ is positive is

$$P_+ = \tfrac{1}{2} + \tfrac{1}{2}\tanh\left(|E_H| \sum_K E_K E_{H-K}\right)/N^{1/2}$$

where the summation is over all values of $K = (h', k', l')$.†

*tanh, the hyperbolic tangent of x, is $\{(e^x - e^{-x})/(e^x + e^{-x})\}$.

†We advance from Eq. (8.1) to Eq. (8.3) by incorporating the commonly used abbreviation that has developed in the literature of direct methods: $H \equiv (h, k, l)$, $K \equiv (h', k', l')$, and hence $H + K \equiv (h + h', k + k', l + l')$. Note that, since $-K$ and K have the same E's (in sign and magnitude) in a centrosymmetric structure, then a relation between H and K, and a relation between H and $-K$ are equivalent.

We will now describe the steps in the determination of a centrosymmetric structure by direct methods. When the list of "triple products" has been prepared the derivations of their signs for centrosymmetric structures requires some initial choices of signs. Initially, in three dimensions, one has a choice of the signs of three reflections for many centrosymmetric space groups; these choices determine which of the several equivalent positions is used for the origin of the unit cell. The choice does not alter the structure. In selecting three origin-fixing reflections (they should not be a "triple product"), it is essential that they be different with respect to the evenness or oddness of their individual indices, and h, k, and l must not all be even. In the example in Figure 8.1c, arbitrary signs were chosen for $F(3\ 3\ 1)$ (odd, odd, odd), $F(9\ 6\ 7)$ (odd, even, odd) and $F(\overline{13}\ 1\ 4)$ (odd, odd, even) at the start.

The reader may ask where negative signs for phases come from. The relationships of signs of reflections with negative values of h, k, and/or l to that of a reflection with all indices positive are listed for each space group in *International Tables,* Volume I. For example in the space group $P2_1/c$, if $k + l$ is odd, $F(hkl) = -F(h\bar{k}l) = -F(\bar{h}k\bar{l})$. Negative signs are introduced into the sign relationships in this way. It is essential to have some negative terms in the calculation of the E-map, because an E-map with all signs positive will give a high peak at the origin, a rarely observed feature in complex structures; in fact this E-map with all signs positive resembles a Patterson function (to be described in the next chapter).

From the list of "triple products" it should be possible to derive for the set of E's a set of signs that have been determined with acceptable probabilities. An example is outlined in Figure 8.1c. If difficulties occur it may be necessary to choose another set of origin-fixing reflections. It may also be necessary to assign symbolic signs (a, b, etc.) to certain reflections and generate the signs of other reflections in terms of these symbols with the hope that eventually the actual signs of these symbols may become clear. This process is referred to as "symbolic addition." For example, in Figure 8.1c it is deduced that the symbolic sign a is positive. If s symbols have been used but their signs cannot be determined in this way, it will be necessary to compute 2^s E-maps. There are several excellent computer programs now available for generating sets of signs by this route of combining the triple product relationship [Eq. (8.1)] with its summed probability [Eqs. (8.3) and 8.4)].

The final stage is the calculation of an E-map, which is an electron density map calculated with E values rather than F values (so that atoms are sharper, corresponding to point atoms). If all has gone well the structure will be clear in this map. Sometimes only part of the structure is revealed in an interpretable way and the rest may be found from successive electron density or difference electron density maps. Sometimes the general orientation and connectivity of the molecule are found but the positioning in the unit cell is wrong because some subsets of signs are in error. This problem is usually recognizable when distances between atoms are calculated and some non-bonded atoms are too close to others. In this case, the development of signs must be done again, this time following some new path such as selecting origin-fixing reflections or assigning symbols to a different set of reflections.

The derivation of signs for a monoclinic centrosymmetric structure is illustrated in Figure 8.1c. Some sign relationships were immediately deduced from a knowledge of the monoclinic space group relationships appropriate for this structure, and others were deduced from them and the arbitrarily chosen signs. This process was continued until the signs of 836 out of the 872 strongest terms were found with no sign ambiguity (although it is usually not necessary to work with this many terms). Part of the resulting Fourier synthesis computed using E values (an E-map) is shown in Figure 8.2.

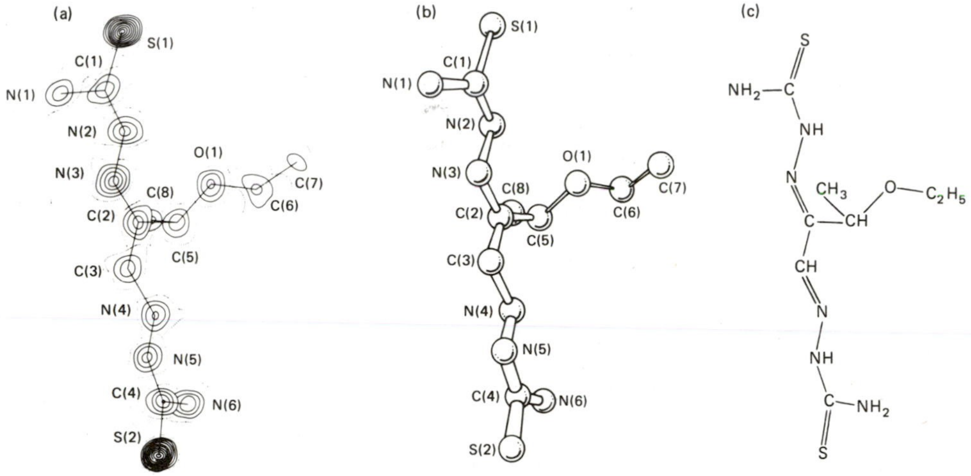

FIGURE 8.2 An Excerpt from an E-Map.

(a) A three-dimensional map calculated with phases derived as in Figure 8.1 and with E values rather than values of $|F|$ as amplitudes. [See *Acta Crystallographica* **B25** (1969), 1620.] (E values are described in the text.) Each peak has been drawn as it appears in the section in which it has the highest value; this is thus a composite map. It is a simple matter to pick out the entire molecule, 2-keto-3-ethoxybutyraldehyde-bis(thiosemicarbazone) from this map; the molecular skeleton and the presumed identity of each atom have been added to the peaks.

(b) A ball-and-stick drawing of the molecule.

(c) The chemical formula of the molecule.

For noncentrosymmetric crystal structures an additional formula may be used to derive approximate values for the phase angle.

$$\phi_H \approx \langle \phi_{H-K} + \phi_K \rangle_K \tag{8.5}$$

where, as before, $H \equiv h, k, l$, $K \equiv h', k', l'$, ϕ is the phase angle of the structure factor, and the brackets refer to an average over all values of K, where $H = (K) + (H - K)$. The so-called tangent formula,

$$\tan \phi_H \approx \frac{\langle |E_K||E_{H-K}| \sin (\phi_K + \phi_{H-K}) \rangle_K}{\langle |E_K||E_{H-K}| \cos (\phi_K + \phi_{H-K}) \rangle_K} \tag{8.6}$$

is used extensively to calculate and also to refine phases for noncentrosymmetric structures. The probability function for noncentrosymmetric structures is more complicated than that given in Eqs. (8.3) and (8.4). It is

$$P(\phi_H) = \exp \left[-4x \cos (\phi_H - \phi_K - \phi_{H-K})\right] / \int_0^{2\pi} \exp(4x \cos \gamma)\, d\gamma \tag{8.7}$$

where $x = |E_H E_K E_{H-K}|/N^{1/2}$ and γ is a dummy variable.

Direct methods for both centrosymmetric and noncentrosymmetric structures have been programmed for many high-speed computers. Since the equations involve probabilistic rather than exact relations, uses of direct methods are most successful when care is taken initially to ensure that the origin-fixing and symbolically assigned phases can be used to determine satisfactorily the phases of a good number of strong reflections so that these may, in turn, be used to place properly an even larger set of reflections. In many structure analyses a reasonable approximate ("trial") structure has been recognizable from an E-map calculated with only 5 or 10 per cent of the observed reflections, although often larger fractions are used, as in the example illustrated in Figures 8.1 and 8.2.

Generally these "direct methods" result in a structure that can be refined (Chapter 11) and so the structure may be considered to be determined. For a variety of reasons, however, such success may be elusive with some structures. There are many possible problems that can arise in using these methods, such as the use of some incorrectly computed E values, a poor choice of origin-fixing reflections, the derivation of too few triple products so that some signs are generated with lower probabilities than one would like, and a preponderance of positive signs for the derived signs so that the resulting E-map has a huge peak at the origin even though there is no heavy atom in the structure. However, with care and experience these problems can usually, although not always, be overcome.

SUMMARY

There are limits to the possible phase angles for individual reflections in both centrosymmetric and noncentrosymmetric structures. This follows from the constraints on the electron density: it must be nonnegative throughout the unit cell and it must contain discrete approximately spherical peaks (atoms). For three intense related reflections in a centrosymmetric structure, the signs are related by

$$s(H) \approx s(K)s(H + K)$$

where s means "sign of" and $H \equiv h, k, l$, $K \equiv h', k', l'$, and $H + K \equiv h + h', k + k', l + l'$. From such relationships it is often possible to derive phases for almost all strong reflections and so to determine an approximation to the structure (a "trial" structure) from the resulting electron density map. Similar methods are available for noncentrosymmetric structures.

The steps in the determination of a structure by "direct methods" consist of:

1. Making a list of E values in decreasing order of magnitude and working with the highest 10 per cent or so.
2. Analysis of the statistical distribution of E values to determine if the structure is centrosymmetric or noncentrosymmetric. This is important if there is an ambiguity in the space group determined from systematically absent reflections.
3. Derivation of triple products among the high E-values.
4. Selection of origin-fixing reflections.
5. Development of signs or phases for as many E values as possible using triple products and probability formulae.
6. Calculation of E-maps and the selection of the structure from the peaks in the map.

All of these steps are now incorporated in computer programs in wide use.

9

The Derivation of Trial Structures

II. Patterson, Heavy Atom and Isomorphous Replacement Methods

The use of high-speed computational methods has greatly simplified structure analysis for small molecules, containing up to about 150 nonhydrogen atoms per molecule. However, the intensity information that we obtain in measuring the diffraction pattern of a crystal can be analyzed in other ways than those just described as "direct methods," and these other methods are of particular use in the determination of certain structures with high symmetry within the asymmetric unit and in the determination of the structures of biological marcromolecules such as proteins and polynucleotides. The two methods described here are the Patterson method and the isomorphous replacement method. Their use can give much insight into the structure being determined. We recommend that the Patterson map of any structure with possible ambiguity from "direct methods" should be determined to see if it is consistent with the proposed trial structure.

THE PATTERSON MAP

It was mentioned earlier, in the discussion of scattering by a group of atoms, that the intensity of a given diffracted beam depended on the relative positions of all the atoms in the unit cell. In the three decades from the mid-1930s to the mid-1960s, the most powerful method of analysis of the intensity distribution in the diffraction pattern was to study the Patterson $|F|^2$-map. The technique is still of great value in unravelling some complex structures, especially those of macromolecules and other molecules containing heavy atoms or in which heavy atoms can readily be substituted.

The Patterson method consists of evaluating a Fourier series for which only the indices and the $|F|^2$ value of each diffracted beam are needed; these quantities are directly derivable from the primary experimental quantities—that is, the directions and intensities of the diffracted beams:

$$P(u,v,w) = \frac{1}{V_c} \sum\sum\sum_{\text{all } h,k,l} |F|^2 \cos 2\pi(hu + kv + lw) \tag{9.1}$$

The Patterson function, P or $P(u,v,w)$, is defined in Eq. (9.1). There is only one Patterson function for a given crystal structure and, for reasons that we explain shortly, a plot of this function is often called a vector map. The function is evaluated at each point u, v, w of a three-dimensional grid that fills a space with the size and shape of the unit cell. No phase information is required for this map, because $|F|^2$ is independent of phase. Appendix 7 gives some useful background information and further details.

The Patterson function, $P(u, v, w)$, at a point u, v, w, calculated by Eq. (9.1), may be shown (Appendix 7) to be the sum of the appearances of the structure when one views it from each atom in turn, a procedure illustrated in Figure 9.1. It may be considered to be obtained by multiplying the electron density at point x,y,z with that at $x + u$, $y + v$, $z + w$ and adding the resulting products for all values of x,y,z.

Thus the Patterson function at a point u,v,w may be thought of as a convolution* of the electron density at all points x,y,z in the unit cell with the electron density at points $x + u$, $y + v$, $z + w$. If any two atoms in the unit cell are separated by a vector (u,v,w), then there will be peak in the Patterson map at (u,v,w). Therefore the orientation and length of every interatomic vector in the structure are represented in the Patterson map.

Conversely, if there is a peak in the Patterson map, at a position related to the origin of the map by a certain vector (with components u, v, w, corresponding to a certain distance from the origin and some particular direction), then *at least one position of that particular vector in the corresponding structure has both ends on atomic positions.* The contributions of individual interatomic interactions to the heights of the peaks in this three-dimensional map are approximately proportional to the values of Z_iZ_j, where Z_i is the atomic number of the atom at one end of the vector and Z_j that of the atom at the other end. In general the value of P at every point u, v, w corresponds to the sum of the situations at the ends of such a vector as it is laid down with its origin at *every possible position* in the structure (remember that a vector is characterized by a certain length and direction,

*See footnote on page 14 and glossary.

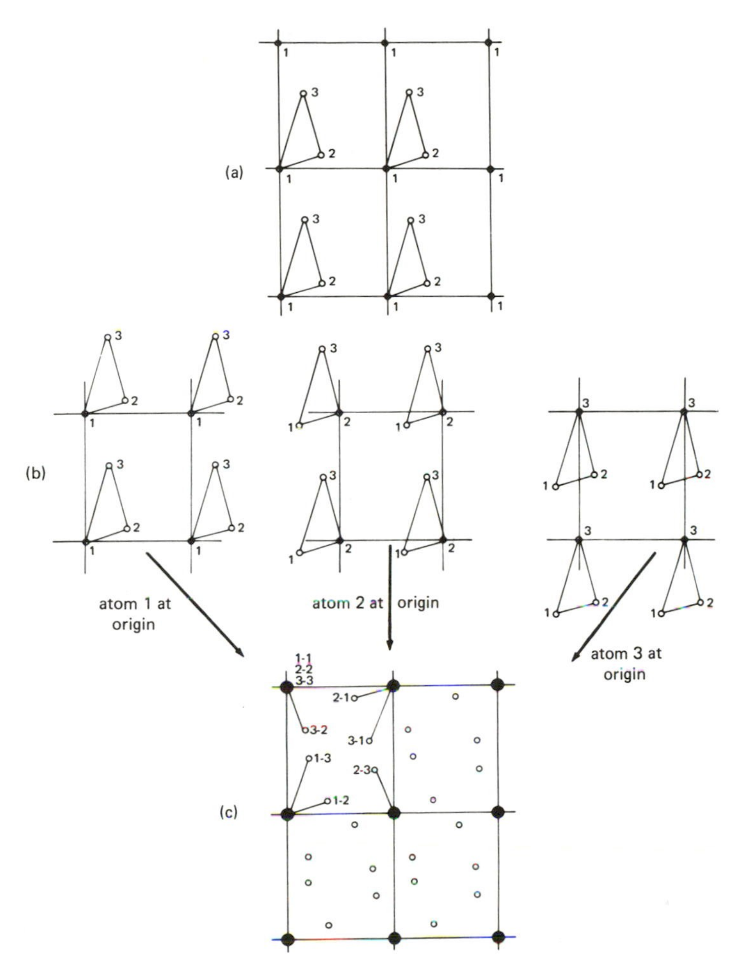

FIGURE 9.1 The Synthesis of a Vector (Patterson) Map for a Known Structure.

(a) The structure contains three atoms. Four unit cells are represented.

(b) The appearance of one unit cell is shown, with portions of others nearby, viewed with respect to each atom in turn, that is, with each atom placed, in turn, at the origin. This shows readily the termini of all interatomic vectors relative to the origin, that is, the view of the structure from atoms 1, 2 and 3 respectively.

(c) The vector map is the sum of all these vectors and represents the positions of the centers of the individual peaks that comprise a Patterson map. Four unit cells are shown. The reader is urged to follow these examples with a piece of tracing paper, laying it down on the original structure and following the steps from (a) to (c). All large filled circles are equivalent origins. Invariably the overlap of the peaks in an actual Patterson map is such that few peaks are separately resolvable and hence peak positions are not easily found.

but its origin may be anywhere).* If there are many pairs of atomic positions related by this vector, or if there are only a few but the atoms involved have high atomic numbers, then the Patterson function will have a high peak at that particular position u, v, w. If the value of the Patterson function at a given position is very low, there is no interatomic vector in the structure with that particular length and direction.

The Patterson map for a one-dimensional structure with identical atoms at $x = \pm\frac{1}{3}$ is shown in Figure 9.2. The values of the function given by Eq. (9.1) are designated $P(u)$, and positions in the one-dimensional map by u.

*H. F. Judson, in *The Eighth Day of Creation* (ref. 129) uses the analogy of a cocktail party in describing the Patterson function. If there are one hundred strangers at the party, there must have been one hundred invitations, and almost five thousand introductions and ten thousand attempts to remember a new name. If the shoes of the guests are nailed to the floor, then handshakes must involve different lengths and directions of arms and different strengths of grip. This analogy may help some readers understand the meaning of the vectors in a Patterson map; they are interatomic vectors of different lengths and directions, with heights proportional to the product of the atomic numbers of the atoms at each end of the vector. If each partygoer could then recount every handshake and the direction, distance, and strength of it, then the location of every guest in the room would be known. Of course one would only use this very complicated method when absolutely necessary.

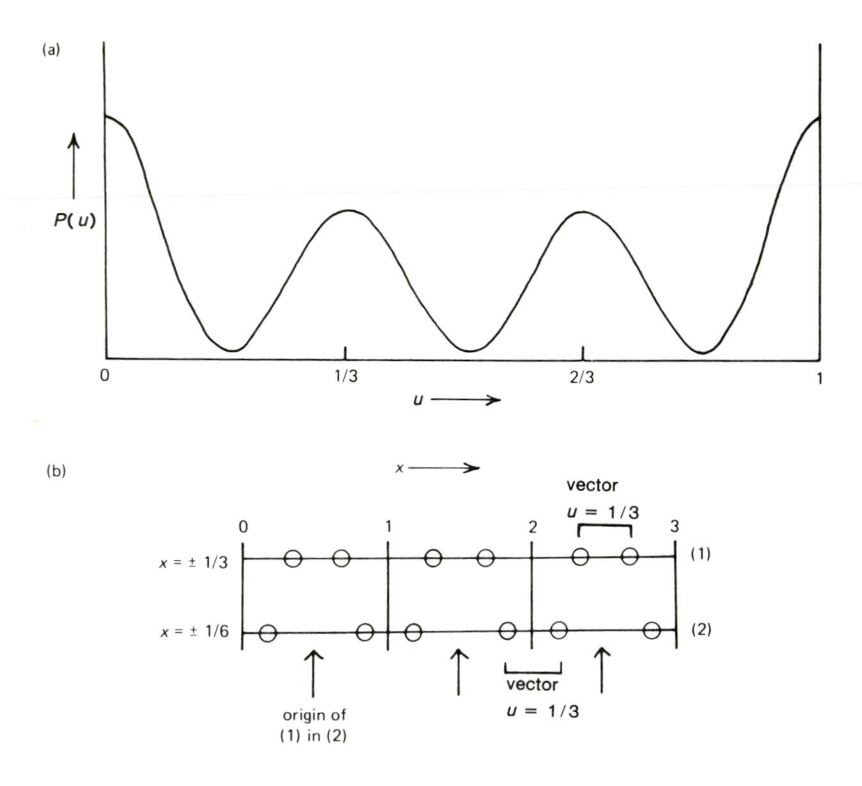

An interesting feature of this map is that the same result would be obtained from a structure in which the atoms were at $x = \pm\frac{1}{6}$. As shown in Figure 9.2, these two structures differ only in that the origin has been changed; the *relative* positions of the atoms are the same in both solutions of the map.

Until the advent of computer-assisted direct methods in the late 1960s, analysis of Patterson maps was the most important method for getting at least a partial trial structure, especially for crystals containing one or a few atoms of atomic number much higher than those of the other atoms present. In principle, for all but the largest structures, a correct trial structure can always be found from the Patterson distribution, but it is often very difficult to unravel the map, especially when the chemical formula of the compound being studied is not known. Some people, however, find it a fascinating mental exercise to try to deduce at least part of a structure from a Patterson map. The usefulness of the map decreases markedly with complicated structures composed of many atoms of about equal atomic number.

The structure for which the Patterson map is shown in Figure 9.3 contains only 13 nonhydrogen atoms in the asymmetric unit, and the complexity of the map is obvious. In this example, similar orientations of the six-membered rings in space group-related molecules give rise to very similar sets of six interatomic vectors, the vectors in each set having nearly the same magnitude and direction, thereby giving a high peak in the Patterson map [see peaks B, C, and D in Figure 9.3a(1) and 9.3a(2)]. Similarly oriented five-membered rings also lead to high peaks (peaks A and E).

The difficulty with Patterson maps is that there are N^2 interatomic vector peaks for a unit cell containing N independent atoms. Since N of these peaks lie at the origin and since the map has a center of symmetry (vector B-A has

---

FIGURE 9.2 The Calculation of a Patterson Map for a One-Dimensional Structure.

(a) The equation of the Patterson function in one dimension is

$$P(u) = \frac{1}{a} \sum_{\substack{\text{all} \\ h}} |F|^2 \cos 2\pi\,(hu)$$

The function plotted is $P(u)$ computed for a one-dimensional structure from the following hypothetical "experimental" data.

h	-3	-2	-1	0	1	2	3
$\|F\|^2$	4	1	1	4	1	1	4

(b) There are two structures consistent with this map, one with atoms at $x = \pm\frac{1}{3}$ and one with atoms at $x = \pm\frac{1}{6}$. As shown, these two structures are related simply by a change of origin.

the opposite direction and the same magnitude* as vector A-B), there are $(N^2 - N)/2$ independent vectors in the map. When N becomes at all large (even as low as 20), the $(N^2 - N)/2$ vector peaks in the Patterson map necessarily overlap one another, since they have about the same width as atomic peaks and occupy a volume equal to that occupied by the N atoms of the structure. For example, with N = 20 there are 20 × 19/2 = 190 Patterson peaks in the same volume that the 20 atomic peaks occupy. With crystals of

*This center of symmetry is evident in Figures 9.1c and 9.3a(1).

FIGURE 9.3 The Analysis of a Patterson Map.

(a) A two-dimensional Patterson map, $P(u,v)$, of an azidopurine is shown in (1) and its interpretation in (2). The peaks in the $P(u,v)$ map that correspond to the multiple superposition of vectors from ring to ring are lettered A to E and are shown in both (1) and (2).

(b) The $P(u,w)$ map for the same structure, indicating the slope of the ring. The contour interval is arbitrary. One molecule is also shown in (2) for comparison with the Patterson map.

From data in *Acta Crystallographica* **B24** (1968), p. 359.

very large molecules, such as proteins, the overlap becomes hopeless to resolve, except for the peaks arising from the interactions between atoms of very high atomic number.

Sometimes a structure contains a complex molecule, with (necessarily) a multitude of vectors but including a group for which all the vectors are

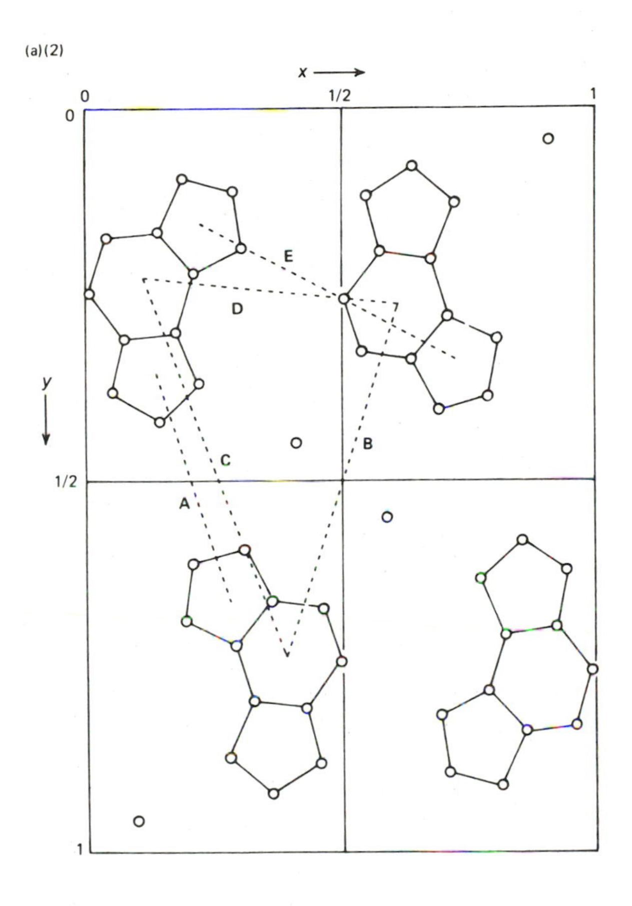

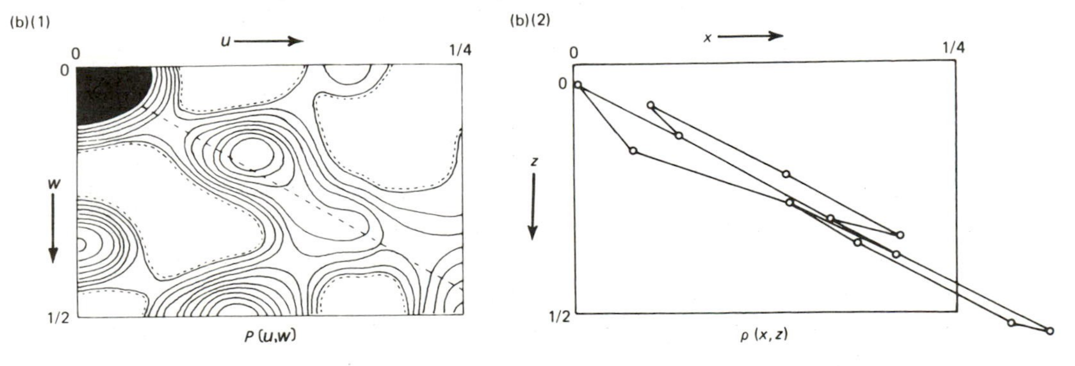

known (relative to one another) rather precisely—for example, a benzene ring in a phenyl derivative. The vector map of this grouping can then be calculated and the arrangement of vectors compared with the arrangement of peaks around the origin of the Patterson map. There will be many more peaks in this region of the Patterson map than those arising from the known structural features alone, but, in at least one relative orientation of the two maps, all peaks in the vector map of the grouping will fall in positive areas of the Patterson map (although they will not necessarily all lie at maxima if the Patterson peaks are composite, as they usually are). This is illustrated in Figure 9.4. The fit of the calculated and observed Patterson maps can be optimized with a computer by making a "rotational search" to examine all possible orientations of one map with respect to the other and to assess the degree of overlap of vectors as a function of the angles through which the Patterson map has been rotated. The maximum overlap normally occurs (except for experimental errors) at or near the proper values of these rotation angles, thus giving the approximate orientation of the group. Then the Patterson map can be searched for vectors between groups in symmetry-related positions, and the exact position of the group in the unit cell can be found and used as part of a trial structure. This method is particularly useful for proteins that have noncrystallographic symmetry.

There are several methods, often quite powerful, for finding the structure corresponding to Patterson maps by transcribing $P(u,v,w)$ upon itself with different relative origins (but always the same orientation). One of the simplest methods of analyzing the Patterson map of a compound containing an atom in a known position is to calculate, graphically or by computer, a vector superposition map. The origin of the Patterson map is put, in turn, at each of the symmetry-related positions of the known atom and the values of $P(u,v,w)$ are noted at all points in the unit cell. The lowest value of P in the different superposed Patterson maps is recorded for each point; the resulting distribution is called a *minimum function.* The principle underlying this approach is that it isolates the vectors arising from the interaction of the known atom with all other atoms in the structure. At points corresponding to the positions of other atoms there will be peaks in the minimum function, each corresponding to a peak in each Patterson map at this position as the origin was moved. A schematic example is illustrated in Figure 9.5. In some of the maps there will be other peaks at this same position, corresponding to other vectors in the structure, but the possible ambiguity that such peaks might introduce is minimized if the lowest value of P in any of the superposed maps is recorded. This method is increasingly powerful when the position of more than one atom is known initially, as in the schematic example illustrated in Figure 9.5.

Sometimes, in crystalline proteins or other macromolecules, there is non-

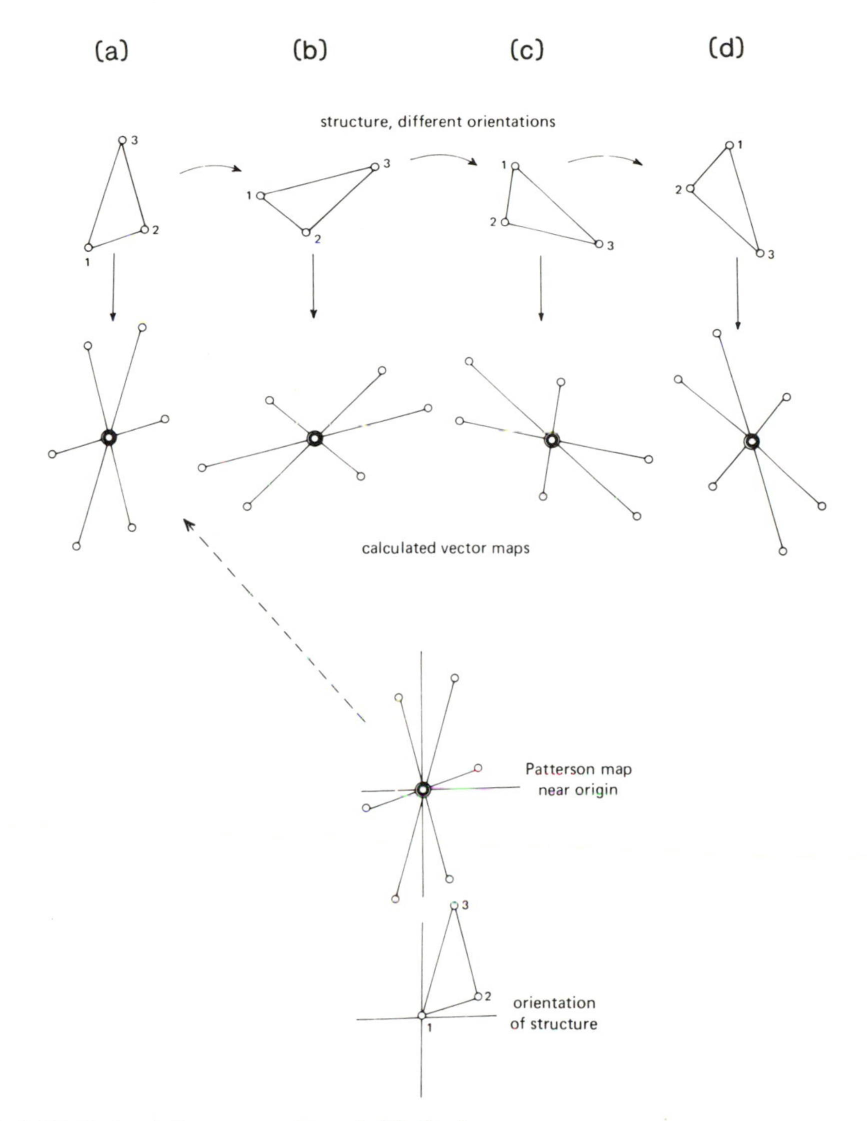

FIGURE 9.4 A Patterson Search Method.

This is a schematic example. If the dimensions of a molecule or part of a molecule in a crystal structure are known, but its orientation (and position) in the unit cell is unknown, the *orientation* may often be found by a comparison of calculated and observed vector maps around the origin. The *position* of the molecule will have to be found in some other way. A comparison of vector maps calculated from trial structures in various orientations with the Patterson map calculated from experimental data indicates that model (a) (above) has the trial structure in its correct orientation. The orientations in (b) to (d) are incorrect.

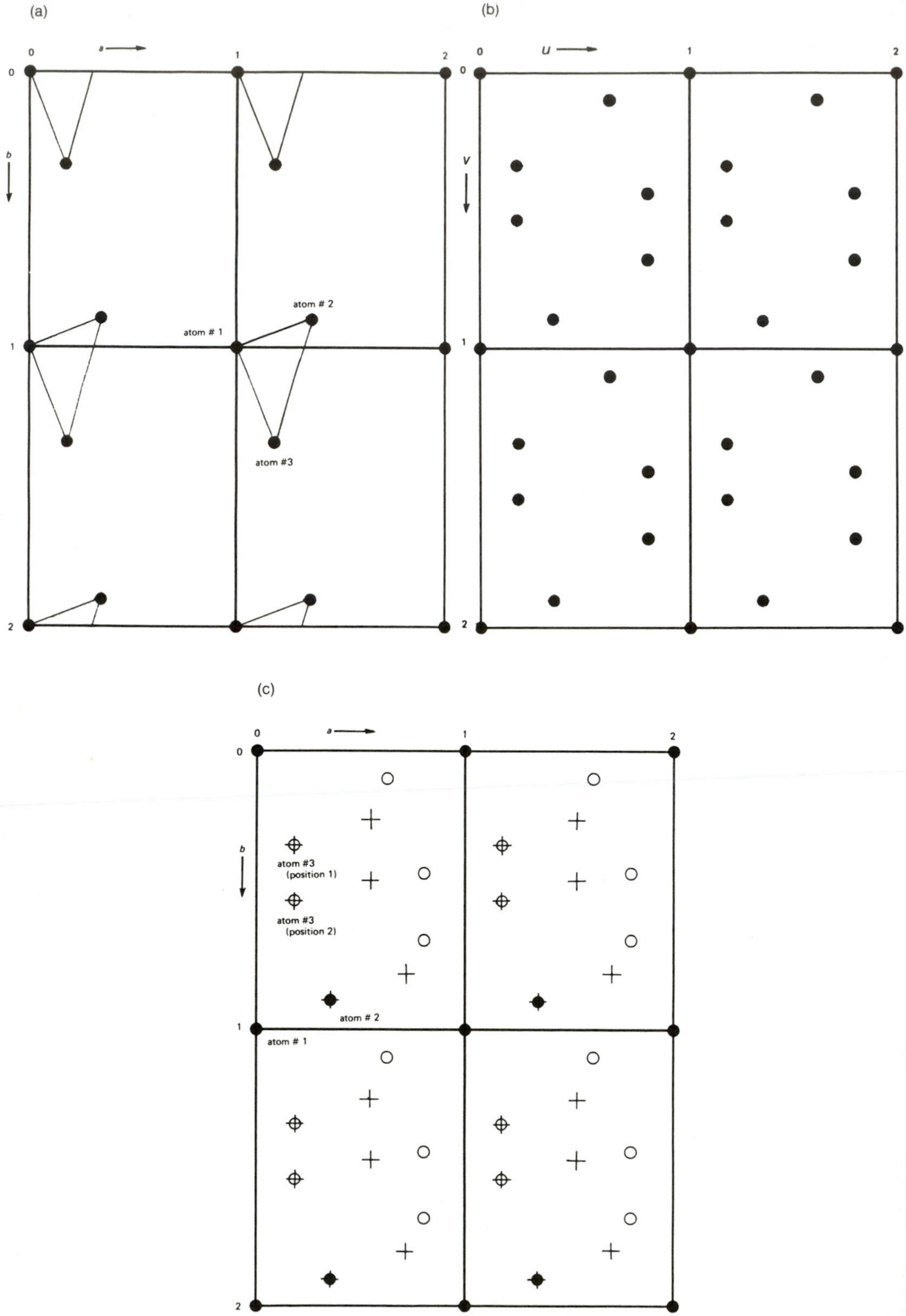

(a)

0 a → 1 2

0

b ↓

atom # 2

atom # 1

1

atom #3

2

(b)

0 u → 1 2

0

v ↓

1

2

(c)

0 a → 1 2

0

b ↓

atom #3
(position 1)

atom #3
(position 2)

atom # 2

1

atom # 1

2

crystallographic symmetry; for example, a dimer of two identical subunits may be contained in the asymmetric unit. Thus there are two identical structures with different positions and orientations in the unit cell. If, however, one copy of the Patterson map of this dimer is rotated on another copy, there will be an orientation of the first relative to the second that gives a high degree of peak overlap. In this way the relative orientations of the two subunits may be determined because the rotations required are directly related to the orientation of the noncrystallographic (local) symmetry element of the dimer, usually a two-fold axis. In a similar way, it may be possible to find the translational components of the noncrystallographic symmetry elements but this is considerably more difficult.

THE HEAVY-ATOM METHOD

In the *heavy-atom method,* one or a few atoms in the structure have atomic number, Z_i, considerably greater than those of the other atoms present. Figure 5.3c shows that if one atom has a much larger atomic scattering factor than the others, then the phase angle for the whole structure will seldom be far from that of the single heavy atom alone, unless, of course, many of the other atoms happen to be in phase with one another, a most improbable circumstance. Heavy atoms thus dominate the scattering of a structure, as illustrated in Figure 5.3c. If the molecule of interest does not contain such an atom, then a derivative, containing, for example, bromine or iodine, can often be prepared, with the hope that the molecular structural features of interest will not be modified in the process.

FIGURE 9.5 Vector Superposition Method.

(a) Crystal structure.

(b) Patterson map.

(c) Vector superposition. A search for the position of a third atom when the positions of the first two (filled circles) are known. The Patterson map illustrated in (b) has been placed with (i) the origin on the position of atom #1 (circles) and (ii) the origin on the position of atom #2 (crosses).
Four unit cells are shown. It can be seen that there are four positions within each unit cell where overlap of Patterson peaks occurs (a circle and cross superposed). Two of these are, necessarily, at the positions of atom #1 (the origin) and atom #2; the other two are possible positions for atom #3, that is, there are two solutions to the vector map at this stage. In practice, this ambiguity is not found when many atoms are present and the method will often show the structure clearly. In higher symmetry space groups, atoms related by crystallographic symmetry operations to atoms at known positions (e.g., heavy atoms) can be used as points to which the origin may be shifted for superposition.

Heavy atoms can usually be located by analysis of a Patterson map, although this depends on how many are present and how heavy they are relative to the other atoms present. In Appendix 8 some data relevant to the Patterson function are given for an organic compound containing cobalt, a derivative of vitamin B_{12} with formula $C_{45}H_{57}O_{14}N_5CoCl \cdot C_3H_6O \cdot 3H_2O$. The appearances of two Patterson projections for this substance are shown in Figure 9.6. In spite of the presence of many other peaks, the cobalt-cobalt peaks are heavier than most of those due to the other vectors and dominate the map. The position of the cobalt atom in the unit cell was thus found from the Patterson map.

Once the heavy atom has been located the assumption is then made that it dominates the diffraction pattern, and the *phase angle for each diffracted beam for the whole structure is approximated by that for the heavy atom.* Figure 9.7 illustrates the application of the heavy-atom method to the structure of the just-mentioned molecule that contained one cobalt atom, one chlorine atom, and about seventy carbon, nitrogen, and oxygen atoms (the structure whose Patterson map is illustrated in Figure 9.6). The first approximation to the electron density was phased with the cobalt atom alone. Peaks in it near the metal atom that were most compatible with known features of molecular geometry were used, together with the metal atom, in a calculation of phase angles for a second approximate electron density map. This process was continued until the entire structure had been found. The combined use of Patterson maps and heavy atom methods made it possible for structures of moderate complexity to be solved in the 1950s and 1960s and, for a while, was the most powerful tool in the analysis of structures of moderate complexity (molecules with, say, 30 to 100 atoms). Direct methods are now more commonly used to solve such structures, these methods having become much more powerful with the greatly increased availability of high-speed, high-capacity computers.

One drawback of the heavy atom method is that when the heavy atom has an atomic number sufficiently high to dominate the vector distribution, it necessarily must dominate the X-ray scattering as well. Consequently, the lighter atoms cannot be located with high precision by any refinement method because their individual contributions to the observed intensities are comparatively slight. If it is desirable to know the structure rather precisely, it is better to use "direct methods" to solve the structure of a compound containing no heavy atom. With many substances, however, the object of the study is a determination of the gross overall molecular structure rather than the fine metric details, so that high precision in the interatomic distances and other aspects of the molecular geometry is less important.

The high scattering power of heavy atoms has been used to help solve the structures of biological macromolecules, with the isomorphous replacement

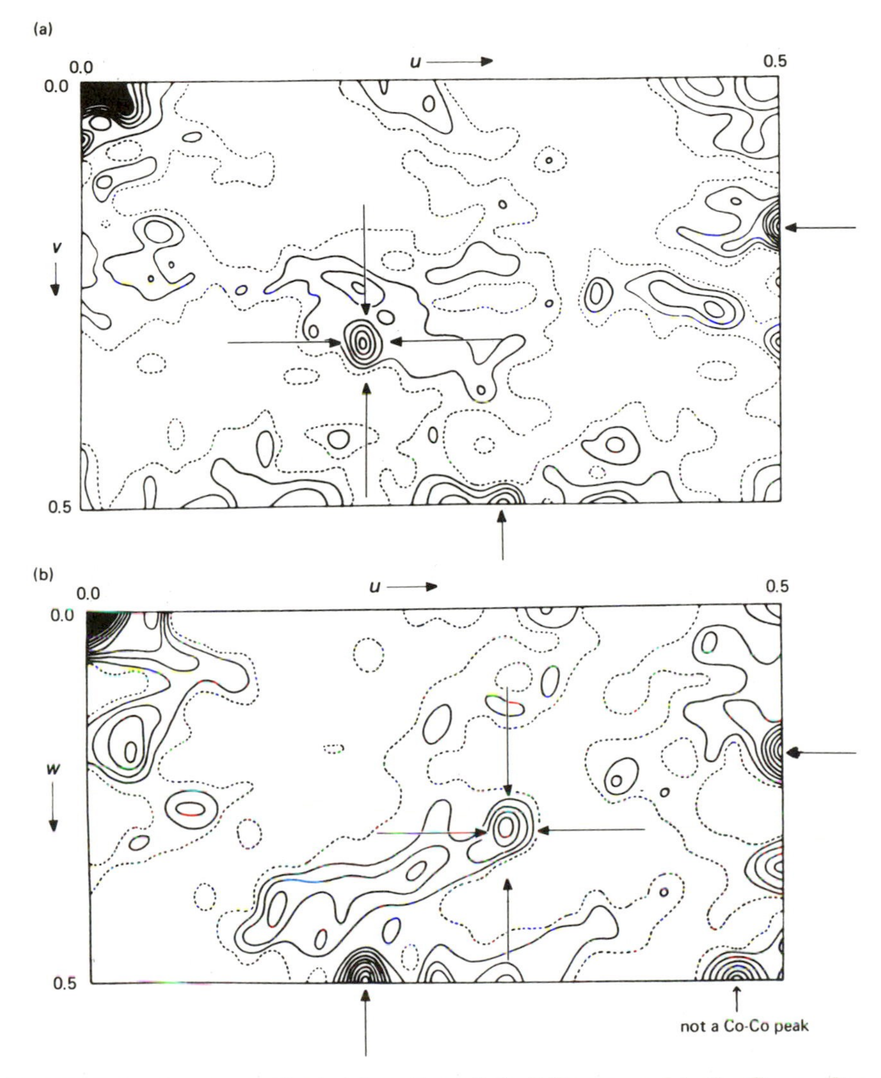

FIGURE 9.6 Patterson Projections for a Cobalt Compound in the Space Group $P2_12_12_1$.

Peaks identified as arising from cobalt-cobalt interactions are indicated by arrows. See Appendix 8 for an analysis of these maps.

(a) $P(u,v)$ Patterson projection down the c-axis. Co-Co peaks appear at 0.00, 0.00; 0.20, 0.32; 0.50, 0.18; 0.30, 0.50.

(b) $P(u,w)$. Patterson projection down the b-axis. Co-Co peaks appear at 0.00, 0.00; 0.30, 0.30; 0.50, 0.20; 0.20, 0.50.

From *Proceedings of the Royal Society,* A **251** (1959), p. 312, fig. 3.

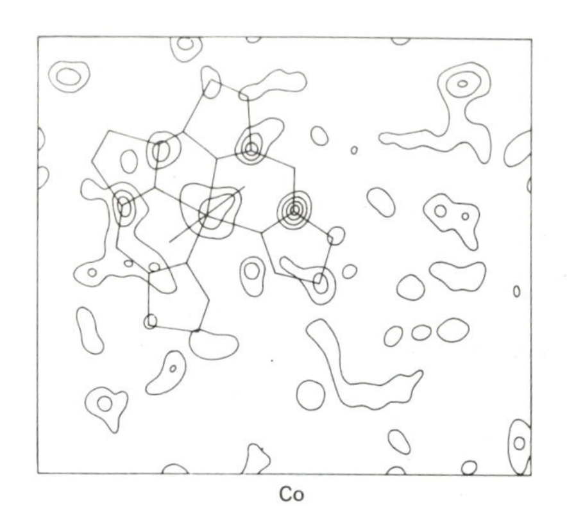

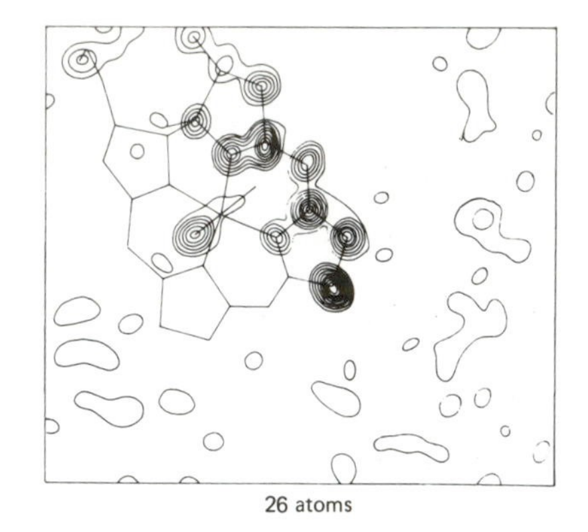

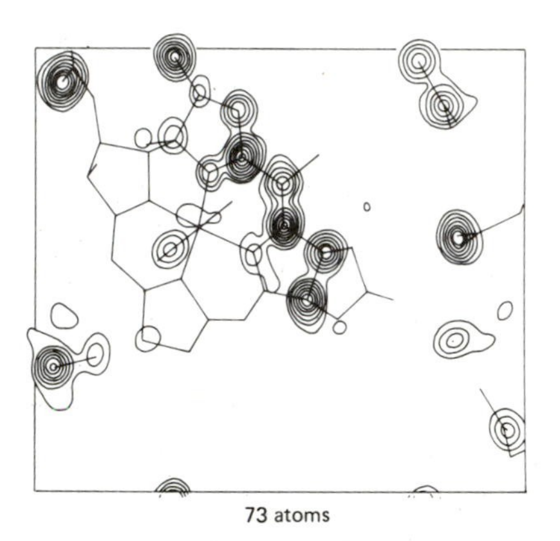

FIGURE 9.7 The Heavy-Atom Method.

One section through a three-dimensional electron density map for a structure with 73 atoms (including various solvent molecules, but not hydrogen atoms) in the asymmetric unit is shown at three different stages of the structure analysis. In the calculation of the first map only a cobalt atom was used to determine phases. For the second map, 26 atoms were used (one Co, 25 C and N), chosen from peaks in the first map. The third map was phased with the positions of all 73 atoms. Most of the features of the map phased on 73 atoms can be found, at least weakly, in the map phased with the heavy atom alone, although in the latter map there are many extra peaks that do not correspond to any real atoms. Note the general reduction in the background density as the true phase angles are approached. Since these are sections of a three-dimensional map, some atoms that lie near but not in the plane of the section are indicated by lower peaks than would represent them if the section passed through their centers. Other atoms implied by the skeletal formula lie so far from this section that no peaks corresponding to them occur here.

From *Proceedings of the Royal Society,* A **251** (1959), p. 328, fig. 14.

method that will be discussed next. The Patterson map of a protein is too complex for direct interpretation but the location of a heavy atom in a protein is possible if data for both the protein and its "heavy atom derivative" are used to determine perturbations to intensities caused by the addition of heavy atoms. For example, if a protein has a molecular weight of 24,000 it contains approximately 2,000 carbon, nitrogen, and oxygen atoms. Then at $\sin\theta = 0°$, the mean value of $\langle |F_P|^2 \rangle = \sum_{j=1}^{2,000} f_c^2 = 98{,}000$. If one uranium atom, atomic number 92 is added, this value of $\langle |F_{PH}|^2 \rangle$ is increased to approximately 106,000, an 8 per cent change in average intensity. With careful experimental techniques differences in intensities of the native protein and the heavy atom derivative can be measured; many of these differences will significantly exceed the average.

If a protein crystal is soaked in a solution of a heavy atom compound (such as a uranyl salt), the heavy atom compound will be distributed throughout the solvent channels in the crystal by diffusion. In some cases the heavy atom will bind to a specific group on the macromolecule and this binding will occur in an ordered arrangement within the macromolecular crystal. The difference between diffracted intensities of the "heavy-atom derivative" of the protein crystal (structure factor F_{PH}) and the diffracted intensities of the "native protein" without any added heavy atom (structure factor F_P) can be used to reveal the position of the heavy atom. The Patterson map (with $||F_{PH}|^2 - |F_P|^2|$ as coefficients) will indicate the position of the heavy atom. With this method of soaking heavy atoms into the native protein it often happens that nonspecific binding of the heavy atom occurs, since there are many binding groups on the surface of a protein. If this does happen, that particular heavy-atom derivative cannot be used. It is necessary to stop the soaking after an appropriate time, determined experimentally, in the hope that only specific binding will occur. The concentration of the heavy-atom salt is often critical. An alternative method, which involves attempting to crystallize protein from solutions containing heavy-atom salts, has not been very satisfactory because the crystals so obtained are often not isomorphous with the native crystal. These crystals must be isomorphous for the use of the isomorphous replacement method that will now be described.

ISOMORPHOUS REPLACEMENT METHOD AND PROTEIN STRUCTURE DETERMINATION

When a pair or a series of isomorphous crystals can be found, isomorphous replacement is a powerful method for the determination of *phase angles*, especially for complex structure for which purely analytical methods (see

Chapter 8) are inadequate. It has provided the basis for the solution of most protein structures determined to date.

Ismorphous crystals are crystals with essentially identical cell dimensions and atomic arrangements but with a variation in the nature of one or more of the atoms present. The alums constitute probably the best-known example of a series of isomorphous crystals. "Potash alum," $KAl(SO_4)_2 \cdot 12H_2O$, grows as colorless octahedra, while "chrome alum," $KCr(SO_4)_2 \cdot 12H_2O$, forms dark lavender crystals of the same shape and structure. The Cr(III) atom in chrome alum is in the same position in the unit cell as the Al(III) atom in potash alum. A common experiment in isomorphism is to grow a crystal of chrome alum suspended from a thread and then to continue to grow it in a solution of potash alum. The result is an octahedral crystal with a dark center surrounded by colorless material.

In general, however, isomorphous pairs (involving isomorphous replacement of one atom by another) are difficult to find for crystals with small unit cells, because variations of atomic size usually cause significant structural changes when substitution is tried. Even with large unit cells, patience and ingenuity are usually needed to find an isomorphous pair for a compound being studied. Perutz and Kendrew searched for years before finding isomorphous derivatives of the proteins hemoglobin and myoglobin. As mentioned earlier it is possible for salts or other compounds of heavy atoms to penetrate the channels between molecules in protein crystals—spaces normally occupied by salt or water molecules—and sometimes an attachment to an active group on the protein occurs. With luck, since the protein is already crystallized, the only change that occurs in the structure (by this isomorphous addition) is a slight distortion in the area of the protein that has reacted with the heavy-atom compound. The existence of isomorphism between a protein and a heavy-atom derivative may be demonstrated by the determination that their unit cell dimensions do not differ by more than about 0.5 per cent, and that there are differences in the diffraction intensity patterns.

In fact, in work with proteins it is necessary to obtain several heavy-atom derivatives, substituted in different places on the molecule. After a Patterson map has been calculated and the position of the heavy atom has been found for each derivative, it is possible to calculate the phases of the reflections directly by a proper consideration of the changes in intensity from one crystal to another. With a crystal as complex as a protein, the labor involved is enormous, for example, measuring the intensities of about 10^4 to 10^5 reflections from each of perhaps four different crystalline materials (with many specimens of each often needed because of radiation damage), and then making appropriate intercomparisons of them all. On the other hand, the rewards are enormous—the entire three-dimensional molecular architecture of the protein molecule, found with minimal chemical assumptions.* Techniques for preparing isomorphous protein derivatives are continually being refined and, as experience has accumulated, the task has become easier. The structures of many proteins are now being studied with great success.

The isomorphous replacement method for calculating phase angles is illustrated in Appendix 9 for a centrosymmetric structure. The atoms or groups of atoms (M and M′) that are interchanged during preparation of the isomorphous pair must first be located, usually from a Patterson map, as described earlier. This allows calculation of their contributions, F_M and $F_{M'}$. If F_M and $F_{M'}$ are positive (they necessarily have the same sign, since their only difference is in the amount of scattering power in the atom or group of atoms), then the overall F values (F_T) must differ in the same way that F_M and $F_{M'}$ differ. Since the absolute magnitudes of these measured values of F are known and the difference equivalent to the change in M can be computed, it is possible to find signs for F_T and $F_{T'}$. The reader should study the numerical example in Appendix 9 in order to understand the principle of the method. The solutions to the equations are in practice inexact, because of experimental errors and also because the remainder of the structure, R, may move slightly during the replacement of one ion or group by another.

With noncentrosymmetric structures, the situation is greatly complicated by the fact that the phase angle may have any value from 0 to 360°. As just noted, even with centrosymmetric crystals, for which only a sign need be determined for each reflection, there may be difficulties because of slight shifts in the unaltered parts of the structure. Proteins, by virtue of their asymmmetry, necessarily crystallize in noncentrosymmetric space groups. When a heavy atom is substituted in a protein the isomorphism is between the unsubstituted protein and a heavy-atom derivative or between several heavy atom derivatives. As a result of the insertion of a bulky heavy atom, the resulting displacement of the remainder of the structure is greatest in the area of the substitution by this heavy atom, a fact that may cause some trouble if this area is near a site of interest, for example, the active site of an enzyme. When a phase angle is to be determined, at least two pairs of isomorphous compounds are required (unless anomalous dispersion is used—see later) and, because there can be errors in phase angles so found, it is better to study even more pairs. The method is illustrated schematically in Figure 9.8. If the heavy-atom position can be found from the Patterson map, then F_H and the phase angle α_H can be computed for a given diffracted beam for each derivative. The construction for graphical determination of the phase angle for the protein (P) is shown in Figure 9.8. For each heavy-atom

*Usually information about the sequence of the amino acids in the protein chain is needed to interpret the electron density maps, especially in poorly resolved regions of the structure.

derivative (PH_1 and PH_2) two possible values* for the phase angle for the protein are found; in the example in Figure 9.8, these are near 53° and 323° for PH_1 (Figure 9.8a) and near 55° and 155° for PH_2 (Figure 9.8b). The phase angle for the free protein for this particular diffracted beam must therefore be near 54°. This process of estimating the phase angle must be repeated for each diffracted beam in the diffraction pattern; usually more than two heavy-atom derivatives (in addition to the free protein) are studied, so that the phases can be more accurately determined.

*With reference to Figure 9.8a, the law of cosines also illustrates the two-fold ambiguity in the phase angle determined from just one heavy-atom derivative.

$$\alpha_P = \alpha_H + \cos^{-1}\{(F_{PH}^2 - F_P^2 - F_H^2)/2F_PF_H\} = \alpha_H \pm \alpha'$$

Thus two values are possible and it is necessary to study several heavy atom derivatives with substituted heavy atoms in different positions in the unit cell; the common value of α_P is determined.

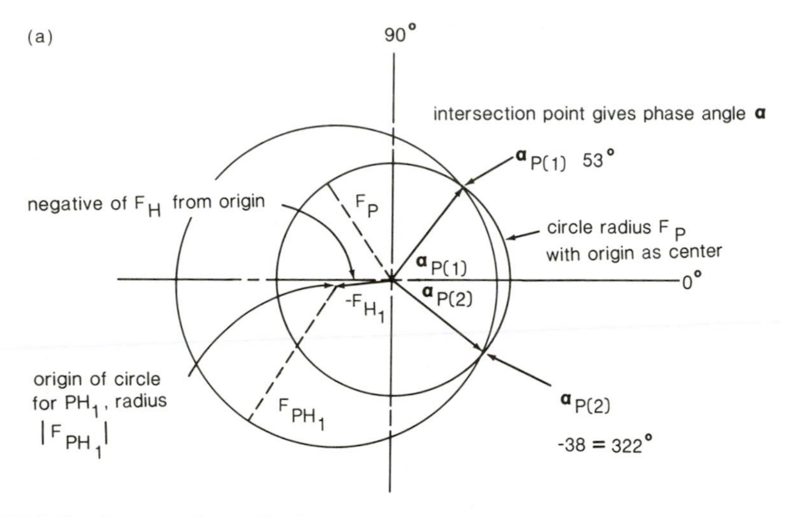

FIGURE 9.8 Isomorphous Replacement for a Noncentrosymmetric Structure. Graphical evaluation of the phase, α_P, of a beam, indices *hkl*, diffracted with a structure factor $|F_P|$ from a protein crystal. The diagrams illustrate the following equation: $\mathbf{F}_P = \mathbf{F}_{PH} - \mathbf{F}_H$.

(a) One heavy-atom derivative is available, with a structure factor $|F_{PH_1}|$ for reflection *hkl*. A circle with radius $|F_P|$ is drawn about the origin. From the position of the heavy atom, determined from a difference Patterson map, it is possible to calculate both the structure amplitude and phase of the heavy-atom contribution ($|F_{H_1}|$, phase angle α_{H_1}). A line of length $|F_{H_1}|$ and phase angle $-\alpha_{H_1}$ (i.e., α_{H_1} + 180° to give $-\mathbf{F}_{H_1}$) is drawn. With the end of this vector as center, a circle with radius $|F_{PH_1}|$ is drawn. It intersects the circle with radius $|F_P|$ at two points, corresponding to two possible phase angles, $\alpha_{P(1)}$ and $\alpha_{P(2)}$, for the native protein.

A measure of the error in phasing is provided by a figure of merit, m. This is the mean cosine of the error in phase angle; it is near unity if the circles used in deriving phases (Figure 9.8) intersect in approximately the same positions. For example, if the figure of merit is 0.8 the phases are in error, on the average, by $\pm 40°$, if it is 0.9 the mean error is $\pm 26°$.

Thus the stages in the determination of the structure of a protein involve the crystallization of the protein, the preparation of heavy-atom derivatives, the measurement of the diffraction patterns of the native protein and its heavy-atom derivatives, the determination of the heavy-atom positions, the computation of phase angles (Figure 9.8), and the computation of an electron density map using native protein data and the phase angles so derived from isomorphous replacement. The map is then interpreted in terms of the known geometry of polypeptide chains so that initially this backbone of the

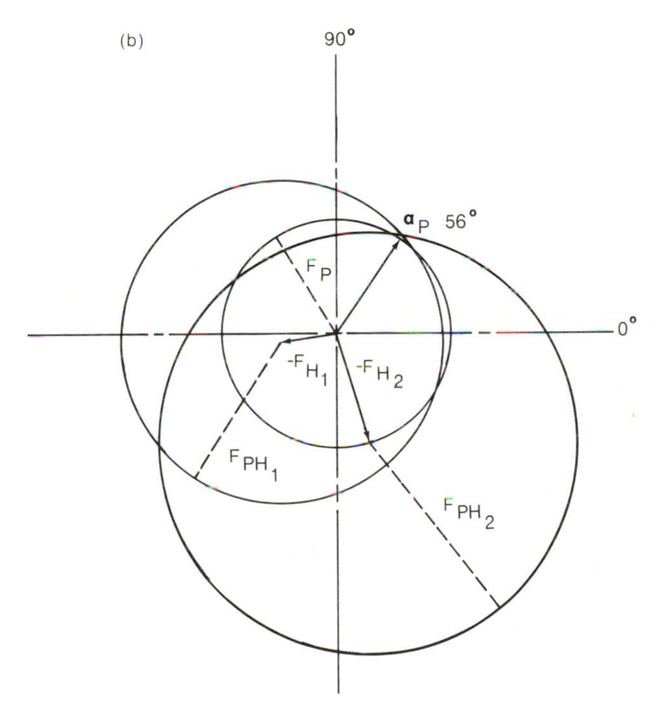

(b) When two or more heavy-atom derivatives are available, then the process described in (a) is repeated and, in favorable circumstances, only one value of the phase angle for the native protein is obtained. Thus, a second derivative is needed to remove the two-fold ambiguity of case (a). This method, of course, depends on accurate measurements of $|F_P|$, $|F_{PH_1}|$ and $|F_{PH_2}|$ and involves the assumption that no other perturbation than the addition of a heavy atom to native protein has occurred. Additional derivatives are sometimes needed to improve the accuracy of the phase angles.

protein is traced through the electron density map. This was formerly done by model building (using a half-silvered mirror in a "Richards box" so that a ball-and-stick model and the electron density map are superimposed and therefore can be visually compared). Nowadays it is more common for this interpretation of the electron density map to be done with the help of computer graphics (as shown in Figure 9.9).

Since the amount of work involved in analysis of protein structure is so great, it is the practice to start with only a small fraction of the diffraction pattern, that observed at small scattering angles. At these low scattering angles the isomorphism is effectively more exact because the differences in positions caused by the substitution do not affect the low-angle reflections as much as they do those at higher angles. However, this means that initially the resolution is poor; no details of the structure can be seen in the corresponding electron density maps. A typical example of the way in which the appearance of a given structure changes as the resolution is improved is shown in Figure 9.10.

The variation of resolving power with scattering angle in structural diffraction studies has direct analogy with the resolution of an ordinary microscope image. If some of the radiation scattered by an object under examination with a microscope escapes rather than being recombined to form an image (this situation is shown in Figure 1.1), the image that is formed will be, to some degree, an imperfect representation of the scattering object. More particularly, fine detail will remain unresolved. Similarly, with X rays, if the diffraction pattern for the customary wavelengths is observed only out to a relatively small scattering angle, the resolution of the corresponding image reconstructed from it will be low. Furthermore, the resolution will be limited by the wavelength chosen even if the entire pattern is observed. As in any process of image formation by recombination of scattered radiation, detail significantly smaller than the wavelength used cannot be distinguished by *any* scheme. On the other hand, the positions of well-resolved objects of known shape can be measured with high precision, and fortunately *all* interatomic distances are well resolved in three dimensions with the usual X rays. Hence usually the positions of the resolved atoms can be measured and the details of molecular geometry calculated quite precisely.

SUMMARY

The Patterson Map The map computed with amplitudes $|F|^2$ and no phase information will give a vectorial representation of the atomic contents of the unit cell. The Patterson function, $P(u,v,w)$, is expressed in the coor-

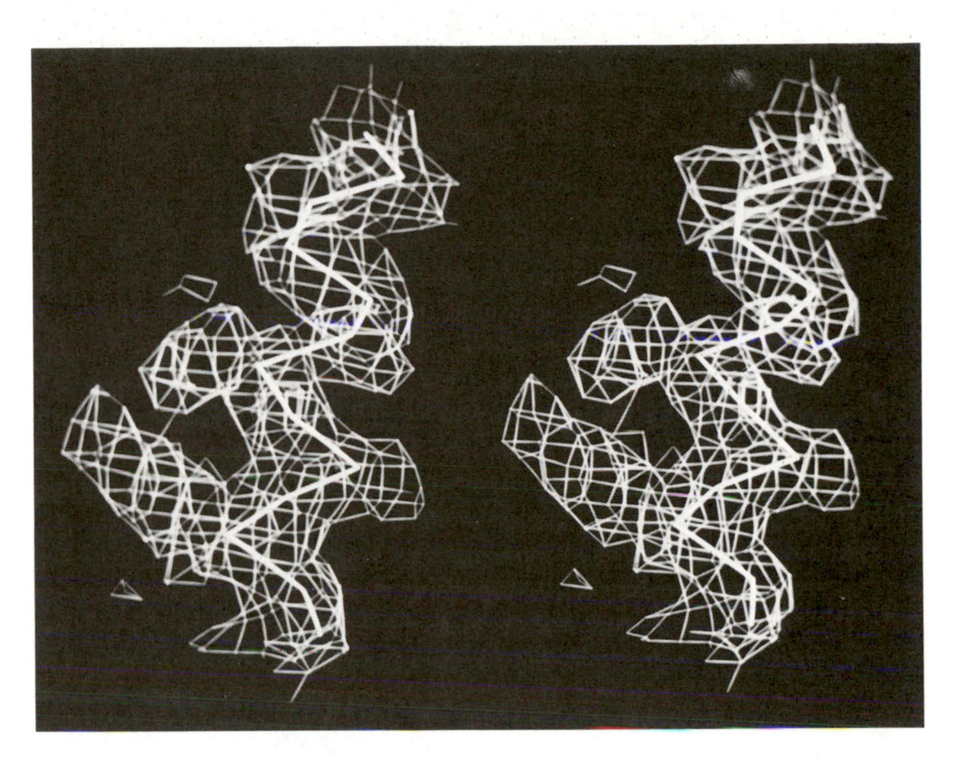

FIGURE 9.9 Protein Backbone Fitting by Computer-based Interactive Graphics.

Areas of high electron density are stored in the computer as three-dimensional information and are represented by cagelike structures on a video screen. Any desired view can be generated. The backbone of the molecule is represented as a series of vectors, each 3.8 Å in length (the distance between α-carbon atoms in a polypeptide chain). Each vector is positioned with one end on an α-carbon atom; the other end of the vector is rotated until it lies in an appropriate area of high electron density. Then coordinates of both ends of the vector are stored in the computer, the process is repeated and the most likely location of the next α-carbon atom is sought. Such vectors are represented in this figure, a stereopair* photographed from a video-screen, as heavy solid lines. In this way the "backbone," that is, the positions of the carbon atoms of the polypeptide backbone of the protein (excluding side chains) may be found.
(Photograph courtesy Dr. H. L. Carrell.)

*Such stereodiagrams can be viewed with stereoglasses or the reader can focus on the two images until an image between them begins to form, and then allow his eyes to relax until the central image becomes three-dimensional. This process requires practice and usually takes 10 seconds or more.

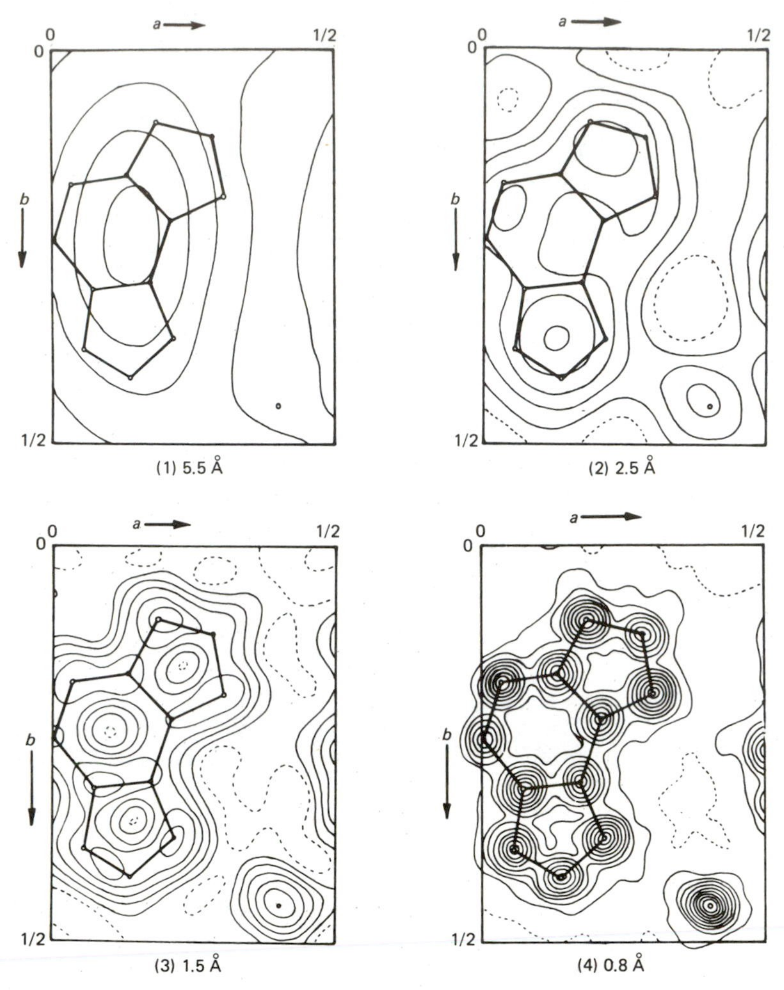

FIGURE 9.10 Different Stages of Resolution for a Given Structure.

The electron density maps shown were calculated after eliminating all observed $|F|$ measured beyond a given 2θ value. The "resolution" obtained is usually expressed in terms of the interplanar spacings $d = \lambda/2 \sin \theta$, corresponding to the maximum observed 2θ values ($\lambda = 1.54$ Å for copper radiation in this example).

	d	*Maximum 2θ*	*Relative number of reflections included in calculation*
(1)	5.5Å	16°	7
(2)	2.5	36	27
(3)	1.5	62	71
(4)	0.8	162	264

In each of the maps, the skeleton of the actual structure from which the data were taken has been superimposed. The first stage (1) (upper left) is typical of dinate system u,v,w in a cell of the same size and shape as that of the crystal. It is calculated by:

$$P(u,v,w) = \frac{1}{V}\sum_{\text{all } h,k,l}\sum\sum |F|^2 \cos 2\pi(hu + kv + lw)$$

The peaks in this map occur at points whose distances from the origin correspond in magnitude and direction with distances between atoms in the crystal, because:

$$P(u,v,w) = V \int\int\int \rho(x,y,z)\rho(x + u, y + v, z + w)\, dx\, dy\, dz$$

Ideally this map can be interpreted in terms of an atomic arrangement. In practice, however, this is only possible if there are comparatively few atoms in the structure or if some are very heavy.

The Heavy-atom Method If one or a few atoms of high atomic number are present, they will dominate the scattering. If these atom(s) can be located from a Patterson map (or in some other way), the phases of the entire structure may be approximated by the phases of the heavy atom(s). In the resulting electron density map, portions of the remainder of the structure will usually be revealed, leading to improved phases and successively better approximations to the structure.

Isomorphous Replacement Method This is the best method for the experimental determination of phase angles and is, at present, the only practical approach for solving very large structures, such as those of proteins. Two crystals are isomorphous if their space groups are the same and their unit cells and atomic arrangements are essentially identical. Suppose that atoms may be added to or replaced in a molecule to produce new crystals isomorphous with those formed by the parent molecule. If the positions of these added or replaced atoms can be found from Patterson maps, their contributions to the phase angle of each reflection can be calculated, and if the atoms are sufficiently heavy, differences in intensities for the two isomorphs can be used to determine the approximate phase angle for each reflection.

---

those encountered early in the determination of a protein structure. For protein structures, a degree of resolution between (2) and (3) is generally as much as is possible. The detail shown in (4) is characteristic of a structure determination with good crystals of low molecular weight compounds with radiation from an X-ray tube with a copper target. These electron density maps may be compared with views of an object through a microscope, each corresponding to a different aperture.

From data in *Acta Crystallographica* **B24** (1968), 359–366.

10

Anomalous Dispersion and Absolute Configuration

The concept of the carbon atom with four bonds extending in a tetrahedral fashion was put forward by van't Hoff and Le Bel in 1874. It coincided with the realization that such an arrangement could be asymmetric if the four substituents were different, as shown in Figure 10.1a. Thus, for any compound containing one such asymmetric carbon atom, there are two isomers of opposite chirality (called enantiomers or enantiomorphs) for which three-dimensional representations of their formulas are related by a mirror plane. Solutions of these enantiomers rotate the plane of polarized light in opposite directions. As discussed in Chapter 7, Pasteur showed that crystals of sodium ammonium tartrate had small asymmetrically located faces and that crystals with these so-called "hemihedral faces" rotated the plane of polarization of light clockwise, while crystals with similar faces in mirror-image positions rotated this plane of polarization counterclockwise. Thus the external form (that is, the morphology) of the crystals illustrated in Figure 10.1b could be used to separate enantiomers.

But even if the chemical formula and the three-dimensional structure of a molecule such as tartaric acid have been determined by standard X-ray diffraction methods, there is an ambiguity about the absolute configuration. Information about the absolute configuration is not contained in the diffraction pattern of the crystal as it is normally measured. Thus, even though the substituents on the asymmetric carbon atoms are known, and even the detailed three-dimensional geometry of the molecule is known, it is not known which of the two enantiomers (mirror-image forms, analogous to those shown in Figure 10.1a) represents the three-dimensional structure of the molecule that has some particular chiral property, such as rotating the plane of polarized light to the right (the dextrorotatory form).

The means of determining the absolute configurations of molecules was provided by X-ray crystallographic studies. It was made possible by the observation that the absorption coefficient of an atom for X-rays shows discontinuities when plotted as a function of the wavelength of the incident X-radiation. These discontinuities, shown in Figure 10.2, are graphically described as "absorption edges." At wavelengths at and below but near the absorption edge, the energy of the X-radiation (inversely proportional to its wavelength) is sufficient either to excite an electron in the strongly absorbing atom to a higher quantum state or to eject the electron completely from the

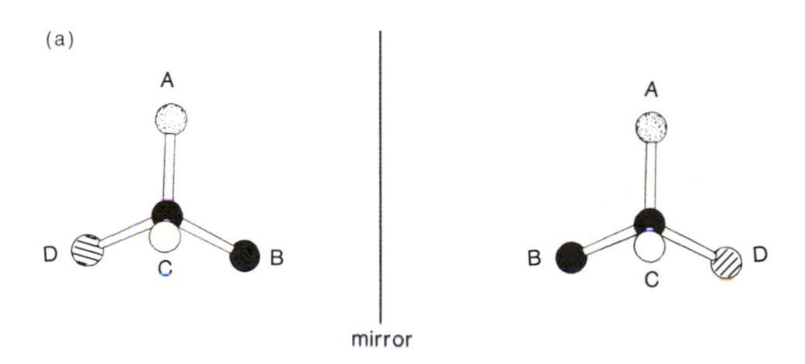

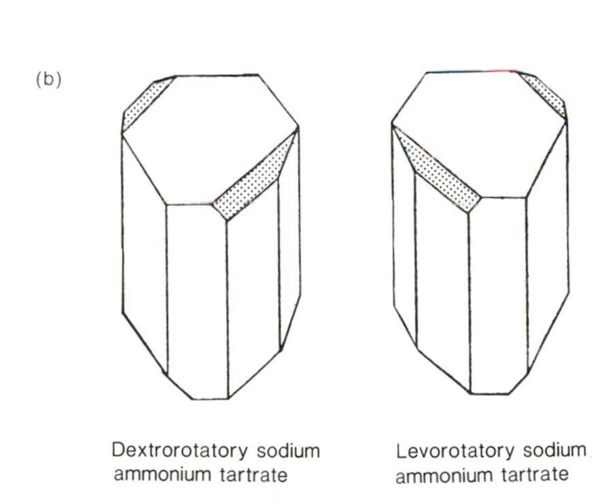

FIGURE 10.1 Absolute Configurations.

(a) The asymmetric carbon atom. If A, B, C, and D attached to the tetrahedral carbon atom are all different, there are two chiral isomers related to each other by a mirror plane. In a similar way the entire structure of a crystal may be chiral.

(b) Hemihedral faces on sodium ammonium tartrate crystals (used by Pasteur to differentiate dextro- from levo-rotatory forms).

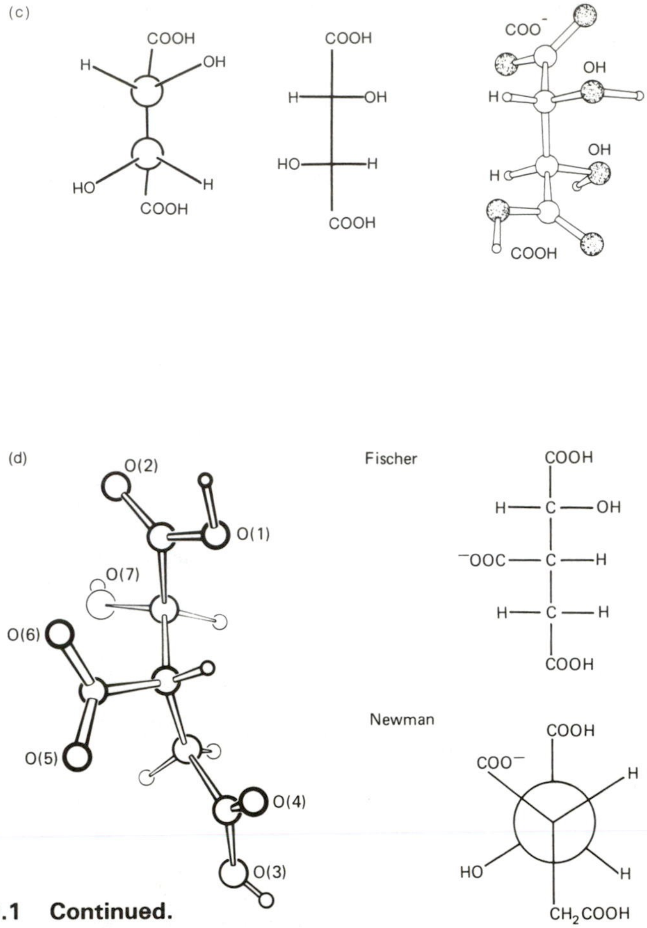

FIGURE 10.1 Continued.

(c) Absolute configuration of (+)-tartaric acid (dextrorotatory tartaric acid). Note that in the actual structure (right) the chain of four carbon atoms has effectively a planar zig-zag arrangement. In the formula on the left, and by convention in all "Fischer formulas," vertical carbon chains are represented as planar but with successive bonds always directed into the page. Thus, in the formulas in the center and left here, the lower half of the molecule has been rotated 180° relative to the upper half as compared to the actual structure. This affects the conformation but not the absolute configuration of the molecule. The conformation of tartaric acid illustrated on the left of (c) is a possible one for this molecule, but it is of higher energy (because bonds are eclipsed) than that shown on the right, the conformation observed in these crystals. Still other conformers may exist in solution or in other crystals.

(d) Absolute configuration of the potassium salt of (+)isocitric acid (isolated from the plant *Bryophyllum calycinum*). Fischer and Newman formulas are shown. The correct designation of this enantiomer is (1*R*:2*S*-1-hydroxy-1,2,3-propanetricarboxylate). The torsion angles are shown in Figure 12.2.

From *Acta Crystallographica* **B24** (1968), p. 585, Fig. 4.

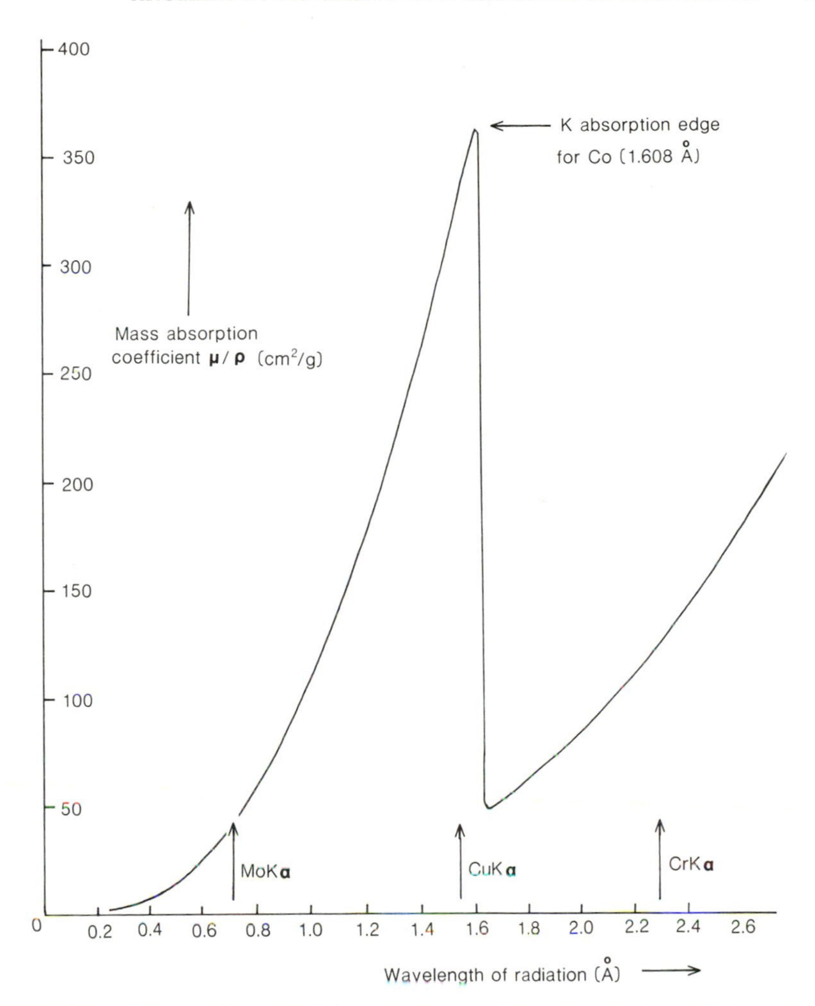

FIGURE 10.2 Absorption of X-Rays of Various Wavelengths by a Cobalt Atom. The mass absorption coefficient for cobalt as a function of wavelength. Note the discontinuity near the absorption edge (1.608 Å) and, elsewhere, the gradual increases as the wavelength of the radiation increases.

atom; this has an effect on the phase change on scattering. The scattering factor for the atom becomes "complex," and the factor f is replaced by

$$f_i + \Delta f_i' + i\Delta f_i'' \tag{10.1}$$

where $\Delta f_i'$ and $\Delta f_i''$ vary with the wavelength of the incident radiation, and $\Delta f_i''$ is largest when this wavelength is near the absorption edge (as for cobalt in a structure studied with copper Kα radiation; see Figure 10.2).

Thus if an atom in the structure absorbs, at least moderately, the X rays being used, then this absorption will result in a *phase change for the X rays*

NO ANOMALOUS DISPERSION

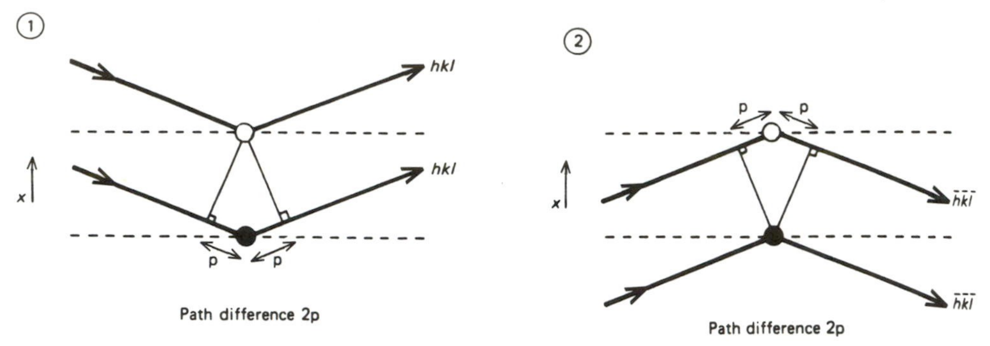

$I_{hkl} = I_{\bar{h}\bar{k}\bar{l}}$ (same path differences on diffraction)

ANOMALOUS DISPERSION

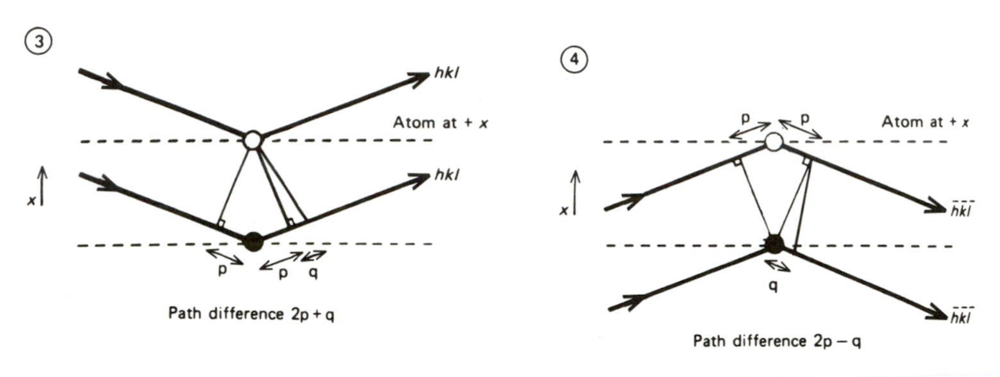

Black atom instantaneously advances wave + q. $I_{hkl} \neq I_{\bar{h}\bar{k}\bar{l}}$ (different path differences on diffraction)

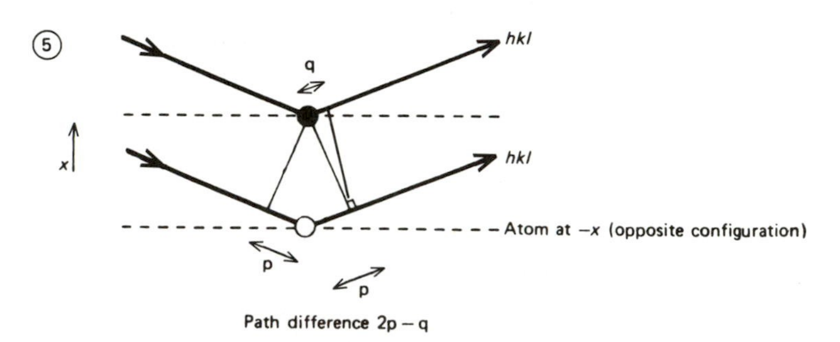

$I_{\bar{h}\bar{k}\bar{l}}$ (atom at $+x$) = I_{hkl} (atom at $-x$)

FIGURE 10.3 Path Differences on Anomalous Scattering.

Effect of anomalous scattering on the path lengths of diffracted X-ray beams. Suppose that for a particular reflection the anomalous scatterer (black circles) causes in effect a path difference, q, in addition to the usual difference of 2p *scattered by that atom, relative to the phase of the X rays scattered by the other atoms* of the structure, the equivalent of advancing or delaying the radiation for a short time [that is, a hesitation ("gulp") before scattering]. This phase change is equivalent in its effect to changing the path length through which the scattered radiation travels, as illustrated schematically in Figure 10.3, and as a result of the altered path lengths there is an effect on the intensities. The effect is found particularly when a heavy atom is present and the wavelength of the X rays used is near an absorption edge for that atom. When there is none of this so-called "anomalous scattering" or "anomalous dispersion," the intensities of the reflections with indices h, k, l and $\bar{h}$, $\bar{k}$, $\bar{l}$ (that is, $-h$, $-k$, $-l$), are the same (Friedel's law); when it is present, the intensities of these two reflections are different because of changes in effective path differences arising from the phase change on absorption by the anomalously scattering atom.

This difference in intensities may alternatively be thought of as a result of the complex nature of f_i [see Eq. (10.1)] so that the absolute value of $\mathbf{F}(hkl)$ is different from that of $\mathbf{F}(-h,-k,-l)$, as illustrated in Figure 10.4. We showed in Eq. (10.1) that if there is an anomalous scatterer in the crystal, f is replaced by $f + \Delta f' + i\Delta f''$. Let $A' = G(f + \Delta f') + A$ and $B' = H(f + \Delta f') + B$, where A and B refer to the rest of the structure and G and H to the anomalous scatterer. As a result since

$$F(hkl) = (A' + iG\Delta f'') + i(B' + iH\Delta f'')$$

$$= (A' - H\Delta f'') + i(B' + G\Delta f'') \qquad (10.2)$$

$$|F(hkl)|^2 = (A' - H\Delta f'')^2 + (B' + G\Delta f'')^2 \qquad (10.3)$$

---

between the radiation scattered by a normal scatterer at this position and by a normal scatterer at some other position (open circles). As shown, the path difference for the hkl reflection with anomalous scattering is 2p + q and for the $\bar{h}\bar{k}\bar{l}$ reflection is 2p − q. If no anomalous scattering had occurred these would be the same—namely, 2p. Since the intensity of a diffracted beam depends on the path difference between waves scattered by the various atoms in the unit cell, the result of anomalous scattering is an intensity difference between hkl and $\bar{h}\bar{k}\bar{l}$. It is possible to compute values of $|F(hkl)|$ and $|F(\bar{h}\bar{k}\bar{l})|$ and see which should be the larger. If for many reflections the relations of the calculated values to the experimentally measured values are the same as those computed for the model, then the model had the correct handedness (configuration); if not, the configuration of the model must be changed. That is, if $|F_o(hkl)| > |F_o(\bar{h}\bar{k}\bar{l})|$, then we must necessarily have $|F_c(hkl)| > |F_c(\bar{h}\bar{k}\bar{l})|$. See Appendix 10.

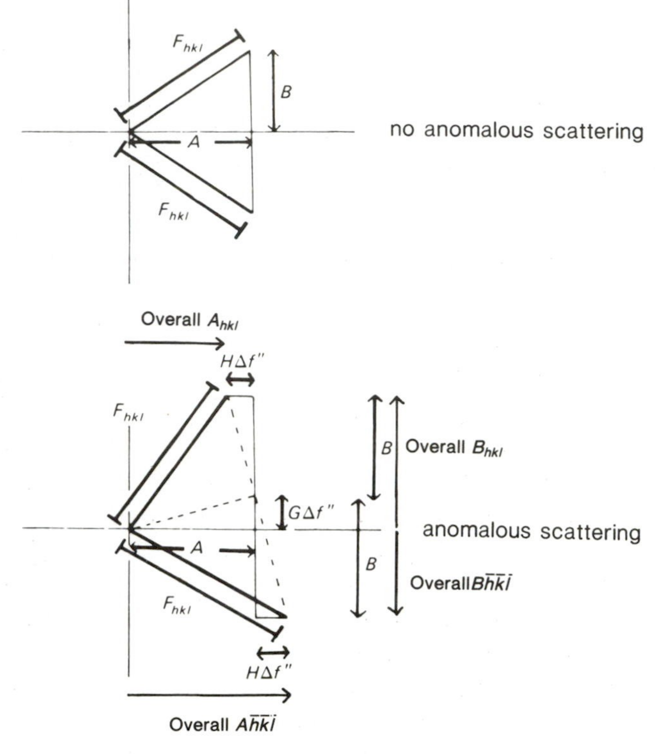

FIGURE 10.4 Effects of Anomalous Scattering on F Values.

The top diagram shows the structure factor vectors for $\mathbf{F}(hkl)$ and $\mathbf{F}(-h, -k, -l)$ in the absence of anomalous scattering and the lower diagram shows the effect of anomalous scattering on such a diagram. When there is anomalous scattering $A(hkl) \neq A(-h, -k, -l)$. The same occurs for B values as shown (see Chapter 5 and especially Figure 5.1 for the fundamentals of such diagrams).

and similarly

$$F(\bar{h}\bar{k}\bar{l}) = (A' + iG\Delta f'') - i(B' + iH\Delta f'')$$

$$= (A' + H\Delta f'') - i(B' - G\Delta f'') \tag{10.4}$$

$$|F(\bar{h}\bar{k}\bar{l})|^2 = (A' + H\Delta f'')^2 + (B' - G\Delta f'')^2 \tag{10.5}$$

Then it follows that

$$|F(hkl)|^2 - |F(\bar{h}\bar{k}\bar{l})|^2 = 4\Delta f''(B'G - A'H) \tag{10.6}$$

Thus, when the incident X-radiation is of a wavelength near that of the absorption edge of an atom in the structure, $|F(hkl)|$ does not equal $|F(-h,-k,-l)|$; in the normal condition the wavelength of the X-radiation

is far from any absorption edge and these two quantities are equal. The magnitude in the difference in $|F|^2$ for the two reflections (called "Friedel pairs") is a function of $\Delta f_i''$ (which depends on how near the absorption edge the incident radiation is) and the positional parameters of both the anomalous scatterer and the rest of the structure.

It is possible from Eq. (10.4) to calculate the expected differences between $|F(hkl)|^2$ and $|F(\overline{h}\overline{k}\overline{l})|^2$ for a given enantiomorph and thus correlate the *observed* difference in these quantities with the absolute configuration of the model used to compute them. In practice, the indices of the reflections are assigned so that h, k, and l are in a right-handed system. Then $|F|^2$ values for pairs of reflections hkl and $\overline{h}\overline{k}\overline{l}$ are measured and the magnitude and sign of their difference is compared with the calculated value of $4\Delta f''(B'G - A'H)$ [see Eq. (10.6)]. G and A' are cosine terms and do not change sign if the "handedness" of the system in which the model is computed is changed. However, B' and H are sine terms, and if the signs of all three coordinates for all the atoms in the model are reversed, then B' and H change sign. Therefore, if $(|F(hkl)|^2 - |F(\overline{h}\overline{k}\overline{l}|^2)$ and $(B'G - A'H)$ have opposite signs, the values of x, y, and z in the model must be replaced by $-x$, $-y$, $-z$ to give the correct model. An example is given in Appendix 10. Since h, k, and l are assigned in a right-handed system, the axes of x, y, z (that is, **a**, **b**, and **c**) must also be in a right-handed system when a model is built or a perspective drawing made. The result of *maintaining the same handedness for the axes in real and reciprocal space* is a three-dimensional representation of the molecule from which the absolute configuration can be seen directly.

In 1930 Koster, Knol, and Prins were able to determine the absolute configuration of a zinc blende (ZnS) crystal.* This contains, in one direction through the crystal (the one perpendicular to the 111 face), pairs of layers of zinc and sulfur atoms separated by a quarter of the spacing in that direction and then another pair one cell translation away, and so on (Figure 10.5). The sense or polarity of that arrangement was determined by the use of radiation (gold, $L\alpha_1$, $\lambda = 1.276$ Å) near the K-absorption edge of zinc (1.283 Å). As a result it was shown that the shiny (1 1 1) faces have layers of sulfur atoms on the surface and the dull (1 1 1) faces have layers of zinc atoms on the surface (see Figure 10.5).

This method was extended, as described above and in Appendix 10, by Bijvoet, Peerdeman, and van Bommel in 1951 to establish the absolute configuration of (+)tartaric acid in crystals of its sodium rubidium double salt using zirconium radiation, which is scattered anomalously by rubidium

*Zinc blende, ZnS, crystallizes in a cubic unit cell, $a = 5.42$ Å, space group $F\overline{4}3m$. The structure contains Zn at (0,0,0), $(0,\frac{1}{2},\frac{1}{2})$,$(\frac{1}{2},0,\frac{1}{2})$, and $(\frac{1}{2},\frac{1}{2},0)$ and sulfur at $(\frac{1}{4},\frac{1}{4},\frac{1}{4})$,$(\frac{1}{4},\frac{3}{4},\frac{3}{4})$,$(\frac{3}{4},\frac{1}{4},\frac{3}{4})$, and $(\frac{3}{4},\frac{3}{4},\frac{1}{4})$.

(a1)

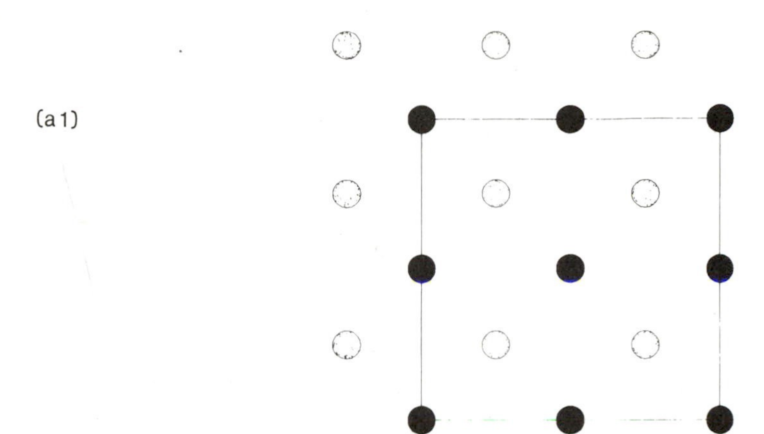

(a2)

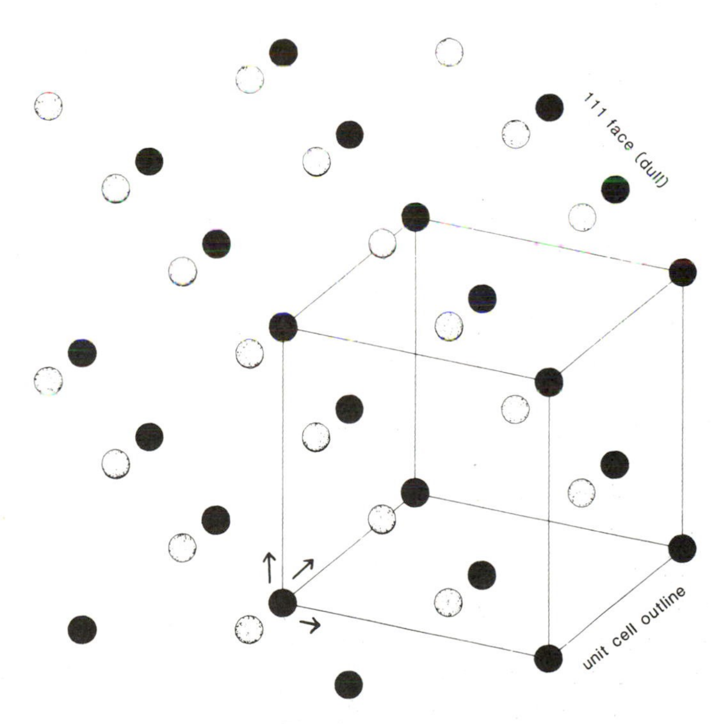

FIGURE 10.5 Polarity Sense of Zinc Blende.

(a) The structure of zinc blende, showing the arrangement of zinc and sulfur atoms. Two views are shown, one down an axis and the other to show the planes of atoms in the [111] direction. Zinc blende is often called sphalerite.

atoms. The absolute configuration was unknown until that time; fortunately that which was found was the one arbitrarily chosen from the two possibilities half a century earlier by Fischer. The absolute configurations of many other molecules have been determined either by X-ray crystallographic methods (see, for example, Figure 10.1d) or by chemical correlation with those compounds for which the absolute configuration had already been established.

It is also possible to use this anomalous scattering effect, in conjunction with isomorphous replacement, to resolve the ambiguity in phase determination by the latter method (see Figure 10.6).

But how can absolute configuration be represented? The R/S system of doing this was introduced by Cahn and Ingold.* They suggested assigning a priority number to the atoms around an asymmetric (carbon) atom so that atoms with higher atomic number have higher priority. If two atoms have the same priority, *their* substituents are considered until differentiation of priorities can be established (otherwise the central atom is not asymmetric). Then the structure is viewed with the atom of lowest priority directly behind the central (carbon) atom and the other substituents are examined. Then if

*See an article in *Angewandte Chemie,* International Edition, **5** (1966), p. 385, by R. S. Cahn, C. Ingold and V. Prelog.

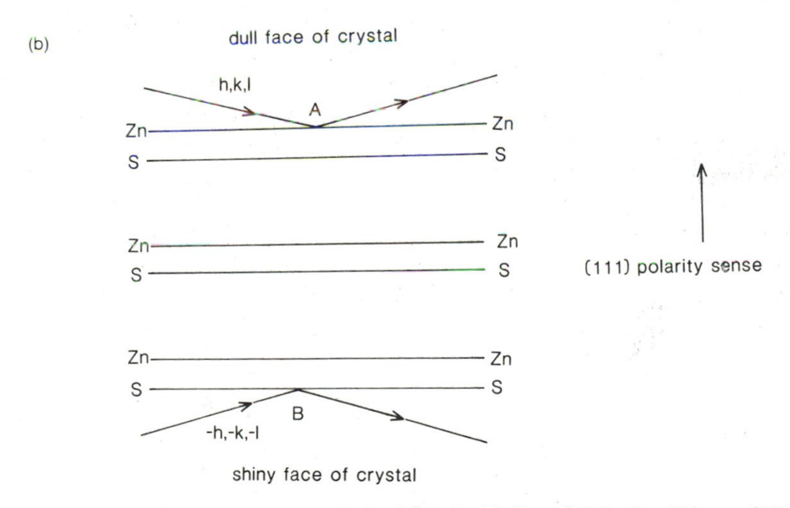

(b) Reflections from the two faces of zinc blende (dull and shiny) will have different relative path differences for the zinc and the sulfur atoms (compare with Figure 10.3). If the radiation is near the absorption edge of zinc, the two types of reflections will have different intensities, allowing one to determine (as did Koster, Knol, and Prins in 1930) that the dull face has zinc atoms on the surface and the shiny face has sulfur atoms on the surface.

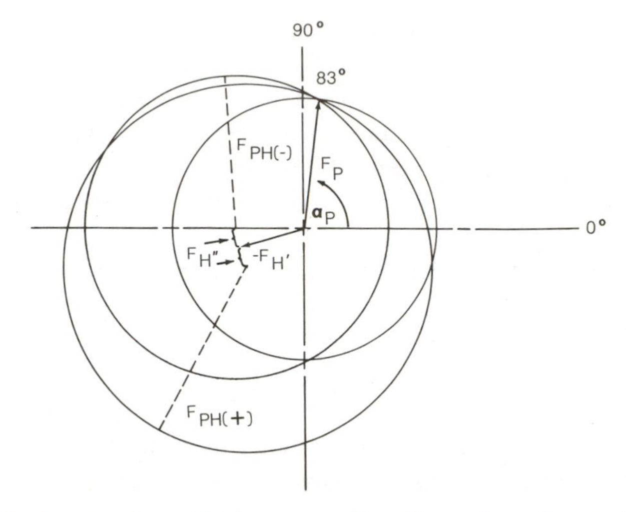

FIGURE 10.6 Isomorphous Replacement Plus Anomalous Scattering.

The effect of anomalous scattering by atom M, introduced to replace another atom, may be used to resolve the ambiguity in phase-angle determination by the isomorphous replacement method. The effect of anomalous scattering (Appendix 9) is to introduce a phase shift, which means in effect, to change the atomic scattering factor of atom M from a "real" quantity, f, to a "complex" one $(f + \Delta f') + i\Delta f''$. Suppose A and B refer to that part of the structure that does not scatter anomalously, and A' and B' to the total structure without any $\Delta f''$ component (i.e., as on p. 141), then

$$A' = G(f + \Delta f') + A$$

and

$$B' = H(f + \Delta f') + B$$

where G and H refer to geometric components for the anomalous scatterer, M. A'' and B'' are components of the structure when anomalous scattering is present and A''_M and B''_M are components for the anomalous scatterer, M, alone. Then

$$A''_M = G(f + \Delta f' + i\Delta f'') = A_M + G\Delta f' + iG\Delta f''$$
$$B''_M = H(f + \Delta f' + i\Delta f'') = B_M + H\Delta f' + iH\Delta f''$$

Then, for the entire structure, including anomalous scattering effects, we have

$$A'' = A + Gf + G\Delta f' + iG\Delta f'' = A' + iG\Delta f''$$
$$B'' = B + Hf + H\Delta f' + iH\Delta f'' = B' + iH\Delta f''$$

As shown in Appendix 10, we have for the entire structure with anomalous scattering (by separating and squaring the "real" and "imaginary" components):

$$|F(hkl)| = \sqrt{(A' - H\Delta f'')^2 + (B' + G\Delta f'')^2}$$
$$|F(\overline{h}\overline{k}\overline{l})| = \sqrt{(A' + H\Delta f'')^2 + (B' - G\Delta f'')^2}$$

the order of the substituents going from highest to lowest priority is clockwise, the central (carbon) atom is designated R (Latin *rectus*, right). If it is anticlockwise, the central (carbon) atom is designated S (Latin *sinister*, left). As a result, once the absolute configuration is established and each asymmetric tetrahedral atom has an R or S designation, sufficient information is provided from these designations to make it possible to build a model with this correct absolute configuration.

The use of anomalous dispersion in structural work has increased recently since the advent of "tunable" synchrotron radiation—that is, X rays whose wavelength may, within certain limits, be selected at will. As a result it is possible to measure the diffraction pattern of a macromolecular crystal with X-radiation of wavelengths near and also far from the absorption edges of any anomalous scatterers present. This, of course, gives information on phases as illustrated in Figure 10.6. The use of anomalous scattering data (that is, measurements for $|F(hkl)|$ and $|F(-h,-k,-l)|$ that have different values) aids in phase determination by isomorphous replacement and helps remove the ambiguity in signs that is shown in Figure 9.8a. This is particularly important if there are experimental problems in obtaining heavy atom derivatives of a macromolecule under study and may make it possible to obtain approximate phases from just one heavy atom derivative.

The effect of anomalous scattering has been used to solve the structure of a small protein, crambin, containing 45 amino acid residues (and which crystallized with 72 water and 4 ethanol molecules per protein molecule). The nearest absorption edge of sulfur is at 5.02 Å, but for Cu $K\alpha$-radiation, wavelength 1.5418 Å, values of $\Delta f'$ and $\Delta f''$ are 0.3 and 0.557, respectively, for sulfur. Pairs of reflections [$|F(hkl)|$ and $|F(-h,-k,-l)|$] were measured to 1.5 Å resolution (the crystals scatter to 0.88 Å resolution); sulfur atom positions were computed from Patterson maps with $|\Delta F|^2$. While it was nec-

---

(See Figure 10.4.) Therefore, $|F(hkl)|$ and $|F(\overline{h}\overline{k}\overline{l})|$ are different; the intensity of each reflection is measured to see which is the greater, as shown in Appendix 10.

A diagram to illustrate the determination of a phase angle for a macromolecule (that is, α_p) by the combination of isomorphous replacement and anomalous scattering is shown. This diagram is constructed in a similar way to Figure 9.8. Circles of radii $|F_{PH(+)}|$ and $|F_{PH(-)}|$ (for reflections of the heavy-atom derivative with indices h,k,l and $-h$, $-k$, $-l$, respectively) are drawn with centers at $-(\mathbf{F}_{H'} + \mathbf{F}_{H''})$ and $-(\mathbf{F}_{H'} - \mathbf{F}_{H''})$, respectively. There are now three circles, radii $|F_P|$, $|F_{PH(+)}|$ and $|F_{PH(-)}|$, and these intersect at a phase angle of α_P (83°). This is probably the phase angle of this reflection, h,k,l, for the native protein.

essary to take into account possible errors in such measurements of the differences of two large numbers, it was, in fact, possible to determine the positions of the three disulfide links (six sulfur atoms). The structure was then determined from an analysis of the Fourier map computed on the heavy-atom parameters of the sulfur atoms together with a partial knowledge of the amino acid sequence.

A modified Patterson map can be used to determine the absolute configuration of a structure provided Bijvoet pairs of reflections have been measured and correctly indexed. The map is computed with a function with $||F(hkl)|^2 + |F(-h,-k,-l)|^2|$ as coefficients and a cosine term; this gives peaks corresponding to Eq. (9.1)—that is, vectors between atoms. Another function, with $||F(hkl)|^2 - |F(-h,-k,-l)|^2|$ as coefficients and a sine term

$$P_s(u,v,w) = \frac{1}{V_c} \sum\sum\sum_{\text{all } hkl} ||F(hkl)|^2 - |F(-h,-k,-l)|^2| \sin 2\pi(hu + kv + lw) \tag{10.7}$$

will have only vectors between anomalously and nonanomalously scattering atoms and these peaks are positive if the vector is from an anomalously scattering atom to a normal atom, and negative if the vector is in the other direction. Thus the absolute configuration of the structure may be determined from such a map.

What effect does anomalous scattering have on the computed electron density, since a term in the scattering factor now has an "imaginary" component? The answer is that the computed electron density must be real, and, to obtain this, any effect of anomalous scattering (which involves a complex scattering factor) must be removed (as described in Appendix 10). This, of course, also removes any means of distinguishing one enantiomorph from the other; such information is contained only in the anomalous scattering data.

SUMMARY

Anomalous Dispersion and Absolute Configuration If an atom in the crystal absorbs appreciably the X rays used, there will be a phase change for the X rays scattered by that atom relative to the phase of the X rays scattered by a nonabsorbing atom at the same site. This phase change on absorption leads to anomalous scattering, and for a noncentrosymmetric structure, results in differences in $|F(hkl)|^2$ and $|F(\bar{h}\bar{k}\bar{l})|^2$ that are not found in the absence of anomalous dispersion. If the structure contains only one enantiomorph of a molecule, its absolute configuration may be determined by a comparison of the signs of the observed and calculated values of $(|F(hkl)|^2 - |F(\bar{h}\bar{k}\bar{l})|^2)$.

Structure Refinement and Structural Information

11

Refinement of the Trial Structure

After approximate positions have been determined for most, if not all, of the atoms, refinement of the structure can be started. In this process the atomic parameters are varied systematically so as to give the best possible agreement of the observed structure factor amplitudes with those calculated for the proposed structure. There are two common refinement techniques, one involving Fourier syntheses and the other a least-squares process. Although they have been shown formally to be nearly equivalent—differing chiefly in the weighting attached to the experimental observations—they differ considerably in manipulative details; we shall discuss them separately here.

Many successive refinement cycles usually are needed before a structure converges to the stage at which the shifts from cycle to cycle in the parameters being refined are negligible with respect to their estimated errors. When least-squares refinement is used, the equations are, as pointed out below, nonlinear in the parameters being refined, which means that the shifts calculated for these parameters are only approximate, as long as the structure is significantly different from the "correct" one. With Fourier refinement methods, the adjustments in the parameters are at best only approximate anyway; final parameter adjustments are now almost always made by least squares, at least for structures not involving macromolecules.

FOURIER METHODS

As indicated earlier (Chapters 8 and 9, especially Figure 9.7 and the accompanying discussion), Fourier methods are commonly used to locate a portion of the structure after some of the atoms have been found—that is, after at least a partial trial structure has been deduced. Initially, this may be only one or a few heavy atoms, or it may comprise an appreciable fraction of the structure if the structure contains atoms with similar scattering powers. Fourier methods are useful also in refining a trial structure and in finding unsuspected atoms or atoms of low scattering power not previously located. Once approximate positions of most of the atoms in the structure are known, the calculated phase angles may be nearly correct. Then an approximate

(a)

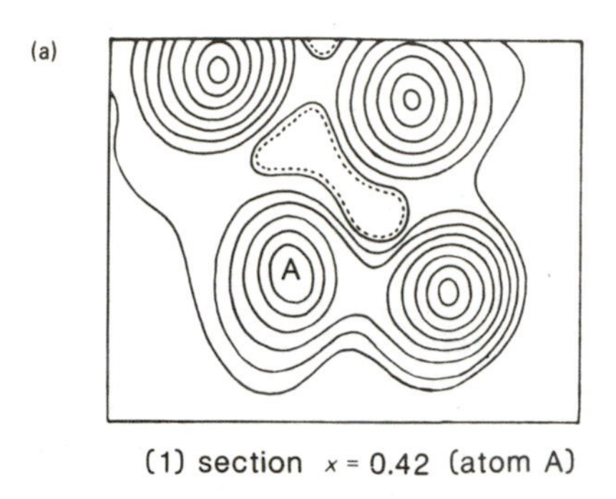

(1) section $x = 0.42$ (atom A)

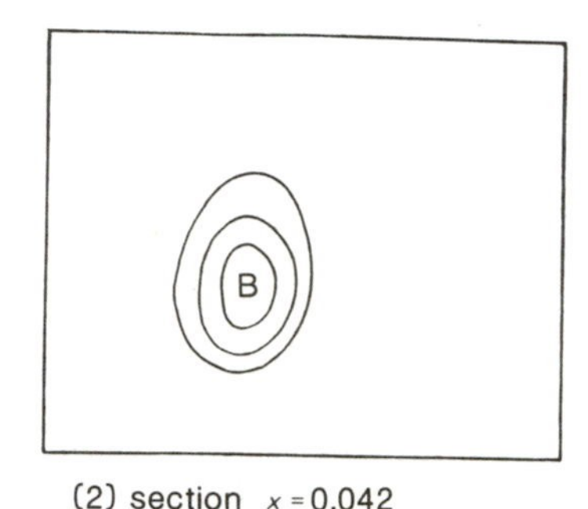

(2) section $x = 0.042$

Atom in incorrect position on $x = 0.42$ (atom no. 24′)

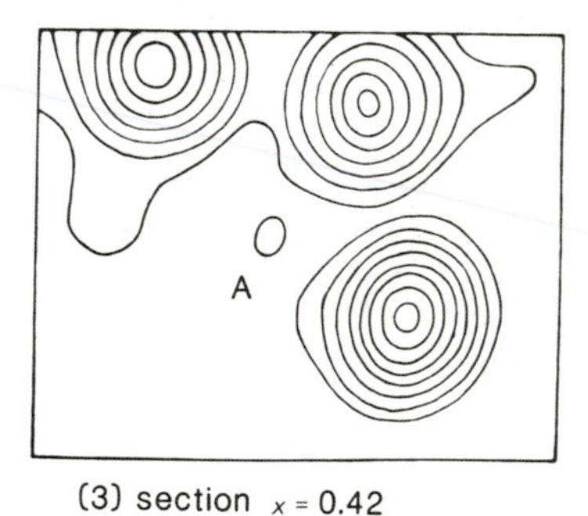

(3) section $x = 0.42$

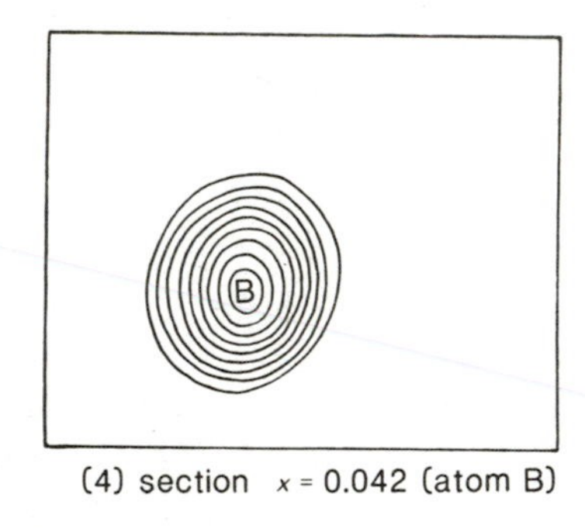

(4) section $x = 0.042$ (atom B)

Atom in correct position on $x = 0.042$ (atom no. 24)

FIGURE 11.1 Fourier Maps Phased with Partially Incorrect Trial Structures.

(a) *The effect of an atom in the wrong position.* This example is from a noncentrosymmetric structure. In (1), one atom, (B), was inadvertently included at the wrong position (marked by an A), in the structure factor calculation. The electron density map phased with this incorrect structure contains a peak at the wrong position, but this peak is lower in electron density than the others near it. A small peak occurs in the correct position, B, shown in (2), although none was introduced there in the phasing. Corresponding sections of a correctly phased map are shown in (3) and (4); the spurious "atom" at A above has disappeared and the correct peak, B, is now a pronounced one.

electron density map calculated with *observed* structure amplitudes and *computed* phase angles will contain a blend of the true structure with the trial structure used to compute the phase angles. If the trial structure contains most of the atoms of the true structure, at or near their proper positions, the resulting electron density map will contain peaks not only near the

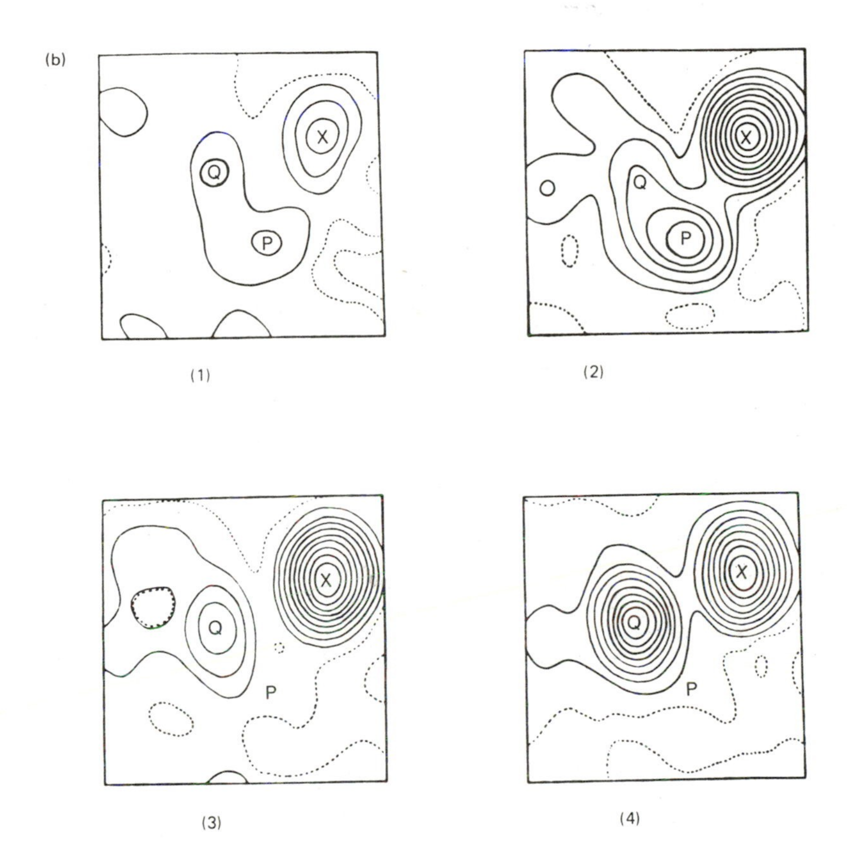

(b) *The effect of an atom near but not at the correct position.* The appearance of a particular section in successive electron density maps is shown as the structure used for phasing becomes more nearly correct. The map (1) is computed from the positions of two heavy atoms and from this the location of atom X was correctly (as it turned out) deduced. But in (2) an atom was incorrectly placed at P; it can be seen that the peak for this atom was elongated in the direction of the correct position, Q. In (3) only atom X (of P, Q, and X) was included in the phasing and peak Q now is more clearly revealed. In (4) the peak at Q is now established as correct. A total of 2, 62, 54, and 68 atoms out of 73 were used in the phasing of maps (1) to (4).

From *Proceedings of the Royal Society,* **A251,** p. 320, figs. 8 and 9.

positions of the atoms in the trial structure but also at other sites representing atoms that were omitted from the trial structure but that are really present. Conversely, if an atom in the trial structure has been incorrectly chosen, the corresponding peak in the electron density map will usually be significantly lower than normal. Finally, if an atom was put into the calculation near, but not at, its correct position, the corresponding peak in the electron density map will usually be biased toward the correct position from the input position. Examples of these effects for a noncentrosymmetric structure are illustrated in Figure 11.1.

In centrosymmetric structures the phase angles are either 0° or 180° and a slight error in the structure may not have any effect on most phase angles. Therefore, a map computed with observed $|F|$ values and computed phase angles may be almost correct even if the model used was slightly in error. However, with noncentrosymmetric structures, for which the phase angles may have any values from 0° to 360°, there will be at least small errors in most of the phases, and consequently the calculated electron density map will be weighted more in the direction of the model used to calculate the phases than with a centrosymmetric structure.

It is usual, when most or all of the trial structure is known, to compute *difference maps* rather than normal electron density maps. For difference maps, the coefficients for the calculation are $(|F_o| - |F_c|)$ and the phase angles are those computed for the trial structure. The difference map is thus the difference of an "observed" and a "calculated" map (both with "calculated" phases). In this map a positive region implies that not enough electrons were put in that area in the trial structure, while a negative region suggests that too many electrons were included in that region in the trial structure. For example, if an atom is included with too high an atomic number, a trough appears at the corresponding position; if it is included (at the correct position) with too low an atomic number, or omitted entirely, a peak appears. Hydrogen atoms are usually located from difference maps calculated from a trial structure that includes all the heavier atoms present (Figure 11.2).

Figure 11.3 shows some examples of further uses of difference maps for refinement of parameters. If an atom has been included near but not at the correct position, the position at which it was input will lie in a negative region, with a positive region in the direction of the correct position. The amount of the shift needed to move the atom to the correct position is indicated by the slope of the contours between the negative and positive regions. If an atom is left out of the trial structure it will appear in the correct position as a peak (provided, of course, that the phase angles used in computing the electron density map, are approximately correct). If the exponent

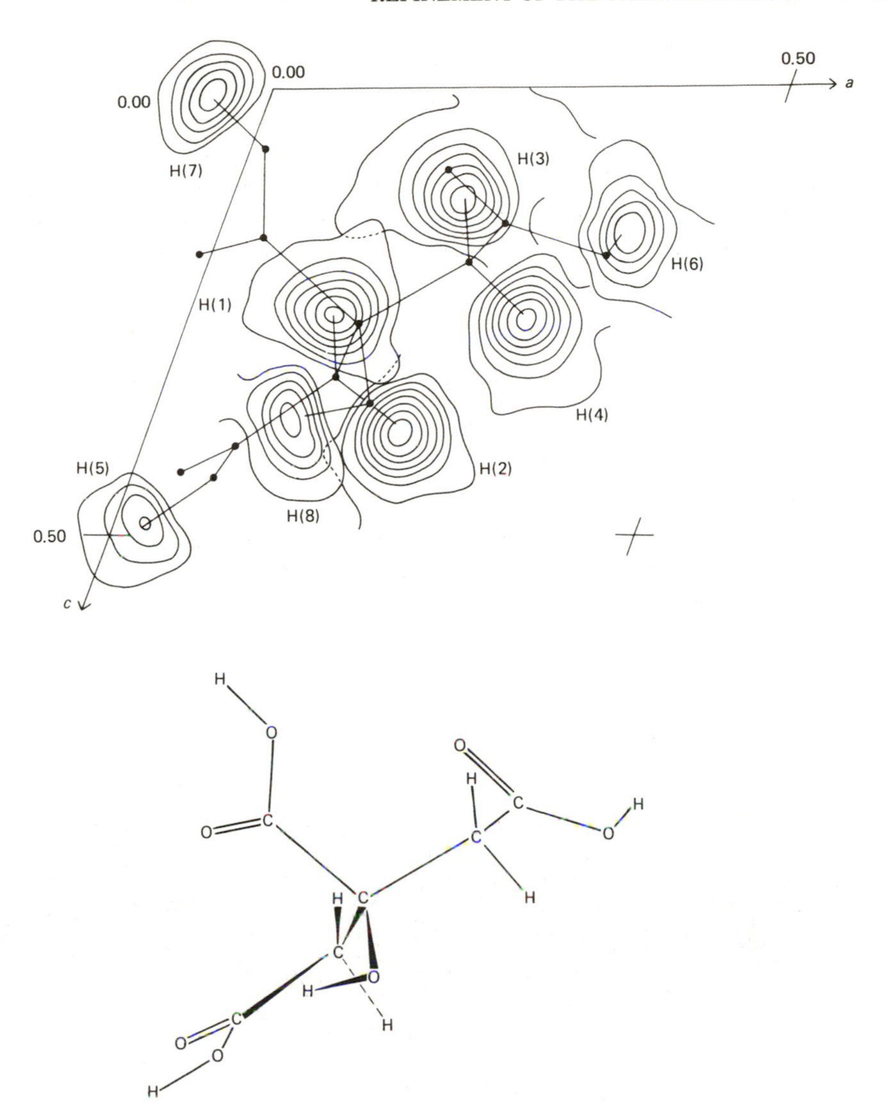

FIGURE 11.2 Hydrogen Atoms Found from a Difference Map.

This is a composite map of sections of a difference map for a monoclinic structure, magnesium citrate decahydrate, viewed down **b**. Eight sections containing hydrogen atoms are shown here. The contour interval is 0.1 e/Å^3; the zero contour is omitted. Solid circles show the final positions of the heavier atoms that were used in the phase-angle calculation. Peaks occur in the map at positions in which not enough electron density has been included in the structure factor calculation, and thus at the positions of hydrogen atoms omitted from the phase-angle calculation. The molecular formula is shown below the map, on the same scale and in the same orientation.

From *Acta Crystallographica* **B25** (1969), p. 1066, fig. 1.

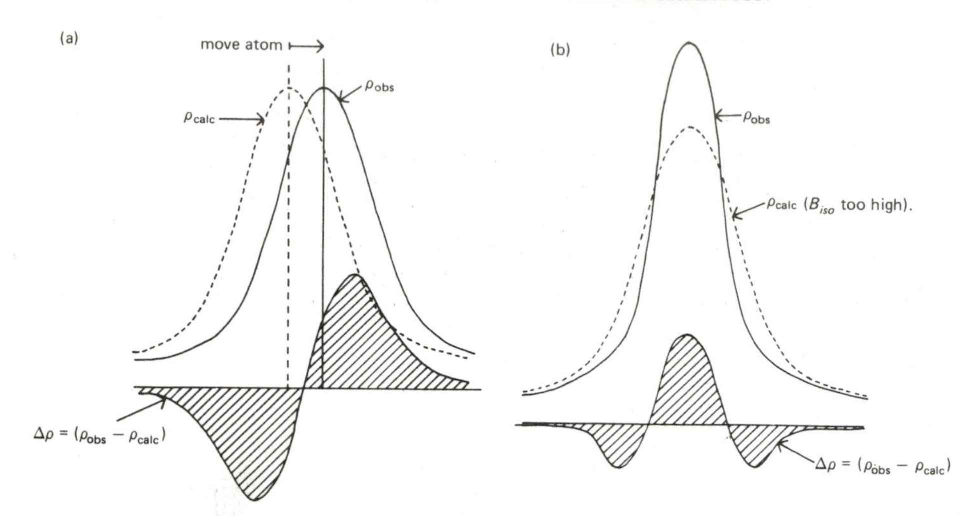

FIGURE 11.3 Refinement by Difference Maps.

A difference map (the difference between the observed and calculated electron density, $\rho_{obs} - \rho_{calc}$) may be used to refine atomic positions and temperature factors. In a difference map a peak (a region of positive electron density) implies that not enough electron density was included in the model at that position, and a trough (a region of negative electron density) implies the opposite.

(a) *An error in the position of an atom.* The peak of ρ_{calc} shows the approximate position used in the calculation of structure factors. The peak in ρ_{obs} is nearer to the correct position. Therefore, the assumed atomic position should be moved (to the right) in the direction of the positive peak in the difference map.

(b) *Incorrect temperature factor of an atom.* If the temperature factor exponent is too high in the model used to phase the map, a peak, surrounded by a region of negative density, occurs at the atomic position, indicating that the exponent should be decreased to give a higher and narrower peak (thus B should be decreased).

in a temperature factor is too small in the calculated trial structure, a trough will appear at the atomic position because the atom has been assumed to be too concentrated; that is, its electrons have been assumed to be confined to a smaller volume than in fact they are, and hence to have too high a density. Similarly, if the vibration parameters are too large in the trial structure, a peak will appear. If the atom vibrates anisotropically, that is, different amounts in different directions, but has been assumed to be isotropic, peaks will occur in directions of greater motion, troughs in directions of lesser motion.

The process of Fourier refinement can be adapted for automatic operation with a high-speed computer. Instead of evaluating the electron density at the points of a fixed lattice, we calculate it, together with its first and second derivatives, at the positions assumed for the atomic centers at this stage. The shifts in the atomic position* and temperature-factor parameters can then be derived from the slopes and the curvatures in different directions. When this *differential-synthesis* method is used, it is normally applied to the difference density. In fact, however, the method is used much less extensively than least-squares refinement, for the latter is somewhat more convenient for computer application and has the advantage of a statistically sounder weighting scheme for the experimental observations.

One of the best criteria of a good structure determination is a flat difference map at the end of the refinement. It is possible to have a good average agreement of $|F_o|$ and $|F_c|$, and thus a low discrepancy index, R, and yet to have many ($|F_o| - |F_c|$) values contributing to a peak or trough in a given area of the map, indicating some error in the structure. Therefore, at the end of every structure determination, a difference map should be calculated and scanned for any peaks.

Another use of difference maps, is in the determination of binding sites of inhibitors (and sometimes poor substrates) on macromolecules.

One question that always arises in discussions of Fourier refinement is: How good must the trial structure be, or how nearly correct must the phases be, for the process to converge? This question cannot be answered precisely; in part, the answer depends on the resolution one is seeking at this stage of refinement. The situation is thus significantly more demanding in the analysis of the structure of a "small" molecule than in the early stages of a protein structure refinement (see Chapter 9), in which one might be using data to a resolution of only 2 Å or more. For an ordinary small-molecule structure analysis, if most of the atoms included are within about 0.3 Å (approximately half their radius) of their correct sites, then a few that are farther away and even one or two that may be wholly spurious can be tolerated. When the initial phases are poor, the first approximations to the electron density will contain much false detail (as illustrated in Figures 9.7 and 11.1b), together with peaks or at near the correct atomic positions. The sorting of the real from the spurious is difficult, especially with noncentrosymmetric structures; experience, chemical information, and a sound knowledge of the principles of structural chemistry are all desirable, and a good deal of caution is essential. A few investigators still follow the formerly common practice of carefully studying contour plots of the approximate electron density or difference den-

*The shift required in x:

$$\Delta x = \frac{-\partial \Delta\rho}{\partial x} \bigg/ \frac{\partial^2 \rho}{\partial x^2} = \frac{-(\text{gradient of difference Fourier at } x_o)}{(\text{curvature of electron density at } x_o)}$$

where x_o is the input position.

sity. A very astute or fortunate crystallographer may be able to recognize portions of a molecule of known structure in a map produced from an extremely poor trial structure, but such perspicacity is uncommon.

THE METHOD OF LEAST SQUARES

The method of least squares, first used by Legendre in 1806, is a common technique for finding the best fit of a *particular assumed model* to a set of experimental data when there are more experimental observations than parameters to be determined. The best parameters for the assumed model are obtained by minimizing the sum of the squares of the deviations between the experimental quantities and the values of the same quantities calculated with the derived parameters. For example, the method is often used to compute the best straight line through a series of points that relate one variable to a second, when it is known that there is an experimental error (assumed random) in the measurement of each point. The equation for a line may be calculated such that the sum of the squares of the deviations from the line is a minimum. (Of course, if the points actually lie on a curve described very well by a nonlinear equation, the method will not tell what this curve is, but will approximate it by a straight line as best it may.) It is possible to "weight" the points; that is, if one measurement is believed to be more accurate than the others, then this measurement may, and indeed should, be given higher weight than the others. The weight assigned to each measurement varies inversely with its precision (more particularly, it is inversely proportional to the square of the estimated standard deviation).

The least-squares method has been extended to the problem of fitting the observed diffraction intensities to calculated ones, and has been for almost three decades by far the most commonly used method of structure refinement, although this practice has not been without serious criticism.* Just as in a least-squares fit of data to a straight line (a two-parameter problem), the observed data are fitted to those calculated for a particular assumed model. If we let $\Delta|F|$ be the difference in the amplitudes of the observed and calculated structure factors, $|F_o| - |F_c|$, and let the standard deviation

*These criticisms are based in part on the fact that the theory of the least-squares method is founded on the assumption that the experimental errors in the data are normally distributed (that is, follow a Gaussian error curve), or at least that the data are from a population with finite second moments. This assumption is largely untested with most data sets. Weighting of the observations may help to alleviate the problem, but it depends on a knowledge of their variance, which is usually assumed rather than experimentally measured. For a discussion of some of these points, see Dunitz's discussion of least-squares methods (ref 1).

of the experimental value of $|F_o(hkl)|$ be $[w(hkl)]^{-1/2}$, then, according to the theory of errors, the best parameters of the model assumed for the structure are those corresponding to the minimum value of the quantity

$$Q = \Sigma w(hkl)[\Delta |F(hkl)|]^2 \qquad (11.1)$$

in which the sum is taken over all independent observations—that is, the unique diffraction maxima. By analysis of the equations* for $|F_c|$, the effects of small changes in the parameters are considered, and changes are found that will tend to reduce the sum in Eq. (11.1) to a minimum. Since even the problem of fitting data to a two-parameter straight line involves much calculation, this method can be used optimally for a many-parameter crystal structure only if a high-speed large-memory computer is available.

The variable parameters that are used in the minimization of Q normally include an overall scale factor for the experimental observations, the atomic position parameters x, y, and z for each atom, j, and the vibration parameters (terms in the exponents of "temperature factors") for each atom, *which may number as many as six*† if the vibration is represented by a general ellipsoid. Occasionally, when disorder is present, occupancy factors (varying from 0 to 1, and perhaps correlated with those of other atoms) may be refined for selected atoms. Thus *in a general case* there may be as many as (9N + 1) or even a few more parameters to be refined for a structure with N independent atoms.

If the total number of parameters to be refined is p, then the minimization of Eq. (11.1) involves setting the derivative of Q with respect to *each* of these parameters equal to zero. This gives p independent simultaneous equations. The derivatives of Q are readily evaluated. Clearly, at least p experimental observations are needed to define the p parameters, but, in fact, since the observations usually have significant experimental uncertainty, it is desirable that the number of observations, m, exceeds the number of variables by an appreciable factor. In most practical cases with three-dimensional X-ray

*Sometimes the equations are formulated with $|F^2|$ rather than $|F|$, so that the equation parallel to Eq. (11.1) then becomes

$$Q = \Sigma\, w(hkl)\, [\Delta\, |F^2(hkl)|]^2$$

Some crystallographers prefer this approach because $|F^2|$ is directly proportional to the observed quantities. We restrict ourselves here to a discussion of refinement using $|F|$; the approach using $|F^2|$ is quite parallel.

†These six vibration parameters, different for each atom, j, are symbolized in various ways (Chapter 12). Here we represent them as b_{11}, b_{22}, b_{33}, b_{12}, b_{23}, and b_{31}, with sometimes an additional subscript to denote the atom, j. As mentioned later, more parameters may be needed to describe the atomic motion in extreme circumstances.

data, m/p is of the order of 5 to 10, so that the equations derived from Eq. (11.1) are greatly overdetermined.

Unfortunately, the equations derived from Eq. (11.1) are by no means linear in the parameters, since they involve trigonometric and exponential functions, whereas the straightforward application of the method of least squares requires a set of linear equations. *If* a reasonable trial structure is available, then it is possible to derive a set of linear equations in which the variables are the *shifts* from the trial parameters, rather than the parameters themselves. This is done by expanding in a Taylor's series about the trial parameters, retaining only the first-derivative terms on the assumption that *the shifts needed are sufficiently small that the terms involving second- and higher-order derivatives are negligible:*

$$\Delta\ |F_c| = \frac{\partial |F_c|}{\partial x_1}\,\Delta x_1 + \frac{\partial |F_c|}{\partial y_1}\,\Delta y_1 + \cdots + \frac{\partial |F_c|}{\partial b_{33,n}}\,\Delta b_{33,n} \qquad (11.2)$$

The validity of this assumption depends on the closeness of the trial structure to the correct structure. If conditions are unfavorable, and Eq. (11.2) is too imprecise, the process may sometimes converge to a false minimum rather than to the minimum corresponding to the correct solution or may not converge at all. Thus this method of refinement also depends for its success on the availability at the start of a reasonably good set of phases—that is, a good *trial structure.*

Since the linearization of the least-squares equations makes them only approximate, several cycles of refinement are needed before convergence is achieved. However, the linear approximation becomes better as the solution is approached because the neglected higher-derivative terms, which involve high powers of the discrepancies between the approximate and true structures, become negligible as these discrepancies become small.

It is often desirable in a least-squares refinement to introduce various constraints on the atomic parameters to make them satisfy some specific criteria, usually geometrical. For example, suppose that the asymmetric unit contains several atoms of relatively high atomic number as well as numerous carbon atoms and other atoms of low atomic number, or suppose that the structure is disordered in some way, or that the available diffraction data are of limited resolution. The individual atomic positions obtained by the usual least-squares process for some of the atoms will then have relatively high standard deviations and the geometrical parameters derived from these positions will be of little significance. If geometrical constraints are introduced—for example, constraining a phenyl ring to be an approximately regular hexagon of certain dimensions, or merely fixing certain bond lengths or bond angles or torsion angles within a particular range of values—the number of parameters to be refined will be significantly reduced and the refinement process accelerated. Such procedures are used widely in refinement of macromolecular structures and are also becoming increasingly common in refinement of smaller structures with, say, 50 or more atoms. It is important in assessing the significance of the geometry reported in a paper to note what, if any, constraints the authors assumed in their work.

If the model used in a least-squares refinement is incorrect or partially incorrect, there are almost always indications that this is so. The discrepancy index, R, may not drop to an acceptable value and the parameters may show certain anomalies. For example, if a false atom has inadvertently been included in the initial trial structure, it may move to a chemically unreasonable position, perhaps too close to another atom, and its temperature factor will increase strikingly to a value far higher than that normally encountered for any real atom. This corresponds physically to a very high vibration amplitude—that is, a smearing of the atom throughout the unit cell, an almost infallible sign that there is no atom in the actual structure at the position assumed in the trial structure.

At the conclusion of any least-squares refinement process, it is always wise to calculate a difference Fourier synthesis. If it is zero everywhere, within experimental error, then the least-squares solution is a reasonable one. If it is not, and the peaks in it are not attributable to light atoms that have been left out of the structure factor calculations or to some other understandable defect of the model, then it is distinctly possible that the least-squares procedure may have converged to a false minimum because the initial approximation (the trial structure) was not sufficiently good. Another plausible trial structure must be sought and refinement tried again.

THE CORRECTNESS OF A STRUCTURE

What assurance is there that the changes suggested by difference maps or least-squares methods are correct, that is, are in the nature of improvements to make the assumed model more nearly resemble the actual distribution of scattering matter in the crystal? In fact, if the experimenter is injudicious or unfortunate, the changes may actually make the model worse, since an image formed with incorrect phases will always contain false detail—for example, peaks that may seem to suggest atoms but that really arise from errors in the phases. If the model is altered in a grossly incorrect way (or if it was inadequate in the first place), the "refinement" process may converge to a quite incorrect solution.

What then are the criteria for assessing the likely correctness of a structure that has been determined by the refinement of approximate phases? There are no certain tests, but the most helpful general criteria are the following. (A number of erroneous structures have been reported because of inadequate attention to these criteria.)

1. The agreement of the individual observed structure factor amplitudes $|F_o|$ with those calculated for the refined model should be comparable to the estimated precision of the experimental measurements of the $|F_o|$. As stressed in Chapter 6, the discrepancy index, R [Eq. (6.6)], is a useful but by no means definitive index of the reliability of a structure analysis.

2. A difference map phased with the final parameters of the refined structure should reveal no fluctuations in electron density greater than those expected on the basis of the estimated precision of the electron density.

3. Any anomalies in the molecular geometry and packing, or other derived quantities—for example, abnormal bond distances and angles, unusually short nonbonded intramolecular or intermolecular distances, and the like—should be scrutinized with the greatest care and regarded with some skepticism. They may be quite genuine, but if so they should be interpretable in terms of some unusual properties of the crystal or the molecules and ions in it.

If writers of crystallographic papers have done their work properly, the information needed for a reader to assess the precision and accuracy of the reported results will be given. The precision of an experimental result, usually expressed in terms of its standard deviation, is a measure of the reproducibility of the observed value if the experiment were to be repeated. The standard deviations of the various observed results—distances, angles, and so on—can be *estimated* by statistical methods, using as a basis the estimated errors of the prime experimental quantities, the intensities and directions of the diffracted radiation and the instrumental parameters of the equipment used. The basic assumption involved in the estimation of standard deviations is that fluctuations in observed quantities are due solely to *random* errors, which implies that the fluctuations are about an average value that agrees with the "true value." However, it is very important to recognize that there may be *systematic* errors, too, arising from failure to correct for various effects, which may be either known—for example, the effect of absorption on the intensities—or unknown—for example, inadequacies of the model because of lack of knowledge of the way in which molecular motion occurs in the crystal. Uncorrected systematic errors can cause the reported values to differ from the "true" values by considerably more than would be estimated on the basis of the precision; that is, the accuracy may be low even if the precision is high. As in any experiment, it is far harder to assess the accuracy than the precision, because many systematic errors are unsuspected; the best way to detect systematic errors is to compare many distinct measurements of the quantity of interest, under different experimental conditions and by different methods if possible.*

If the distribution of errors is normal, statistical tables can be used to assess the probability that one observation or derived quantity is "significantly" different from another—that is, that the difference arises not merely from random errors but rather is one that further sufficiently precise measurements could verify. If two measurements differ from one another by twice the estimated standard deviation (e.s.d.) of either, the probability is about 5 per cent that the difference between them represents a random fluctuation; if they differ by 2.7 times the e.s.d., the probability is only about 1 per cent that the difference represents a random fluctuation—in other words, there is about 99 per cent probability that they represent two distinct values, which further precise measurements would verify as being different. It is a matter of taste what one accepts as being "significantly different"; some people accept the 2 e.s.d. (or "95 per cent confidence") level, while those who are more conservative may choose the 2.7 e.s.d. (or "99 per cent confidence") level, or an even higher one. Because systematic errors are so difficult to eliminate, the e.s.d.'s calculated on the assumption that only random errors are present are usually quite optimistic as estimates of the *"accuracy"* of the results, however valuable they may be as measures of *precision.* Hence, in comparing results from different studies—for example, in comparing two bond lengths, or in trying to decide whether a bond angle is significantly different from that expected on the basis of some theoretical model—it is usually sound not to regard the difference as significant unless it is at least three or more times the e.s.d. For example, if a bond length is measured to be 1.560 Å with an e.s.d. of 0.007 Å, it is probably not significantly different from one measured to be 1.542 Å.

There are several known sources of systematic errors in even the more precise crystal structure analyses published to date. Most of these effects are under study in various laboratories and some of the most careful recent studies take them into account. They include:

1. Scattering factor curves (uncorrected for thermal motion) are normally assumed to be spherically symmetrical, which is clearly not correct for bonded atoms. Extensive studies (both theoretical and experimental) of this asymmetry, which is detectable only in the most precise work, are now under way.

*A classic example of this approach led to the discovery of the noble gases by Rayleigh and Ramsey, through a comparison of highly precise measurements of the density of nitrogen prepared from various pure nitrogen compounds with that of a sample obtained by fractionation of liquid air.

2. The motions of some molecules in crystals are very complicated, and the usual ellipsoidal approximation for the motion of each atom may be a considerable oversimplification, especially if the motion is appreciable. Furthermore, in some crystals the correlated motions of molecules in different unit cells—so-called "lattice vibrations"—may give rise to appreciable "thermal diffuse scattering" (e.g., streaks extending out from the usual Bragg diffraction peaks). Correction must be made for such effects in the most precise work.

3. Many errors that can in principle be eliminated—for example, those arising from absorption or instrumental effects—may not have been properly taken into account.

4. Sometimes the diffracted beam is rediffracted in the crystal when two planes are in a position to "reflect" simultaneously. This can give rise to significant errors in measurements of intensities.

Failure to correct for systematic errors may occur because the errors are regarded as minor and the corrections too complicated to be worthwhile, because an appropriate method of correction is not known, or because the source of error is overlooked. A critical reader will seek to discover what the author has done about known sources of systematic errors. Of course, it takes experience to assess the likely effects of having ignored some of them. Because of the ever-present possibility of systematic errors in even the most careful work, it is usually unwise to regard measured interatomic distances in crystals as more accurate than to the nearest 0.01 Å, although the stated precision may be as low as 0.001 Å. An exception is the now relatively unusual circumstance that the distance involves no parameters at all other than the unit-cell dimensions, for example, the Na^+ to Cl^- distance in NaCl or the C-C distance in diamond, each of which can be measured accurately to better than 0.001 Å at any given temperature. However, even when an interatomic distance is known with high precision and apparent accuracy, it must always be remembered that it represents only the distance between the average positions of the atoms as they vibrate in the crystal. For substances such as rock salt, the root-mean-square amplitude of vibration of the atoms at room temperature is 0.08 Å, and for organic molecules it is larger by a factor of two or three.

SUMMARY

Since there are so many measured reflections (50 to 100 or more per atom in precise structure determinations), the "trial structure" parameters, representing atomic positions and extents of vibration, may be refined to obtain the best possible fit of observed and calculated structure factors.

Fourier Methods Either electron density or difference electron density maps may be calculated, the latter being especially useful in the later stages of refinement. A peak in a difference map represents too little scattering matter in the trial structure, a trough too much. The model is adjusted appropriately to give as flat a final difference map as possible; this map should ideally be zero everywhere, but fluctuations occur as a result of experimental uncertainties or inadequacies of the model used.

Least Squares In any crystal structure analysis there are many more observations than parameters to be determined. The best parameters corresponding to some assumed model of the structure are found by minimizing the sum of the squares of the discrepancies between the observed values of $|F|$ (or $|F|^2$) and those calculated for an appropriate trial structure (or a partially refined version of it). This method of refinement has only been practicable for sets of three-dimensional data since the advent of high-speed computers.

The Correctness of a Structure All the following criteria should be applied:

1. The agreement of individual structure factor amplitudes with those calculated for the refined model should be consistent with the estimated precision of the experimental measurements of the observations.
2. A difference map, phased with final parameters for the refined structure, should reveal no fluctuations in electron density greater than those expected on the basis of the estimated precision of the electron density.
3. Any anomalies in molecular geometry or packing should be scrutinized with great care and regarded with some skepticism.

12

Structural Parameters: Analysis of Results

The results of an X-ray structure analysis are coordinates of the individual atoms in each unit cell, and vibration parameters (exponential terms in the "temperature factors") that may be interpreted as indicative of molecular motion. We now review the ways in which these atomic parameters can be used to gain an understanding of the structure.

Calculation of Molecular Geometry When molecules crystallize in an orthorhombic, tetragonal, or cubic unit cell it is reasonably easy to build a model using the unit-cell dimensions and fractional coordinates, because all the interaxial angles are 90°. However the situation is more complicated if the unit cell contains oblique axes and it is often simpler to convert the fractional crystal coordinates to orthogonal coordinates. The equations for doing this for bond lengths, interbond angles, and torsion angles are presented in Appendix 11. If the reader wishes to compute interatomic distances directly, this is also possible if one knows the cell dimensions (a, b, c, α, β, γ), the fractional atomic coordinates (x,y,z for each atom), and the space group. For example, the square of the distance between two points (x_1,y_1,z_1) and (x_2,y_2,z_2) is

$$l^2 = [(x_1 - x_2)a]^2 + [(y_1 - y_2)b]^2 + [(z_1 - z_2)c]^2 - [2ab\cos\gamma(x_1 - x_2)(y_1 - y_2)] - [2ac\cos\beta(x_1 - x_2)(z_1 - z_2)] - [2bc\cos\alpha\,(y_1 - y_2)(z_1 - z_2)] \quad (12.1)$$

If the lengths of three bonds AB = l_1, AC = l_2, BC = l_3 are known, then the angle B-A-C = δ may be calculated with the law of cosines,

$$\cos\delta = \frac{l_1^2 + l_2^2 - l_3^2}{2l_1l_2} \quad (12.2)$$

Some examples from early structure determinations are illustrated in Figure 12.1. Many that were studied at first were those that crystallized in cubic unit cells because the structure analysis of cubic systems was the most straightforward. For example, sodium chloride, NaCl, crystallizes at room temperature in the space group *Fm*3*m*, a cubic space group, with cell edge $a = 5.6402(2)$ Å (the 2 in parentheses is an estimated standard deviation in the last place quoted). Since this structure* involves a sodium ion at the origin ($x = y = z = 0.0$) and a chloride ion at $\frac{1}{2}, 0, 0$ it can be readily deduced that the nearest distance between cations and anions is 2.82 Å (Figure 12.1a). Potassium chloride has a similar structure in a unit cell with $a = 6.2931(2)$ Å and therefore a $K^+ \ldots Cl^-$ distance of 3.15 Å. Cesium chloride has a cubic unit cell with a cesium ion at the origin and a chloride ion in the center of the cell at $x = y = z = \frac{1}{2}$ to give a primitive unit cell *(not a body-centered unit cell because atoms at the origin and the center of the unit cell are different)*. Thus the space group is primitive, *Pm*3*m*. Since the unit-cell edge is $a = 4.120(2)$ Å, the $Cs^+ \ldots Cl^-$ distance is $4.120 \times (\sqrt{3})/2 = 3.57$ Å. Approximate ionic radii for many common ions in crystals have been derived from data such as these; there is always an element of arbitrariness in assigning radii, and no set is completely consistent because ions are not "hard spheres," their effective radii varying somewhat with environment. Radii estimated in different ways may vary† as much as 0.2 Å or even more. Typical values are: Na^+, 0.95–1.17 Å; K^+, 1.33–1.49 Å; Cl^-, 1.64–1.81 Å; F^-, 1.16–1.36 Å.

Iron pyrite, FeS_2, also crystallizes‡ in a cubic unit cell, space group *Pa*3, $a = 5.4175(5)$ Å, with an iron atom at the origin and a sulfur atom at x,x,x, where $x = 0.39$. Thus the Fe—S bond is 3.05 Å and S—S bonds are about 2.06 Å in length (Figure 12.1b).

Diamond, shown in Figure 12.1c, crystallizes§ in a cubic unit cell, $a = 3.5597$ Å, space group *Fd*3*m*, with eight carbon atoms per unit cell. The nearest neighbor to an atom at the origin is the atom at $x = y = z = \frac{1}{4}$ so that the C—C distance is $3.5597 \times (\sqrt{3})/4 = 1.541$ Å, the C—C—C bond angle is 109.5°, and the C—C—C—C torsion angles are 60° or 180° depending on which carbon atom is chosen as the fourth (see equations in Appendix 11).

Of course, much more complicated structures than those illustrated in Figure 12.1 are now being studied and the amount of information on bond lengths and the environments of various chemical groupings is escalating. Data on the results of X-ray and neutron diffraction studies on more than

*$4Na^+$ at 0, 0, 0; 0, $\frac{1}{2}$, $\frac{1}{2}$; $\frac{1}{2}$, 0, $\frac{1}{2}$; $\frac{1}{2}$, $\frac{1}{2}$, 0 and $4Cl^-$ at $\frac{1}{2}$, 0, 0; 0, $\frac{1}{2}$, 0; 0, 0, $\frac{1}{2}$; $\frac{1}{2}$, $\frac{1}{2}$, $\frac{1}{2}$.

†See the discussion by Dunitz, ref. 1, p. 413.

‡4Fe at 0, 0, 0; 0, $\frac{1}{2}$, $\frac{1}{2}$; $\frac{1}{2}$, 0, $\frac{1}{2}$; $\frac{1}{2}$, $\frac{1}{2}$, 0 (as Na^+ in NaCl); 8S at $\pm(x, x, x; \frac{1}{2} + x, \frac{1}{2} - x, -x; -x, \frac{1}{2} + x, \frac{1}{2} - x; \frac{1}{2} - x, -x, \frac{1}{2} + x$; where $x = 0.39$).

§C at 0, 0, 0; 0, $\frac{1}{2}$, $\frac{1}{2}$; $\frac{1}{2}$, 0, $\frac{1}{2}$; $\frac{1}{2}$, $\frac{1}{2}$, 0; $\frac{1}{4}$, $\frac{1}{4}$, $\frac{1}{4}$; $\frac{1}{4}$, $\frac{3}{4}$, $\frac{3}{4}$; $\frac{3}{4}$, $\frac{1}{4}$, $\frac{3}{4}$; $\frac{3}{4}$, $\frac{3}{4}$, $\frac{1}{4}$.

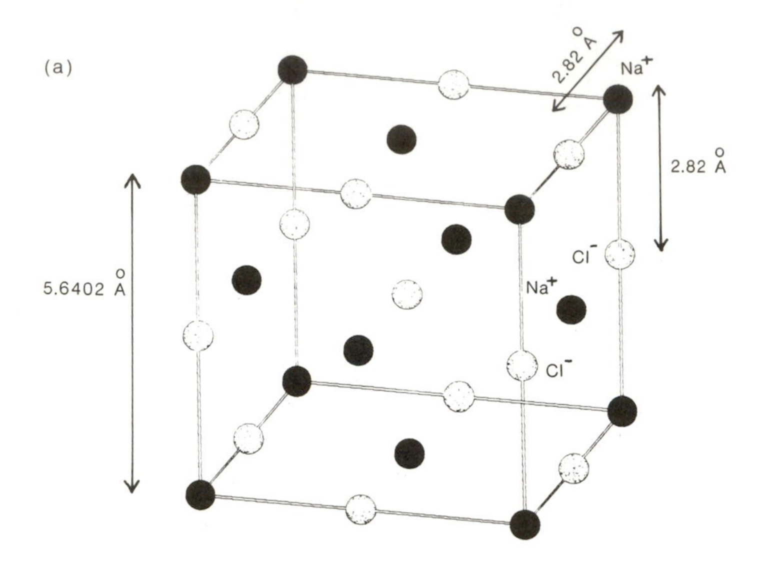
(a)
2.82 Å
Na⁺
2.82 Å
5.6402 Å
Cl⁻
Na⁺
Cl⁻

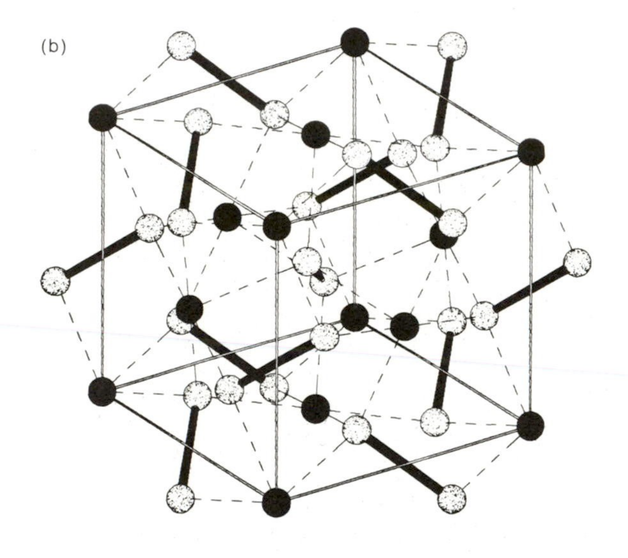
(b)

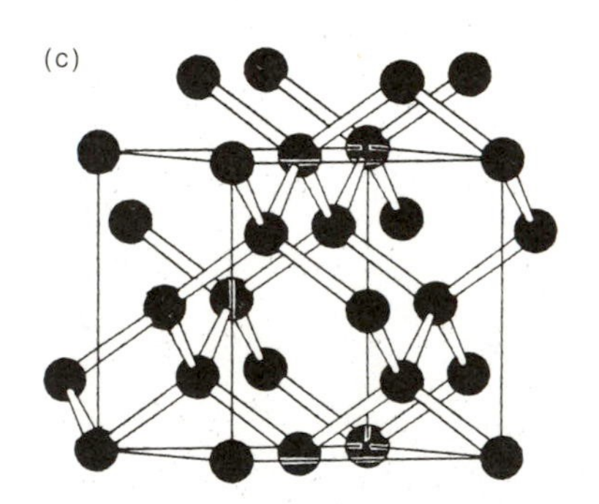
(c)

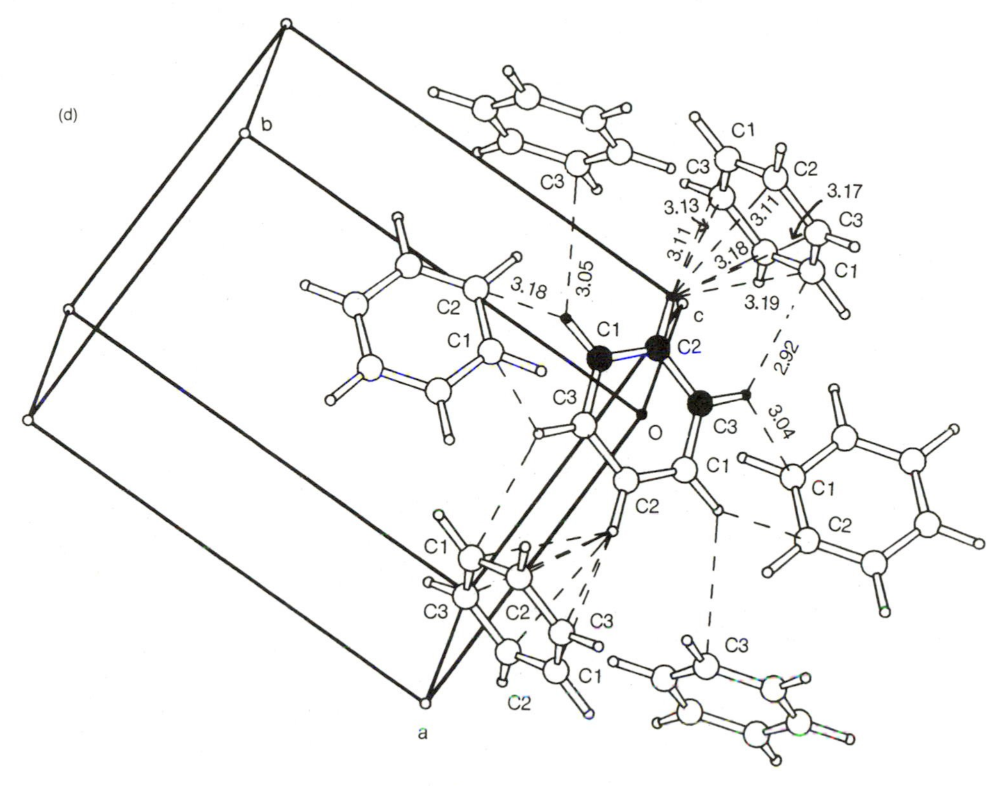

FIGURE 12.1 Geometry of Some Simple Structures.

(a) Sodium chloride (Na^+ black, Cl^- open circles).
(b) Iron pyrite, FeS_2 (Fe black, S stippled).
(c) Diamond.
(d) Benzene, space group *Pbca*, $a = 7.44$, $b = 9.55$, $c = 6.92$ Å. Atoms at $\pm\{x, y, z; \frac{1}{2} + x, \frac{1}{2} - y, -z; -x, \frac{1}{2} + y; \frac{1}{2} - z; \frac{1}{2} - x, -y, \frac{1}{2} + z\}$

Atom	*x*	*y*	*z*
C(1)	−0.0569	0.1387	−0.0054
C(2)	−0.1335	0.0460	0.1264
C(3)	−0.0774	−0.0925	0.1295
H(1)	−0.0976	0.2447	−0.0177
H(2)	−0.2409	0.0794	0.2218
H(3)	−0.1371	−0.1631	0.2312

The asymmetric unit is indicated by black atoms.

See *Proceedings of the Royal Society,* A **279** (1964), p. 98, London.

40,000 structures of small and medium-sized molecules containing at least one carbon atom are available on the Cambridge Crystallographic Data File maintained at Cambridge, England. Data files on inorganic structures, on proteins and on metals also are available (see Bibliography).

Molecular Conformations The torsion angles, τ, in a molecular structure are frequently of interest. These are a measure of the amount of twist about a bond and are defined, for a bonded series of four atoms (A—B—C—D), as the angle of rotation about a bond B—C needed to make the projection of the line B—A coincide with the projection of the line C—D, when viewed along the B—C direction. The positive sense is clockwise for this rotation. Thus the torsion angle is a representation of the structure viewed so that the atom C is completely obscured by atom B, as illustrated in Figure 12.2. A chain of methylene ($-CH_2-$) groups will generally have a staggered conformation so that angles are 180° for the C—C—C—C torsion angle and 60° for C—C—C—H or H—C—C—H. The torsion angle is actually independent of the direction of view; that is, the A—B—C—D torsion angle equals the D—C—B—A torsion angle (Figure 12.2a, cf. (3) and (4)). However, for a pair of enantiomers (mirror images) the torsion angles of equivalent sets of atoms have opposite signs (Figure 12.2a, cf. (4) and (5)). Details of the calculation of torsion angles are given in Appendix 11.

Many studies of molecular conformations involve lists of torsion angles because these angles are useful in indicating similarities (or significant variations) in conformation (e.g., for sugars and for steroids). The conformations observed in such compounds often result from interactions between hydrogen atoms in the structure.

The *least-squares plane* through a group of atoms in a molecule also is a commonly computed quantity* and can be used as a point of reference in describing the rest of the molecule. It is thus useful when the shapes of molecules are being compared.

Intermolecular Distances If one wishes to determine intermolecular distances (i.e., distances between atoms in different molecules), then the space-group symmetry will have to be used, although if there are several independent molecules in the asymmetric unit, some intermolecular distances can be calculated without taking the space group into account. Such calculations are particularly useful for investigating the presence of hydrogen bonds and also, incidentally, for checking whether two molecules are unusually close (an indication either of an unexpected intermolecular interaction or of an incorrect structure). For example, if the compound crystallizes in the space group

*See the discussion by Dunitz, ref. 1.

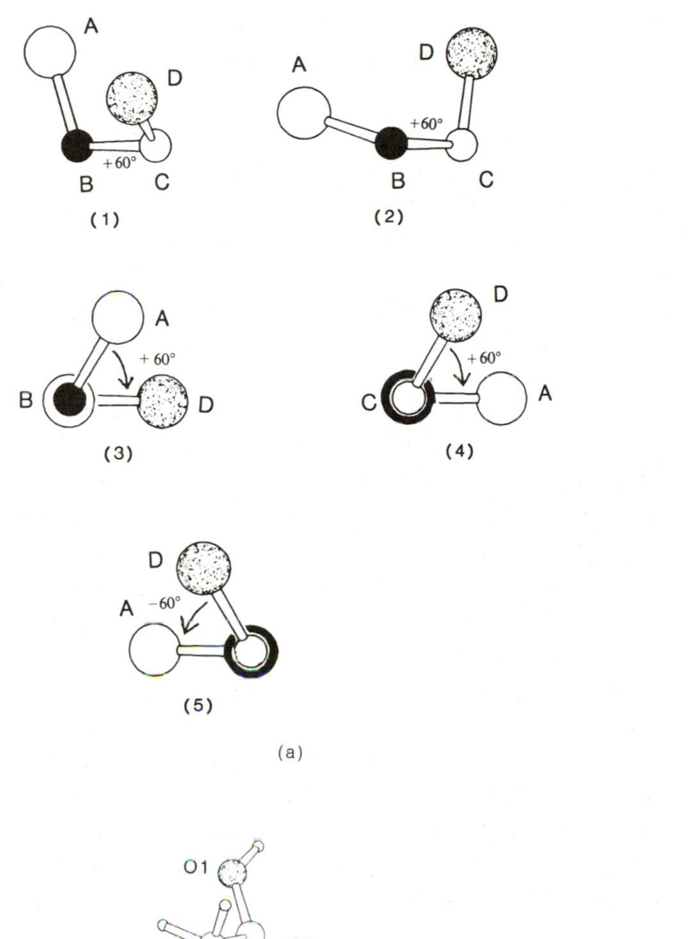

FIGURE 12.2 Torsion Angles.

(a) Four views of a set of four atoms, A—B—C—D, with a torsion angle of 60° about the B—C bond. The fifth view shows that mirror images have torsion angles with equal magnitudes and opposite signs.

(b) The isocitrate ion (see Figure 10.1 (d)) showing some relevant torsion angles.

$P2_12_12_1$, then, using Eq. (12.1) and the information in Figure 7.2, the distance may, for instance, be computed between one atom at x_1,y_1,z_1 and another at $\frac{1}{2} - x_2$, $1 - y_2$, $\frac{1}{2} + z_2$ (where x_1, y_1, z_1, and x_2, y_2, z_2 are the coordinates of atoms in one molecule). Systematic calculations of distances and angles are now done almost entirely by computer programs, which search for all distances (intramolecular and intermolecular) within a selected range (in Å) around each atom in a chosen molecule.

Benzene, for example, has been studied in the crystalline state at −55°C by neutron diffraction (it is a liquid at room temperature). The structure is illustrated in Figure 12.1d. Crystals are orthorhombic, space group *Pbca* with cell dimensions $a = 7.44$, $b = 9.55$, and $c = 6.92$ Å, and with half of a molecule in the asymmetric unit; coordinates are listed in Figure 12.1d. The average C−C bond is calculated to be 1.390 Å and average C−H bonds are 1.07 Å in length. In addition, one hydrogen atom of one molecule points toward the π-electron system of the aromatic ring of a neighboring molecule with minimum H . . . . C intermolecular distances of 3.04 Å. This kind of C−H . . . π-electron interaction occurs in many crystal structures of aromatic compounds.

Precision Of course, all the quantities listed (bond lengths, interbond angles, torsion angles, and least-squares planes) have errors. These result from the fact that the X-ray measurements necessarily contain experimental errors. Further, the atomic scattering model used is not an exact representation of electron density, merely the sum of ellipsoidal electron density around each atomic nucleus; also, there are errors in the cell dimensions. Estimates of random errors may be made from least-squares refinements, and their values can be used to assess the estimated standard deviations in bond lengths, bond angles, and torsion angles. As indicated earlier, unsuspected systematic errors also may be present.

The applications of estimated standard deviations* have been discussed in Chapter 11. It is always necessary to quote an estimated standard deviation with any computed geometrical quantity. The e.s.d. of a bond length is a function both of the precision in measurement of $|F|$ values (reflected in the R value) and of the relative atomic numbers of the various atoms in the structure. For example, the e.s.d. of a C−C bond in a structure containing only carbon and hydrogen atoms may be 0.002 Å for an R value of 0.05, but can increase to 0.02 Å or more for a structure with $R = 0.05$ that contains a heavy atom. E.s.d.s of the coordinates of the heavy atom may be very small, but, because the carbon atoms contribute relatively little to the scattering

*Dunitz has an extensive discussion of calculations of e.s.d.s of derived quantities, including the need for taking correlations between different parameters into account (ref. 1).

(see Figure 5.2b and the discussion of the heavy atom method in Chapter 9), their e.s.d.s are much larger.

Atomic and Molecular Motion The extent of vibration of each atom in a structure also can be measured, and can be represented by exponential parameters in the temperature factors.* However, before deriving these values it is important that absorption and other factors that affect the intensity distribution be taken into account; otherwise the parameters in the temperature factors will not be a true representation of atomic vibrations, and they may not be anyway if there is *static* disorder.

The effect of the vibration of atoms in crystals on the scattering of X rays by these atoms has been discussed in Figure 5.2 and the accompanying text. The simplest assumption that can be made is that the motion of each atom is the same in all directions, that is, that the motion is isotropic. The decrease of scattering intensity that results from this motion then depends only on the scattering angle and not on the particular orientation of the crystal with respect to the incident X-ray beam. As indicated in Figure 5.2c, such isotropic motion causes an exponential decrease of the effective atomic scattering factors as the scattering angle, 2θ, increases. The scattering factor for an atom at rest, f, is replaced by

$$f e^{-B_{iso}[(\sin^2\theta)/\lambda^2]} \tag{12.3}$$

B_{iso} is related to the average of the square of the amplitude of vibration, $\langle u^2 \rangle$, by $B_{iso} = 8\pi^2 \langle u^2 \rangle \cong 79 \langle u^2 \rangle$. For a typical B value of around 4 Å^2 (for an atom in an organic molecule at room temperature), this means that $\langle u^2 \rangle$ is about 0.05 Å^2, and the root-mean-square vibration amplitude, $\langle u^2 \rangle^{1/2}$, is then around 0.22 Å. At liquid nitrogen temperatures (near 100K), B values are typically reduced by a factor of 2 or 3 from those at room temperature, and the r.m.s. amplitude will then be of the order of 0.15 Å.

However, it is clear that the approximation of isotropic motion is a poor one for atoms in most crystals, because the environments of these atoms are far from isotropic. The increasing availability of high-speed computers during the last three decades has made it worthwhile to attempt to collect precise intensity data and to analyze these data for relatively subtle effects, such as more complicated patterns of atomic and molecular motion. The next simplest approximation after isotropic motion is to assume that the motion is ellipsoidal—that is, that it can be described by the six parameters of a gen-

*The name "temperature factor" has persisted to denote the constants in the exponential factors in Eqs. (12.3) and (12.4), despite the fact that it has long been recognized that some such correction would be needed even near 0 K because of the persistence of vibrations at low temperatures, and that a static disorder may simulate a dynamic one if studies are made only at a single temperature.

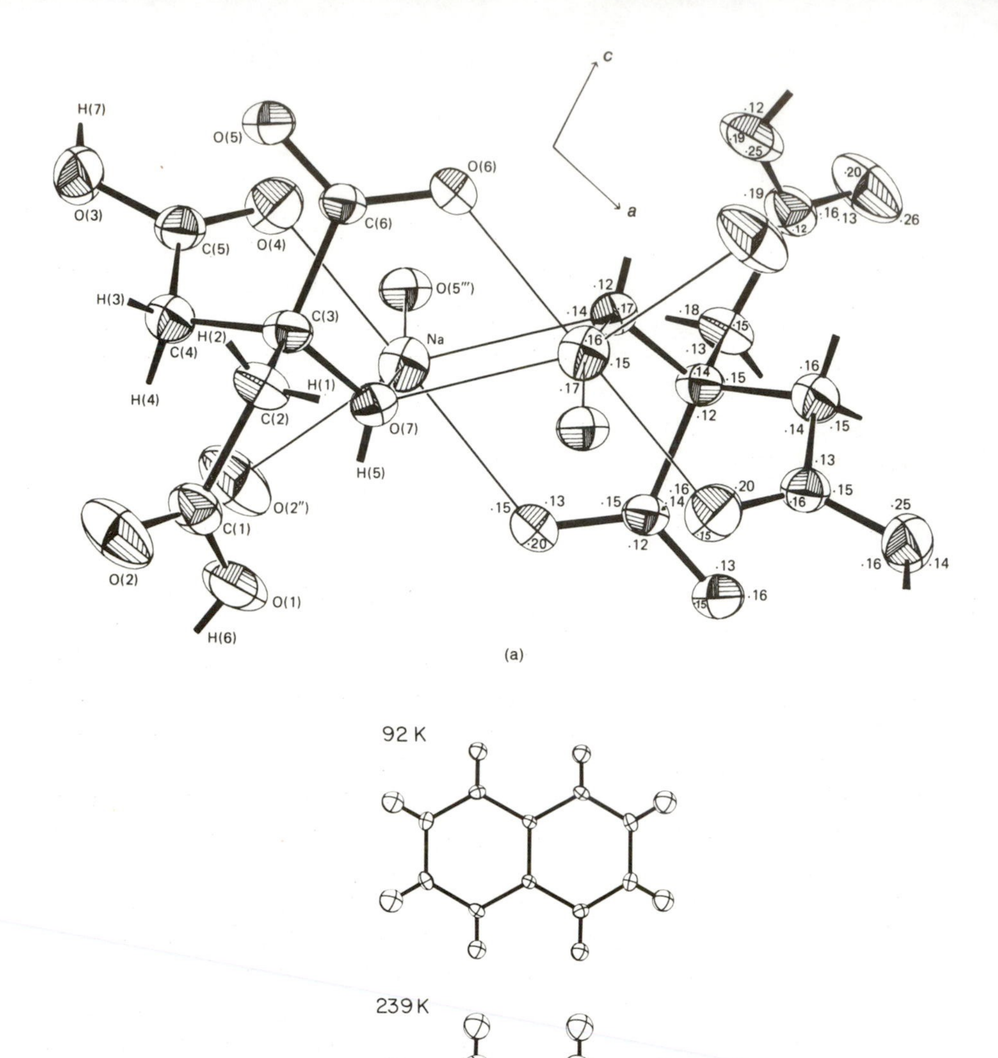

FIGURE 12.3 Molecular Motion.

(a) The anisotropic motion of atoms is usually described by "thermal ellipsoids," as in this example, taken from a study of the structure of sodium dihydrogen citrate. Two complete dihydrogen citrate ions and two sodium ions are shown, grouped around a center of symmetry; two atoms [O(5) and O(2)] of each of two other dihydrogen citrate ions are also shown. The thick lines represent covalent bonds; the thin ones denote coordination interactions of oxygen atoms with the sodium ion.

From *Acta Crystallographica* **19**, (1965), p. 564, fig. 2.

eral ellipsoid rather than the single parameter characteristic of a sphere. Three of these six parameters may be considered to define the lengths of three mutually perpendicular axes describing the amount of motion in these directions, and three to define the orientation of these ellipsoidal axes relative to the crystal axes. Figure 12.3(a) illustrates this representation of atomic motion in a portion of the structure of sodium dihydrogen citrate. It is noteworthy that both the degree of anisotropy and the extent of atomic motion itself vary in different parts of the citrate ion, being greatest for some of the peripheral atoms. The usual way of taking this kind of ellipsoidal motion into account in the structure factor equations is by means of an anisotropic exponential factor analogous to that in Eq. (12.3), with six anisotropic vibration parameters, b_{ij}, as multipliers of the indices for each reflection hkl in the exponent, thus:

$$e^{-(b_{11}h^2+b_{22}k^2+b_{33}l^2+b_{12}hk+b_{23}kl+b_{31}hl)} \quad (12.4)$$

Increasingly, anisotropic vibration parameters are reported as components of a symmetric tensor, U, rather than as b values because the latter are dimensionless and their magnitudes cannot be related to vibration amplitudes without taking the cell dimensions into account. The relation between U_{ij} and b_{ij} values is simple:

$$U_{ii} = b_{ii}/2\pi^2 a_i^{*2}, \qquad U_{ij} = b_{ij}/4\pi^2 a_i^* a_j^* \qquad (i \neq j) \quad (12.5)$$

The mean square vibration amplitude in any direction, specified by the cosines, l, of the angles this direction makes with reciprocal axes, is given by

$$\langle u^2 \rangle = U_{11}l_1^2 + U_{22}l_2^2 + U_{33}l_3^2 + 2U_{12}l_1l_2 + 2U_{23}l_2l_3 + 2U_{31}l_3l_1 \quad (12.6)$$

The anisotropic vibration parameters, b_{ij} or U_{ij}, differ from atom to atom in a structure. The effect of temperature is illustrated in Figure 12.3(b).

This ellipsoidal description of atomic motion is a convenient one for computation, unlike more complex models that may be more realistic physically,

---

The three numbers near each of the ellipsoids in the right half of the drawing indicate the root-mean-square displacements (Å) along the three principal axes of that ellipsoid. The anisotropy of the motion is very evident for some of the atoms, especially for those at the ends of the ion; for these peripheral atoms, the motion is always greatest in directions perpendicular to the bonding direction. This result is just what one would expect, and thus is evidence for the reality of this interpretation of the diffraction data.

(b) Two views of naphthalene, measured at 92K (upper diagram) and 239K (lower diagram). Note the smaller root-mean-square displacements at the lower temperature.

See *Acta Crystallographica* **B38** (1982), 2218–2228. (Photograph courtesy C.P. Brock and J.D. Dunitz)

and it has proved adequate for most structure analyses to date. At the present stage of crystal diffraction techniques, only rarely is an attempt made to analyze atomic motion in terms of a model more complex than that of an ellipsoid, although it is clear that the motions of atoms in crystals may frequently be more complicated; for example, the atoms may move along arcs rather than straight lines, or under the influence of an anharmonic potential function that is steeper on one side of the equilibrium position than on the other. Analysis of such motion requires the best possible data. In many analyses, even the admittedly oversimplified ellipsoidal model may be at times unjustified, because the intensity data simply are not of sufficient quality. "Vibration parameters" so derived are sometimes, at least in part, artificial; for example, appreciable uncorrected absorption errors in a crystal of irregular shape may be compensated for by spurious anisotropy of motion of some atoms in the structure. However, by suitable choice of radiation and crystal size and shape, such absorption errors can be minimized or corrected for, and the reality of derived anisotropies of atomic motion in many structures has been firmly established.

Rigid-body Motion Some molecules may be regarded as nearly rigid bodies, which implies that when they move the relative positions of all atoms (and consequently all interatomic distances) remain constant. The motion may thus be considered to be motion of the molecule as a whole. This is clearly only an approximation because there are always "internal" vibrations—motion of an atom in the molecule relative to its neighbors—but in many crystals the overall motion of the molecules (or ions) is far greater than the internal vibrations. Analysis of the individual anisotropic thermal parameters of molecules in crystals sometimes reveals striking patterns of molecular motion, which can frequently be correlated with the shape of the molecule and the nature of its surroundings in the crystal. The molecular motion may, in general, be described in terms of three components: a translational motion (vibration along a straight-line path), a librational motion (vibration along an arc), and a combination of translation and libration that may be regarded as vibration along a helical path. Some molecules that are not completely rigid may be considered to be composed of segments that are themselves rigid, coupled together in a nonrigid way—for example, molecules such as biphenyl and its derivatives with appreciable torsional oscillation about the inter-ring bond, or torsional oscillation of the methyl groups in durene (1,2,4,5-tetramethylbenzene). Methods have been developed for analysis of internal torsional motion and similar motions in many molecules, and it has been possible to obtain, from diffraction data, rough estimates of force constants for and barriers to such motions.

One important consequence of librational motion is that *intra*molecular distances appear to be somewhat foreshortened, especially for distances that are perpendicular to axes about which there is appreciable librational motion. Approximate corrections to *intra*molecular distances are not hard to make if the pattern of motion is known, but with molecules that are not rigid, the corrections are not themselves precise and consequently the corrected distances cannot be. This is an example of a systematic error that can make the accuracy of a derived result considerably poorer than would be implied by a statistical analysis based on the assumption that only random errors were present. Only wide limits can usually be put on *inter*molecular distances if there is appreciable molecular motion because the correlation (if any) of the motion of one molecule with that of its neighbors is unknown.

Neutron Diffraction In many ways neutron and X-ray diffraction complement each other, since they involve different phenomena. Neutrons are scattered by nuclei (or any unpaired electrons present, the magnetic moment of the electron interacting with that of the neutron). Although there have been a few studies of the distribution of unpaired electrons (e.g., in certain orbitals of selected transition metal ions), such applications have been rare, and in most crystal diffraction studies with neutrons, all electrons are paired and the scattering of the neutrons is essentially by the nuclei present. X rays, on the other hand, are scattered almost entirely by the electrons in atoms. Hence, if the center of gravity of the electron distribution in an atom does not coincide with the position of the nucleus, atomic positions determined by the two methods will differ. Such differences are particularly noticeable for the positions of hydrogen atoms, unless X-ray data have been collected to an unusually high angle corresponding to $\sin \theta/\lambda$ of near 1.2, nearly twice as great as usual (and thus corresponding to nearly eight times as many data, if all reflections are collected). Bonds involving hydrogen atoms derived from X-ray data usually appear shorter by about 0.1 Å because the bound hydrogen atom has no spherically symmetrical inner shell of electrons, its lone electron being displaced toward the atom to which the hydrogen atom is bonded.

The amount of scattering by nuclei does not vary much (or in any regular way) with atomic number. This fact may be used to clear up some ambiguities in an X-ray study. Typical scattering factor data for X rays and neutrons are listed in Appendix 12. Hydrogen has a negative* scattering factor for neutrons (as shown in Figure 12.4) while deuterium has a positive one, both quite high, so that these two isotopes may readily be distinguished; as

*If a nucleus has a negative scattering factor, the radiation scattered by that nucleus differs in phase by 180° ($\cos 180° = -1$) from the radiation that would be scattered from a nucleus that has a positive scattering factor and is situated at the same position.

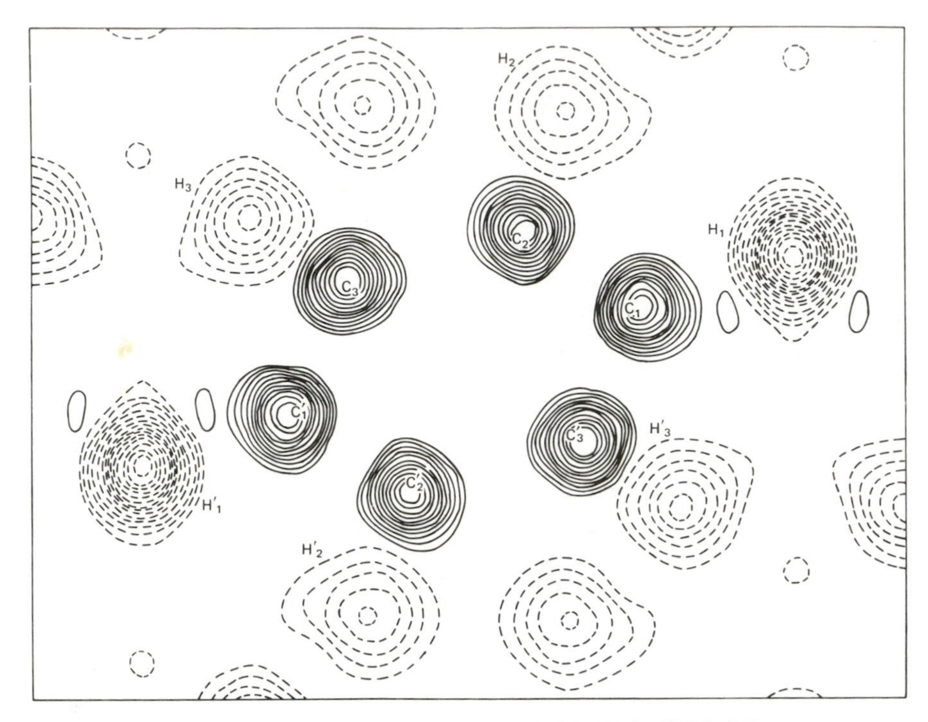

FIGURE 12.4 Projection of the Neutron Scattering Density for Benzene.
Positive density (mainly at carbon atom positions) is indicated by full lines; negative density (mainly at hydrogen atom positions) by broken lines. The unlabeled hydrogen atoms are parts of other benzene molecules. The ring plane is not perpendicular to the direction of the projection; thus the ring does not appear as a regular hexagon. The deeper trough at H_1 and H_1' results from the fact that there are two hydrogen atoms superimposed on each other at these positions in this projection.
(Figure courtesy of Dr. G. E. Bacon.)

far as X rays are concerned, they are identical. Neutron diffraction can thus be useful in studying the structures of reaction products that have been labelled with deuterium. It is also possible with neutrons to distinguish atoms with nearly the same atomic number that cannot readily be distinguished with X rays (for example, Fe, Co, and Ni), because their scattering power for neutrons may be very different. Atomic positions for hydrogen or deuterium may be determined as accurately as those for uranium and many other heavy atoms. This is a particularly important advantage of neutron diffraction studies. There may also be anomalous scattering with neutrons, as with X rays. Since nuclei are extremely small relative to the usual neutron wavelengths, which are about 1 Å, the intensity of neutrons scattered from a stationary nucleus would not decrease markedly at high angles, as it would for X rays. Atomic vibrations, even at low temperature, will, however, cause a decrease of intensity at high angles, as with X rays (Figure 5.2).

One disadvantage of neutron diffraction is that larger crystals are needed than for X-ray structure analysis in order to get sufficient diffraction intensity with the neutron flux available from the present reactors. In order to collect data on myoglobin a crystal with minimum dimensions of 2 mm was needed. One advantage of neutrons is that they do not cause as much radiation damage as do X rays.

The combined use of neutron and X-ray diffraction to solve a biochemical problem is illustrated by the analysis of the structure of lithium glycolate.* Deuterated glycolic acid, HO—CHD—COOH was prepared biochemically and the structure of the lithium salt determined by X-ray diffraction methods. Since hydrogen and deuterium have the same atomic number they could not be distinguished by this method. Crystals of the lithium salt were prepared using lithium hydroxide enriched with the isotope of atomic weight 6. It was then possible to determine the absolute configuration of the lithium salt by neutron diffraction because the scattering amplitude of ^{6}Li is anomalous to neutrons $(0.18 + 0.025i) \times 10^{-12}$ cm and the scattering amplitudes of hydrogen and deuterium (-0.378 and $+0.65$ ($\times 10^{-12}$ cm), respectively) are so different.

Very High Resolution Studies The disposition of the electron density in a molecule is of particular interest to chemists since it provides information on what keeps the atoms together in a molecule. Some measure of this may be obtained by X-ray diffraction studies. The valence-electron scattering is mainly concentrated in reflections with low $\sin\theta/\lambda$ values. In order to view the valence-electron density by means of difference electron density maps, it is necessary to obtain precise and unbiased positional and temperature parameters; this requires high-order data for which the spherical atom approximation is more closely valid. It is usual to measure diffraction patterns to values of $\sin\theta/\lambda$ of 0.65 Å^{-1}, since this is the limit accessible to the most commonly used X-radiation, copper Kα-radiation, wavelength 1.5418 Å. When data are measured to the maximum scattering angles for MoKα-radiation (λ = 0.7107 Å) or, even better, AgKα-radiation (λ = 0.5609 Å), and especially when measurements are made at low temperatures, many more data result and the structure perceived in the X-ray experiment—that is, the electron density—is seen at much higher resolution; atoms are therefore located with higher precision. It is possible to look at bonding effects by

*C. K. Johnson, E. J. Gabe, M. R. Taylor, and I. A. Rose. *Journal of the American Chemical Society,* **87** (1965), pp. 1802–1804.

computing either an "X−X" map (a difference map using atomic positions from an analysis of only the high-order X-ray diffraction data) or an "X−N" map (a difference map using atomic positions from a neutron diffraction analysis, and hence atomic nuclear positions). These are known as "deformation density maps" since they really represent the difference between the total electron density as observed with X rays and the electron density corresponding to a superposition of spherical atoms (each with the experimental thermal motion), which is the deformation of the electron density when atoms form a molecule. Peaks in this map in positions expected for bonding electrons and lone pair electrons are generally found. For several molecules that have been studied (e.g., oxalic acid*) quite good agreement exists between the experimental deformation density and a theoretical one, provided the theoretical model is sufficiently sophisticated (i.e., an extended basis set is used). In addition the lack of apparent electron density in N−O and C−F bonds in deformation density maps, though at first surprising, is in accordance with the deformation density map definition and is predicted by theory. The future of this area of analysis is bright.

SUMMARY

Molecular Geometry This may be computed from the unit-cell dimensions and symmetry and the values of x, y, and z for each atom that have been derived from electron density maps or by least-squares methods. Bond lengths, interbond angles, torsion angles, least-squares planes through groups of atoms and the angles between such planes give much useful chemical information.

Atomic and Molecular Motion The fall-off in intensity with increasing scattering angle becomes more pronounced with increasing vibrations of atoms. Atomic vibration itself becomes greater as the temperature of the specimen rises. For spherically symmetrical motion, the reduction in intensity is simply represented by an exponential function, $e^{(-2B_{iso}\sin^2\theta/\lambda^2)}$. Thermal motion is frequently represented by more sophisticated models.

Neutron Diffraction Neutrons are scattered by atomic nuclei (or by unpaired electrons); X rays are scattered significantly only by the electrons in atoms. Scattering factors for neutrons do not vary systematically with atomic number or atomic weight. Neutron diffraction studies can often clear up ambiguities in X-ray work. The two methods are thus to some degree complementary.

*See papers by P. Coppens, T. M. Sabine, R. G. Delaplane, J. A. Ibers and others, *Acta Crystallographica* **B25** (1969) 2423–2460 and **A40** (1984) 184–195.

13

Micro- and Noncrystalline Materials

As discussed earlier, the crystalline state is characterized by a high degree of internal order. There are two types of order that we will discuss here. One is *chemical order,* which consists of the connectivity (bond lengths and bond angles) and stoichiometry in organic and many inorganic molecules, or just stoichiometry in minerals, metals and other such materials. Some degree of chemical ordering exists for any molecule consisting of more than one atom and the molecular structure of chemically simple gas molecules can be determined by gaseous electron diffraction or by high resolution infrared spectroscopy. The second type of order to be discussed is *geometrical order,* which is the regular arrangement of entities in space such as in cubes, cylinders, coiled coils, and many other arrangements. For a compound to be crystalline it is necessary for the geometrical order of the individual entities (which must each have the same overall conformation) to extend *indefinitely* (that is, apparently infinitely) in three dimensions such that a three-dimensional repeat unit can be defined from diffraction data. Single crystals of quartz, diamond, silicon, or potassium dihydrogen phosphate can be grown to be as large as six or more inches across. Imagine how many atoms or ions must be identically arranged to create such macroscopic perfection!

Sometimes, however, this geometrical order does not extend very far, and microarrays of molecules or ions, while themselves ordered, are disordered with respect to each other on a macroscopic scale. In such a case the three-dimensional order does not extend far enough to give a sharp diffraction pattern. The crystal quality is then described as "poor" or the crystal is considered to be *microcrystalline,* as in the naturally occurring clay minerals.

On the other hand, in certain solid materials the spatial extent of geomet-

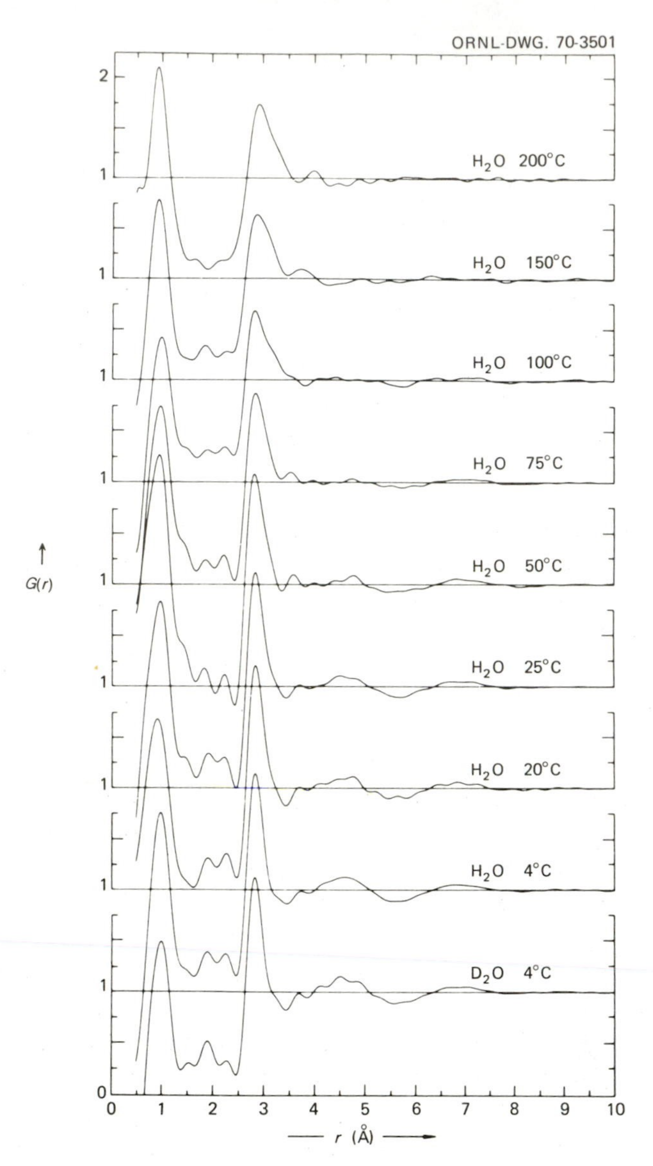

FIGURE 13.1 Radial Distribution Functions.

Radial distribution curves obtained by X-ray diffraction studies on liquid water at temperatures from 4°C to 200°C are shown. Sample pressures were atmospheric up to 100°C; above 100°C, the pressure was equal to the vapor pressure. The vertical coordinate, $G(r)$, for the curves represents a normalized radial distribution function; that is, it gives information on the number of neighbor atoms or molecules at a distance r from an average atom or molecule in the system compared to that expected for a liquid without distinct structure.

The peak near 1 Å represents the intramolecular O—H interaction and that at 2.9 Å represents interactions between oxygen atoms of neighboring water molecules. A sequence of broad peaks follows, notably those near 4.5 Å and 7 Å, that may be attributed to preferred distances of separation for second and higher coordination shells. At distances large compared to atomic dimensions,

rical order may be less than three-dimensional, and this reduced order gives rise to interesting properties. For example, the geometrical order may only exist in two dimensions; this is the case for mica and graphite, which consist of planar structures with much weaker forces between the layers so that cleavage and slippage are readily observed. In a similar way certain biological structures such as membranes and micelles have less than three-dimensional order. Sometimes the geometrical order can be increased by external forces; for example, "liquid crystals" can be *temporarily* aligned in three dimensions by externally applied forces such as electric or magnetic fields (hence their use in liquid crystal displays in watches, computers, and other instruments). Even less geometrical order is shown by fibers such as silk, hair, and some long-chain polymers that have essentially only one-dimensional order.

Many times there is no evident geometric order beyond the immediate near-neighbor environment of the fundamental building unit. This is characteristic of liquids, glasses, and rubbers whose spherically symmetrical diffraction patterns indicate that in no direction in space is there geometric order extensive enough to define a period. In such *amorphous* materials the only regularities seen in the diffraction pattern are those due to recurring bond distances. Thus diffraction patterns from amorphous materials provide information about interatomic distances only when a particular distance stands out from the average of all—usually because it is heavily weighted either by frequent occurrence or by involvement of atoms with scattering factors that are large relative to those of the other atoms present, but occasionally simply because it is unique, with no other distances of comparable magnitude occurring in the sample.

Liquid Diffraction Careful diffraction studies of liquids have provided much valuable structural information on time-averaged interatomic distances; these are spherically symmetrical in space and therefore are generally represented by radial distribution functions, that is radially averaged electron density maps. Examples, calculated from the diffraction patterns of water at various temperatures, are shown in Figure 13.1. These show the

---

and also with increasing temperature, the values of $G(r)$ merge to unity—that is, to the value for a structureless liquid.

In liquid water the average coordination in the first shell represents about 4.4 molecules (independent of temperature), compared to exactly 4 molecules in ice, in support of the idea that the increase in density upon melting of ice is due to a small increase in the average coordination number in the first coordination shell. Other details in the distribution curves are compatible with an approximately tetrahedral coordination of molecules, as found in ice.

The curves were kindly provided by Dr. A. H. Narten from Oak Ridge National Laboratory Report 4378, June 1970.

expected interatomic distances (O—H, O . . . O and H . . . O) and the effects of neighboring molecules, which change as the temperature is raised.

Glass Diffraction Traditional glass, used throughout history to construct containers, windows and ornaments, is made by fusion of a mixture of lime, silica, and soda and subsequent blowing or pressing of the product into the desired shape. Such glass is, of course, solid at ordinary temperatures. Glass stemware made from it is often referred to as "crystal" in spite of the fact that it is not crystalline. Its diffraction pattern has a halolike appearance, resembling the diffraction pattern of a liquid; this demonstrates clearly that it is not crystalline and that there is no well-defined geometrical order within it. The best model to date of such glass is that of random chains, nets and three-dimensional arrays of SiO_4 tetrahedra, linked through oxygen atoms, with appropriately situated cations. Many attempts have been made to fit models with different kinds of short-range order to the observed diffraction patterns and to other quantitative physical and chemical data available on various glasses in an effort to define more precisely what might be meant by "the structure of glass."

In contrast to traditional glasses that are the products of fusion and can be "thawed" and reworked without crystallizing, there are now known to be many other glass-forming composition systems and, as a result, several ways of generating glasses and other amorphous materials. Each of these gives rise to properties that are useful. For example, amorphous metal films can be made by "splat cooling"—that is, a jet of liquid metal is directed onto a cold surface and therefore is cooled to a solid so rapidly from the melt that it has been deprived of the time required for crystal organization. Another industrial example is provided by the use of a chemical reaction in the gas phase to generate an extremely fluffy amorphous "soot" that may be sintered and compressed to three-dimensional solidity without crystallizing. Optical waveguide-laser communication technology depends, in large measure, on the purity, composition control, and perfection of such processes, achievable by starting with pure gases, such as silicon tetrafluoride and oxygen, reacting them to form a condensed phase of pure silica "soot" where, presumably, the surface is both highly energetic and unique such that particles "join" under pressure without melting (sintering) to form a continuum; such sintering without melting precludes the possibility of any crystallization. A third example is provided by glass-ceramics, which, although noncrystalline as formed, cannot be heated to the softening point because they undergo crystallization from the solid state; this crystallization must be controlled carefully in order to obtain a glass-ceramic with the desired physical properties.

Fiber Diffraction The diffraction patterns in Figure 3.7 show the effect on the diffraction pattern of partial but incomplete internal order. Figure 3.7d displays quite effectively the result of one-dimensional internal order (characteristic of certain fibers), with elongated streaks instead of spots on the photograph. Many fibers are composed of units with helical structures, with some order along the axis of the helix but often little order in the packing of adjacent helical units. DNA, certain fibrous proteins, and many other natural and synthetic materials have such structures. Some X-ray photographs of DNA are shown in Figure 13.2a and b; note that the fiber axis is vertical in Figure 13.2, but horizontal in Figure 3.7d.

The coordinates of the atoms in a helical structure are best described by cylindrical polar coordinates, and the scattering factor of a cylindrical system is most appropriately represented in terms of Bessel functions. A zeroth-order Bessel function is high near the origin and then dies away like a ripple in a pond, while higher order Bessel functions are zero at the origin and then rise to a peak at a distance proportional to their order and then die away, again like a ripple. These Bessel functions are used in calculating the Fourier transform of a helix, which describes the scattering pattern of the helix. The "cross" that is so striking in Figure 13.2b is characteristic of helical diffraction patterns. Because the helix is periodic along the axial direction, layer lines are formed; they are visible in both Figures 13.2a and 13.2b. Two chief pieces of information may be derived from such a photograph as that in Figure 13.2b. These are the distance between "equivalent" units of the helical structure (for example, the base pairs in DNA) and the distance along the helix needed for one complete turn. From these two data the pitch of the helix can be deduced. Often little else can be determined with certainty although it may be possible to select or eliminate certain models ("trial structures"). Further details may be found in the books by Wilson (ref. 192) and Holmes and Blow (ref. 15).

The diffraction pattern of a crystalline dodecameric fragment of DNA is shown in Figure 13.2c; the crystal structure that corresponds to this diffraction pattern is shown in Figure 13.2d. Note that Figure 13.2c represents a sampling of the diffraction pattern in Figure 13.2b.

Small-angle Scattering Structural features that are large compared to the wavelength of the radiation being used cause significant scattering only at small angles (Figures 3.1 and 5.2). "Small-angle scattering" at angles, 2θ, no larger than a few degrees, is thus used to measure long-range structure. For example, for a biological macromolecule it may be used to measure the radius of gyration and to study the hydration of the macromolecule. It has

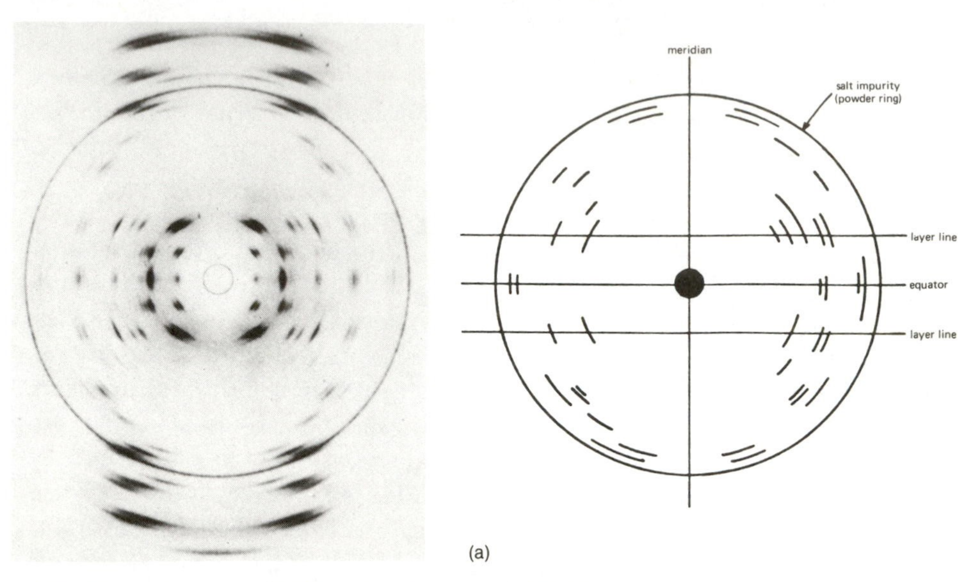

(a)

FIGURE 13.2 Some Diffraction Patterns of DNA.

Diffraction patterns of two common forms of DNA and of a synthetic analog. Each diffraction photograph has been taken with the fiber axis vertical.

(a) Structure A is the "crystalline" form of DNA found at relatively low humidity (75 per cent). The fibers pack with fair regularity to form a lattice and the diffraction pattern resembles that of a single crystal, although the spots are drawn out into arcs as a result of the random orientation of different crystallites in the specimen. The left diagram is a reproduction of a photograph of the A-form of DNA; the right diagram is a schematic transcription of some predominant features of the photograph, with added indications of what some of these features are. Note the formation of layer lines. The powder ring is due to a salt impurity.
(Photograph courtesy of Dr. H. R. Wilson.)

(b) B-DNA, the diffraction pattern of which is illustrated in the upper diagram, is another form of DNA, in which the individual molecules are packed together less regularly. This fibrous noncrystalline form is that for which Watson and Crick first proposed their famous DNA structure. The fibers are randomly oriented around the fiber axis, and a helical diffraction pattern with a characteristic cross is obtained.
(Photograph courtesy of Dr. R. Langridge.)

(c) 20° Precession photograph of the $0kl$ zone of the diffraction pattern from crystals of a dodecameric polynucleotide, CGCGAATTCGCG, which represents a portion of DNA, studied at room temperature. The crystals are orthorhombic, space group $P2_12_12_1$ with cell dimensions $a = 24.87$, $b = 40.39$, $c = 66.20$ Å. The c^* axis is vertical, and appears, by comparison with Figure 13.2(b), to suggest that the c axis is also the axis of a B-type double helix. Note that reflections with $l = 18$ to 20 are intense, indicating some feature that repeats every 66.2/(18 to 20)Å = 3.3 to 3.7 Å; this is the stacking of the bases. Strong reflections occur in a cross pattern at about 25 to 30° to the vertical axis. This corresponds to the pitch angle of the phosphate backbone in the double helix. The structure is shown in (d). This photograph is a sampling of the fiber diffraction photograph shown in Figure 13.2(b) and should be compared with it; both are

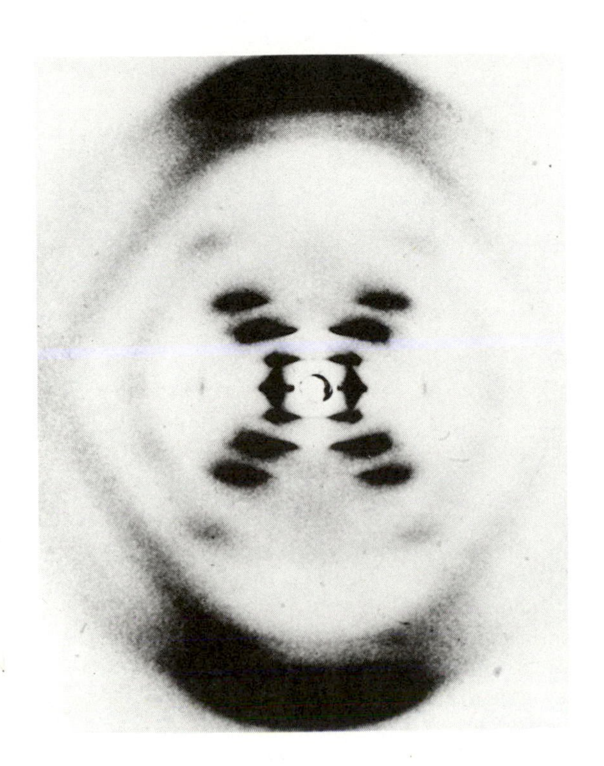

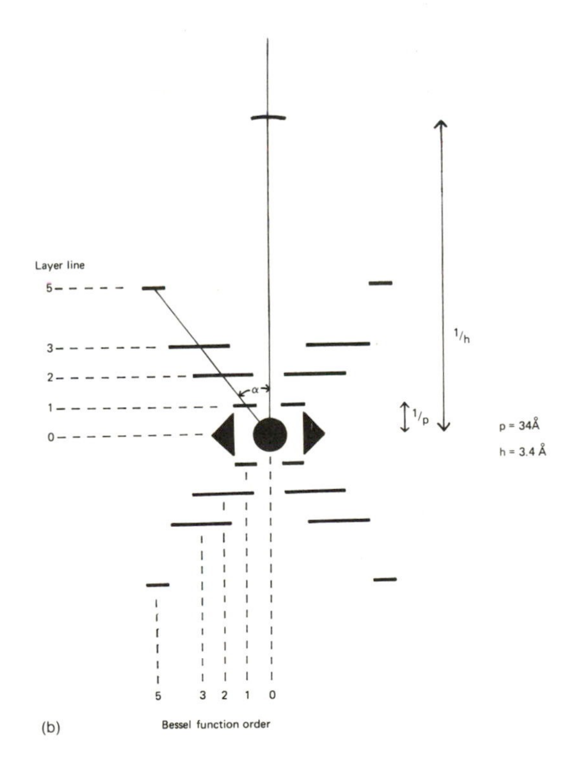
Layer line
5
3
2
1
0
α
1/h
1/p
p = 34Å
h = 3.4 Å
5
3
2
1
0
(b)
Bessel function order

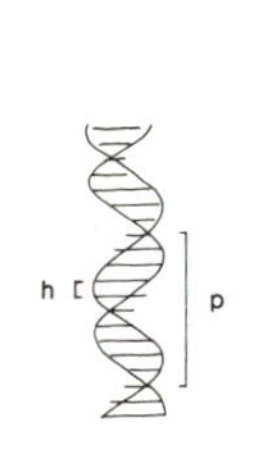
h
p

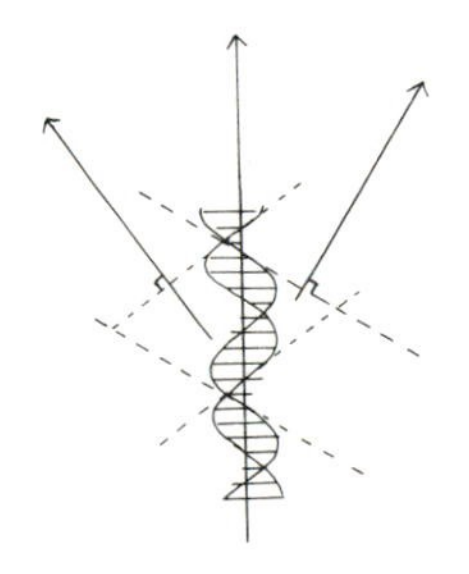

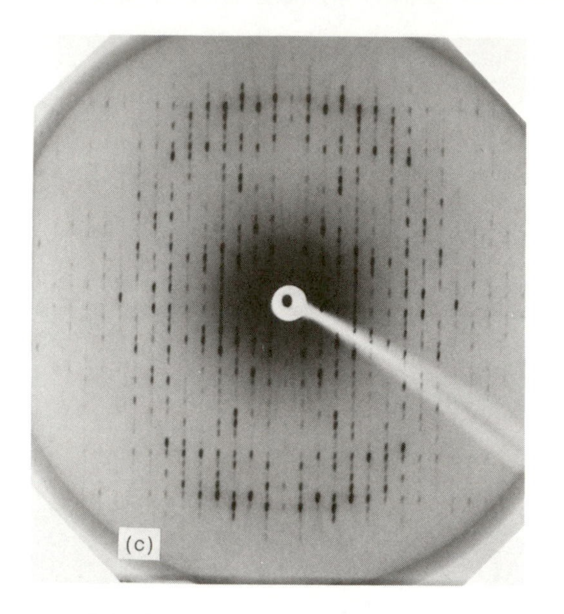

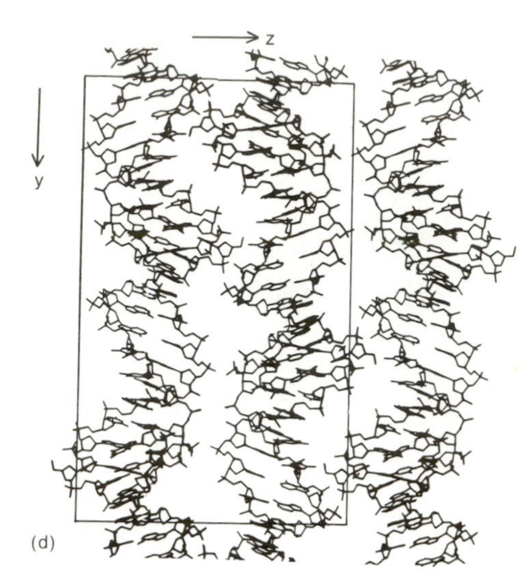

FIGURE 13.2 Continued

in the same orientation. Photographs of crystals or fibers of A-DNA do not show a cross so clearly because the bases are stacked at an angle to the helix axis and the reinforcement is less perfect.

(d) The structure of the polynucleotide CGCGAATTCGCG in the orientation to give the diffraction pattern shown in Figure 13.2(c). (Photograph courtesy of Richard E. Dickerson.)

R. E. Dickerson, M. L. Kopka, and P. Pjura (1985) in *Biological Macromolecules and Assemblies.* Volume 2. Nucleic Acid and Interactive Proteins. (Ed. F. Jurnak and A. McPherson.) Ch. 2, pp. 38–126. New York: John Wiley and Sons, Inc.
Wing, R. M., Drew, H.R., Takano, T., Broka, C., Tanaka, S., Itakura, K., and Dickerson, R. E. (1980). *Nature* 287 755–758.

been widely applied to the study of liquids, polymers, liquid crystals, and biological membranes.

Powder Diffraction The diffraction pattern of a powder (packed in a capillary tube) may be considered that of a single crystal but with the pattern of the crystal in all possible orientations (as are the crystallites in the capillary tube). Powder diffraction is an extremely powerful tool for the identification of crystalline phases and for the qualitative and quantitative analyses of mixtures. It is used for analysis of unit-cell parameters as a function of temperature and pressure and to determine phase diagrams (diagrams showing the stable phases present as a function of temperature, pressure and composition). Compilations of common powder diffraction patterns are available (by computer or in a book); they are maintained by the Joint Committee for Powder Diffraction Standards, Swarthmore, PA, USA. A

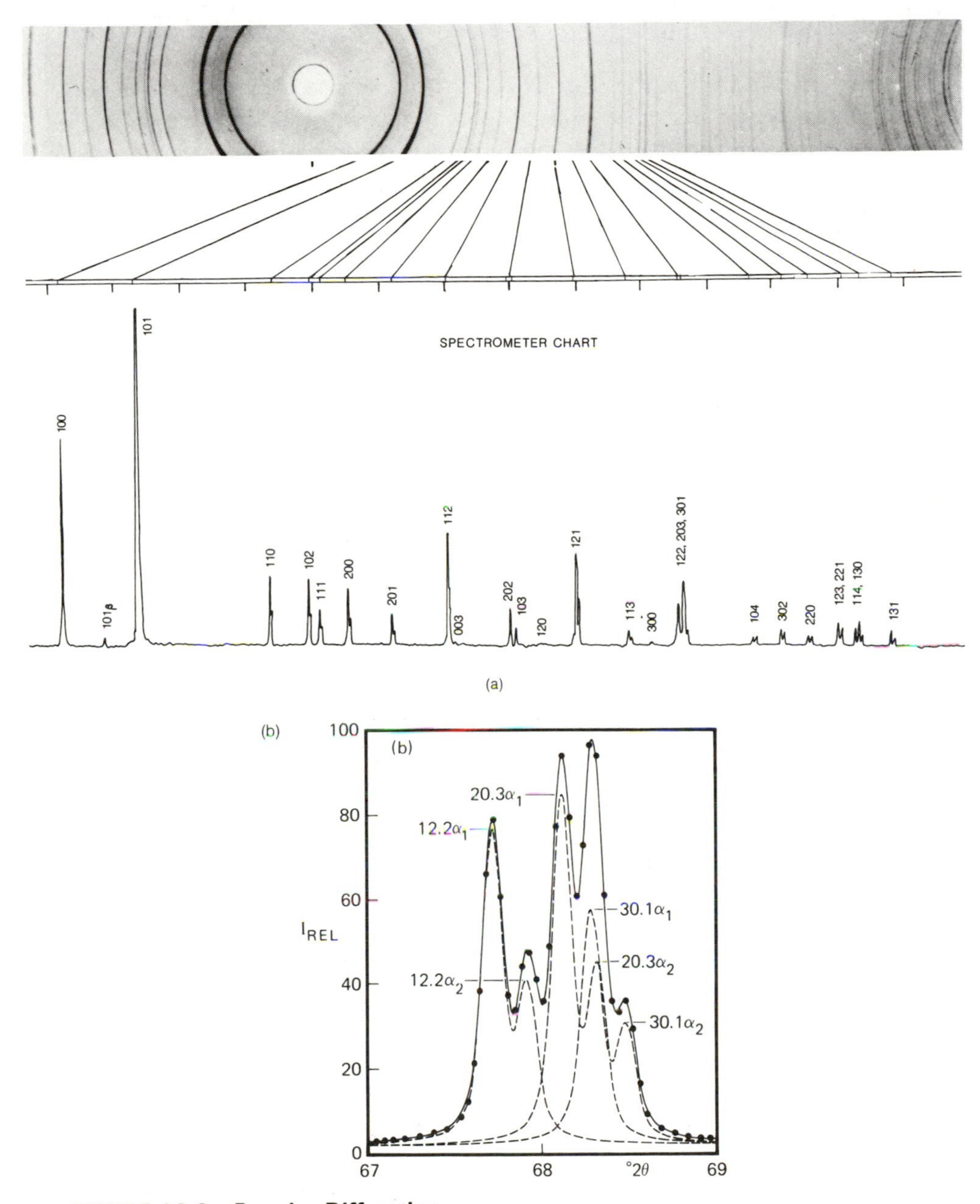

FIGURE 13.3 Powder Diffraction.

(a) Comparison of an 11.46-cm diameter powder camera film (upper photograph) with a scanned diffractometer pattern of quartz (with copper Kα-radiation) 12.2, 20.3, and 30.1 represent, respectively, the 122, 203, and 301 reflections for this crystal.

(b) Profile fitting of a portion of the diffraction pattern of quartz. Dots are experimental points from step-scanning and dashed lines are the individual results for each reflection. The sum is represented by a solid line. Note the separation of the α_1 and α_2 wavelengths of the radiation.
(Photographs and diagram courtesy of Dr. William Parrish.)

comparison of the powder diffraction pattern obtained experimentally with the highest diffracted intensities of some powder diffraction patterns in the file, a search that can be done by computer, will often reveal the chemical composition of a powder. Thus, the method is of great importance industrially. For example, the composition of particles in an industrial smoke stack may be determined by analysis of the diffraction pattern.

Powder methods may even be used for simple structural studies. There are now sophisticated methods, originally introduced by H. M. Rietveld in 1967, for the adjustment of parameters to give the best fit with an experimental neutron powder diffraction pattern. The technique is now also used for X-ray structures and, for simple structures, can give precise unit-cell dimensions and atomic coordinates and temperature factors. This is, of course, of great value when suitably large crystals cannot be grown. Other useful information can also come from careful powder diffraction studies. For example, an analysis of profile broadening (Figure 13.3) can lead to an estimate of average crystallite sizes in the specimen.

SUMMARY

Studies of Structures That Are Not Fully Crystalline The diffraction patterns of liquids and glasses are spherically symmetrical and only radial information can be obtained. However, from substances exhibiting partial order more information may be derived. For example, for a helical structure, the pitch of the helix and the repeat distance along it can be deduced.

Powder Diffraction The diffraction pattern of a powder also gives only radial information since the powder contains crystallites in all possible orientations. Powder diffraction is used for the identification of crystalline phases and for the qualitative and quantitative analyses of mixtures.

14

Outline of a Crystal Structure Determination

The stages in a crystal structure analysis by diffraction methods are summarized in Figure 14.1a for a substance with a molecular weight less than about 2000. The principal steps are:

1. Obtain or grow suitable single crystals—sometimes a tedious and difficult process. The ideal crystal is 0.2–0.3 mm in diameter. (Somewhat larger specimens are needed for neutron diffraction.) Various solvents, and perhaps various different derivatives, may have to be tried before suitable specimens are obtained.

2. Obtain the unit-cell dimensions and the space group. This can usually be done in a day, more or less, barring complications. The cell dimensions are obtained by measurements of spacings in the recorded diffraction pattern, these spacings being reciprocally related to the dimensions of the crystal lattice. The space group can be deduced from the symmetry of, and the systematic absences in, the diffraction pattern.

3. After the density of the crystal has been measured, which is easily done to 0.2 per cent or so by flotation, the formula weight of the contents of the unit cell can be determined to equally high precision. If it has been established how much (if any) solvent of crystallization is present, it is then possible to make precise statements about the formula weight of the primary constituent of the crystal. With some hygroscopic or unstable materials, analysis for solvent is difficult. If the molecular weight of the main constituent is unknown, and the solvent content is also unknown, these two quantities can only be established after a complete structure determination.

4. Decide whether to proceed with a complete structure determination. The investment in effort may be considerable and there is always the chance

FIGURE 14.1 The Course of a Structure Determination by Single-Crystal X-Ray Diffraction.

(a) Flow diagram for determination of small structures (10^2 or fewer atoms per asymmetric unit).

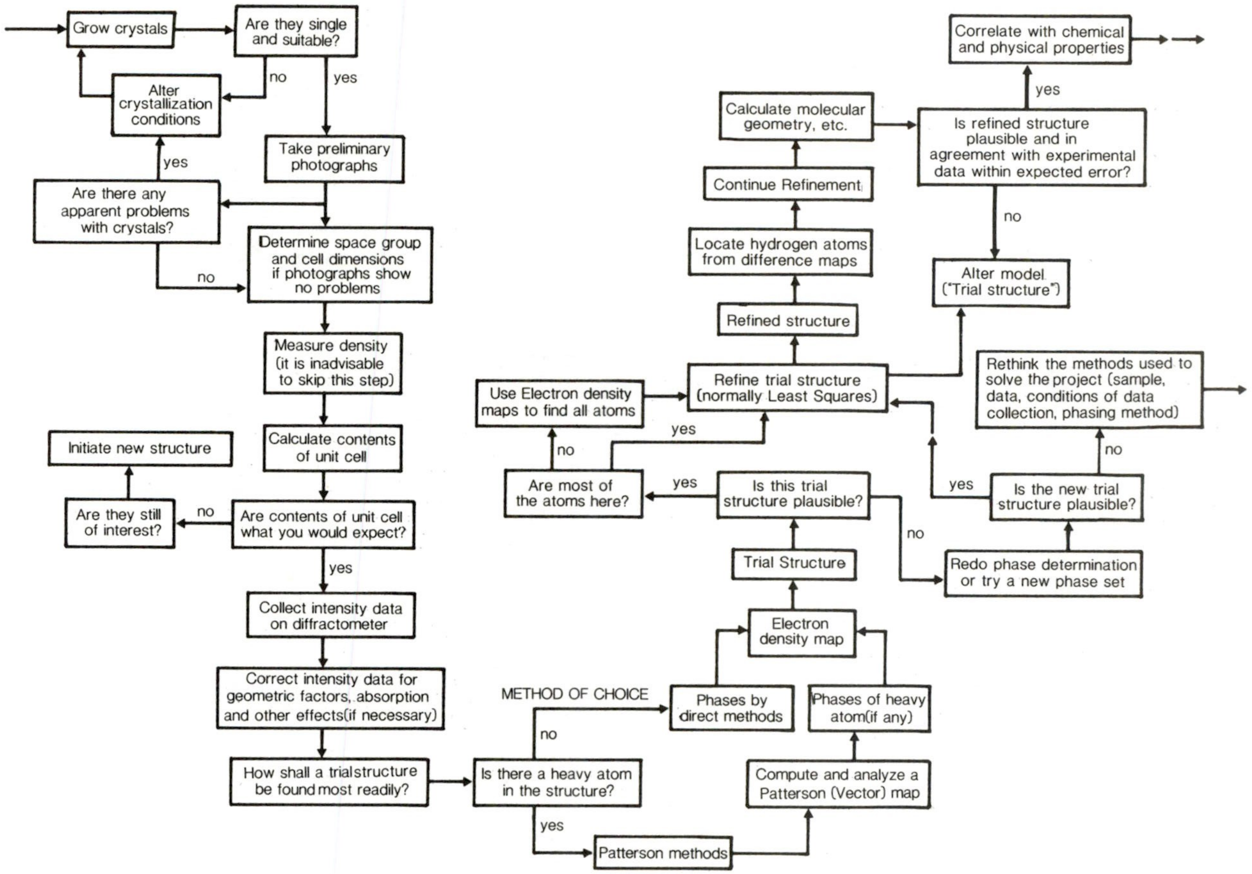

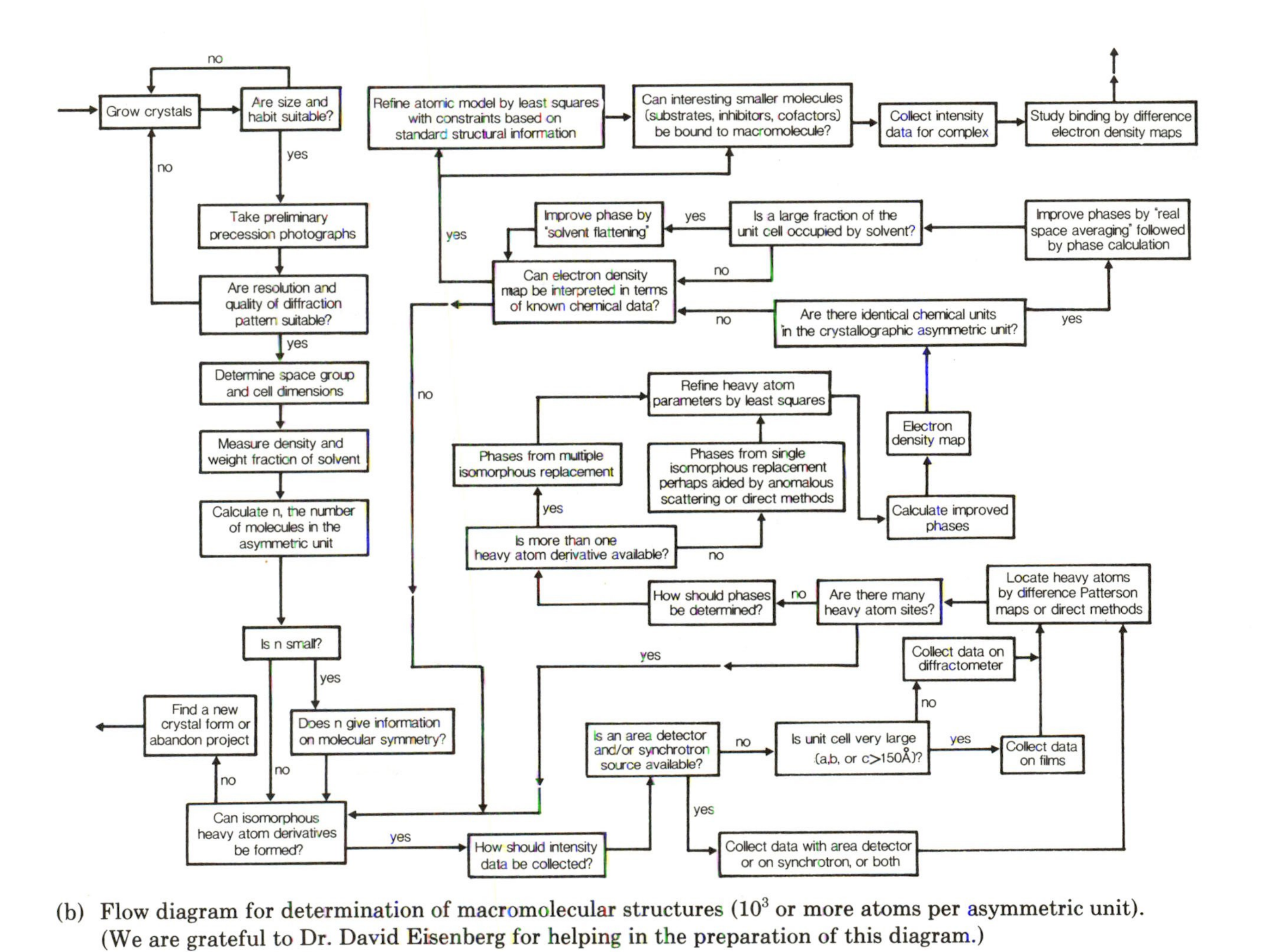

(b) Flow diagram for determination of macromolecular structures (10^3 or more atoms per asymmetric unit). (We are grateful to Dr. David Eisenberg for helping in the preparation of this diagram.)

that the structure will prove unsolvable. Hence one must try to weigh properly the relevant factors, among which are:

(i) Quite obviously, the intrinsic interest of the structure.

(ii) Whether the diffraction pattern gives evidence of twinning, disorder, or other difficulties that will make the analysis, even if possible, at best of limited value. This will depend in part on the type of information sought.

If the answer to (ii) is unfavorable, another crystal specimen or polymorph must be sought. However, under happier circumstances, one can proceed.

5. The next stage is then to record, usually now with a counter-diffractometer, most of the accessible diffraction maxima. If they are recorded photographically, all the intensities must be measured either visually (now rare) or with a densitometer. In any case, the intensities must be appropriately correlated, averaged, and multiplied by various geometrical factors to convert them to relative values of $|F|$. For a typical molecular structure, there may be between 10^3 and 10^4 unique diffraction maxima to be measured, or even more with a very large molecule. The normal time involved in the collection and estimation of these intensity data is from a day or two to a week or two, the exact amount depending on the equipment available and the experience and other concurrent obligations of the experimenter. Without an automatic diffractometer, the time involved can be significantly greater. The data processing is done with a computer as are all subsequent steps, appreciably reducing the necessary time involved.

6. Attempt to get a "trial structure" or approximate phases. The normal procedure is to try some of the direct methods, with the aid of an appropriate computer program, or to calculate a three-dimensional Patterson map and hope to find the heavy atom(s) present (if any) or some recognizable portion of the molecule. Meanwhile, collection of data on other compounds that may prove easier to solve if this one is unusually stubborn should be going on; every laboratory has its collection of unsolved structures, some of which yield to new and improved methods or brighter minds that come along, and a few of which persist indomitably against all challengers.

7. A satisfactory trial structure is one that is chemically plausible and for which there is good agreement between observed and calculated structure factors. It must then be refined, as discussed earlier. The resulting structure should have an R index [Chapter 6, Eq. (6.6)] consistent with the precision of the data that were collected, and should meet the criteria discussed earlier under the heading "The Correctness of a Structure" (Chapter 11).

8. One byproduct of a complete and successful structure analysis of an optically active material can be a determination of its absolute configuration, provided that it contains an atom that absorbs sufficiently the X rays being used. This technique has been applied successfully to many organic natural products and was previously discussed and illustrated (Chapter 10).

When a macromolecule is crystallized, somewhat different techniques are used to determine its structure (Figure 14.1b) and these stages have been discussed. The principal steps are:

1. Suitable single crystals are grown by vapor diffusion of solvent or related methods (Chapter 2 and Figure 2.1).

2. The unit-cell dimensions, space group, and density are determined. These will indicate if the analysis is feasible or not. Sometimes a subunit of an enzyme or other large macromolecule is the asymmetric unit; then the structure analysis becomes more feasible because only a fraction of the atomic positions need be determined. On the other hand, it sometimes happens that several molecules comprise the asymmetric unit. This is not always unfortunate, because the resulting additional symmetry in the Patterson function may provide valuable help in solving the structure.

3. The next stage is to assess the degree of order in the crystal under study. This is determined by the measurable reflections at highest $\sin \theta/\lambda$ values. It must then be decided whether the ultimate resolution will be sufficient to provide information about the detailed structure. If the resolution is poor, one must try to grow better crystals or look for another source of the biological macromolecule (e.g., different animal or bacterium).

4. Next, attempts are made to get heavy-atom derivatives by soaking macromolecule crystals in solutions of heavy-atom salts. If this works the project will succeed, but many soaking experiments may be needed before good derivatives (more than one is usually essential) are obtained.

5. Then diffraction data for native protein and heavy-atom derivatives must be measured, either with a diffractometer (if the unit cell is small) or by film methods or area detectors (if the unit cell is large). Data on anomalous scattering, if relevant, may also be collected [i.e., $|F(hkl)|$ and $|F(\overline{hkl})|$]. In general, for convenience, the low-resolution data are measured first, and later higher-order data are collected if the isomorphism is exact enough for the heavy-atom data to be useful.

6. Heavy atoms are then located from difference Patterson maps (sometimes by direct methods). If there are too many indicated positions, another heavy-atom derivative may need to be used.

7. If there are only a few heavy-atom sites (or, better yet, only one), then the phases can be determined. Because of the ambiguity in each phase determined for one heavy-atom derivative, it is necessary to use two or more different heavy atom derivatives, with anomalous scattering data if possible.

8. The heavy-atom parameters are refined by least-squares methods, improved phases are derived, and an electron density map is computed.

9. The next stage is the interpretation of the electron density map; this may be difficult, especially if the resolution is low. Computer graphics are now used for this step. If the molecule under study is a protein, it is usual first to "trace the chain" of the polypeptide backbone. If the macromolecule

is a nucleic acid, the phosphate groups and the bases are sought from the electron density map. The determination of side chain coordinates for the protein follows from higher-resolution data and from a knowledge of the amino acid sequence of the protein. Without sequence information the analysis of the electron density map is difficult unless phasing is good to atomic resolution.

10. For an enzyme the question of the location of the active site of the catalytic process then arises. This may often be found by soaking into native crystals either inhibitors, poor substrates (if the substrate is too good, reaction may readily occur), or cofactors. Then diffraction data are measured and a difference electron density map is computed using phases from the native protein. In this way the site of attachment of a substrate can be found and it may be inferred that this is the active site of the enzyme.

CONCLUDING REMARKS

We have attempted to present enough about the details of structure determination so that an attentive reader can appreciate how the method works. As mentioned earlier, a glossary and bibliography have been included so that those interested may delve further into the subject. We will summarize by answering our initial questions.

Why Use Crystals and Not Liquids or Gases? A crystal has a precise internal order and gives a diffraction pattern that can be analyzed in terms of the shape and contents of one repeating unit, the unit cell. This internal order is lacking in liquids and gases and for these only radial information may be derived. Such information may be of use in distinguishing structures but the directional quality of any structural information is lost.

Why Use X Rays (or Neutrons) and Not Other Radiation? These radiations are scattered by the components of atoms and have wavelengths that are of the same order of magnitude as the distances between atoms in a crystal (approximately 10^{-10} m). Hence they lead to diffraction effects on a scale convenient for observation.

What Experimental Measurements Are Needed? The unit-cell dimensions and the density of the crystal, and the indices and intensities of all observable "reflections" in the diffraction pattern.

What Are the Stages in a Typical Structure Determination? These stages have been described above in detail for both small molecules and macromolecules, and further information may be obtained from references in the bibliography. The stages involve preparation of a crystal, the indexing and measurement of intensities in the diffraction pattern, the determination of a "trial structure," and the refinement of this structure.

Why Is the Process of Structure Analysis Often Lengthy and Complex? Because 50 to 100 distinct intensity measurements are needed per atom in the asymmetric unit for a resolution of 0.75 Å, because the determination of a trial structure may be difficult, because the refinement requires much computation, and because in the end so much structural information is obtained that analysis of it takes time. Many structures are readily or even automatically solved, while others, tackled by the same competent crystallographer, may take months or years to solve. It is hard for the noncrystallographer who may have been led to believe that the determination of structure is now almost automatic, to comprehend this "never-never land" in which the crystallographer occasionally finds himself while trying to arrive at a trial structure for certain crystals.

Why Is It Necessary to "Refine" the Approximate Structure That Is First Obtained? Because the initially estimated phases may give a poor image of the scattering matter. Since the least-squares equations are not linear, many cycles of refinement are usually necessary. By refinement one can tell whether the approximate structure is correct and obtain the best possible atomic positions consistent with the experimental data and the assumed structural model.

How Can One Assess the Reliability of a Structure Analysis? By checking the estimated standard deviations of the derived results, and by considering measures of the agreement of the values of $|F_o|$ with the values of $|F_c|$, by the absence of any unexplained peaks in a final difference map, and by the chemical reasonableness of the resulting structure.

We hope we have made it possible for you to read accounts of X-ray structure analyses with some appreciation of the scope and the limitations of the work described. Perhaps you are even interested enough to want to try the techniques yourself. If so, we trust that this introduction serves as a useful background and reference. But also we hope that you realize that there is more to the crystallographer's discipline than just diffraction methods. When the crystal structure is known, it is a first step in the interpretation of physical properties, chemical reactivity, or biological function in terms of three-dimensional structure.

APPENDIXES

APPENDIX 1 Some Information About Crystal Systems and Crystal Lattices

There are seven crystal systems defined by the minimum symmetry of the unit cell. It is conventional to label the edges of the unit cell a, b, c and the angles between them α, β, γ, with α the angle between **b** and **c**, β that between **a** and **c**, and γ that between **a** and **b**. If the lattice has six-fold symmetry, sometimes four axes of reference are used. These are x, y, u, z, where x, y, and u lie in one plane inclined at 120° to each other and with z perpendicular to them. The indices of reflections are then $hkil$ with the necessary condition that $i = -(h + k)$. We use the simpler cell here. For the seven crystal systems the minimum symmetry and the diffraction symmetry are:

	Minimum point group symmetry of a crystal in this system	Diffraction symmetry (Laue symmetry)
1. Triclinic	None.	$\bar{1}$
2. Monoclinic	Two-fold axis parallel to **b**.	$2/m$
3. Orthorhombic	Three mutually perpendicular two-fold axes.	mmm
4. Tetragonal	Four-fold axis parallel to **c**.	$4/m$ or $4/mmm$
5. Trigonal/ rhombohedral	Three-fold axis parallel to $(\mathbf{a} + \mathbf{b} + \mathbf{c})$.	$\bar{3}$ or $\bar{3}m$
6. Hexagonal	Six-fold axis parallel to **c**.	$6/m$ or $6/mmm$
7. Cubic	Three-fold axes along the cube diagonals.	$m3$ or $m3m$

Diagrams of the unit cells are shown below, together with symmetry-imposed restrictions on the unit-cell dimensions.

Diagrams of unit cells	Crystal systems	Rotational symmetry elements and cell-dimension restrictions
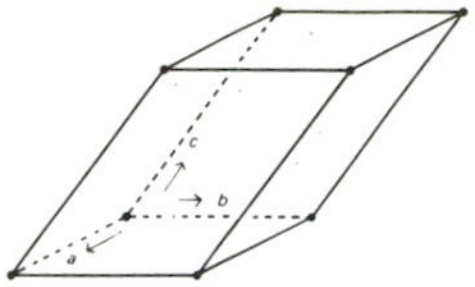	Triclinic	No rotational symmetry. No restrictions on axial ratios or angles.
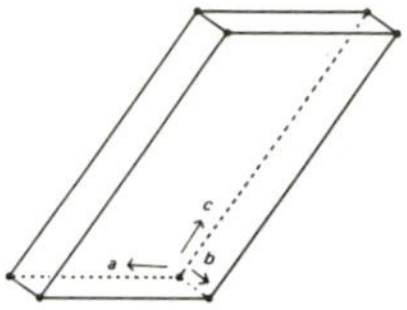	Monoclinic	**b** chosen along the two-fold rotation axis.[a] Angles made by **b** with **a** and by **b** with **c** must be 90°.
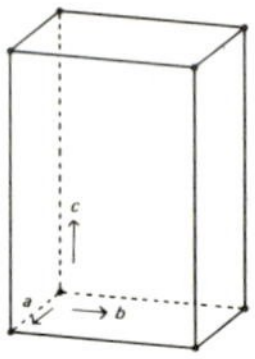	Orthorhombic	Three mutually perpendicular two-fold rotation axes chosen as **a**, **b**, **c** coordinate axes. No restrictions on axial ratios. All three angles must be 90°.
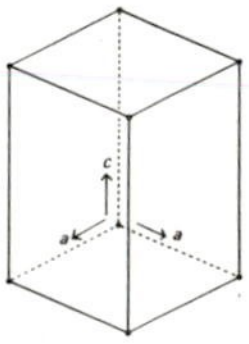	Tetragonal	Four-fold rotation axis chosen as **c**. Two-fold rotation axes perpendicular to **c**. Lengths of **a** and **b** identical. All angles must be 90°.
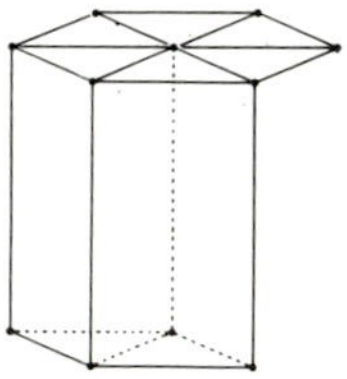	Hexagonal[b]	**c** is chosen along the six-fold axis. Two-fold rotation axes perpendicular to **c**. Angle between **a** and **b** must be 120°, other two angles must be 90°.
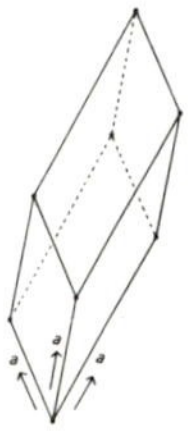	Rhombohedral	Three-fold rotation axis along one body diagonal of unit cell. This makes all three axial lengths necessarily the same and all three interaxial angles also necessarily equal. There is no restriction on the value of the interaxial angle, α.

Diagrams of unit cells	Crystal systems	Rotational symmetry elements and cell-dimension restrictions
	Cubic	Three-fold rotation axes along all four body-diagonals of unit cell. Four-fold axes parallel to each crystal axis. Two-fold axes are also present. All axial lengths are identical by symmetry. All angles must be 90°.
	Face-centered (F) and body-centered (I) cubic	Symmetry at each lattice point is the same as for simple cubic. F has four lattice points per unit cell, the extra three being at face centers. I has two points per unit cell, the extra one being at the center of the cell.

[a]This means that if the cell is rotated 360°/2 = 180° about an axis parallel to **b** the cell so obtained is indistinguishable from the original.

[b]The six-fold axis present in hexagonal lattices is perhaps not evident from the shape of the unit cell, because the inclusion of the cell edges as solid lines in the diagram obscures the symmetry. If only the lattice points are shown in a layer normal to the unique **c**-axis (one cell is outlined here in dashed lines), the six-fold symmetry is apparent (ignoring the dotted lines). There is a six-fold rotation axis perpendicular to the plane of the paper at every lattice point; this is indicated by the dashed lines drawn from one lattice point.

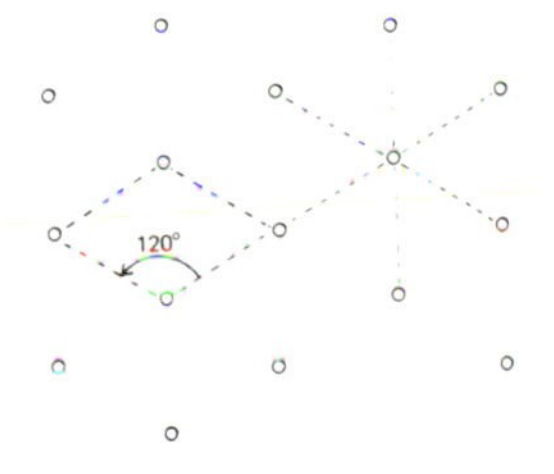

There are five additional Bravais lattices that are obtained by adding face-centering and body-centering to certain of the seven space lattices just listed. Face centering involves a lattice point at the center of opposite pairs of faces and is designated F if all faces are centered and A, B, or C if only one pair of faces is centered. In body-centered unit cells there is a lattice point at the center of the unit cell; a body-centered cell is designated I. These centerings cause additional systematic absences in the measured reflections as follows:

A $(k + l)$ odd
B $(l + h)$ odd
C $(h + k)$ odd
F $(h + k)$, $(k + l)$, $(l + h)$ all odd
I $(h + k + l)$ odd

The 14 Bravais lattices are then:

Triclinic	*P*
Monoclinic	*P C*
Orthorhombic	*P C F I*
Tetragonal	*P I*
Hexagonal	*P*
Rhombohedral	*P*
Cubic	*P F I*

(*C* in monoclinic can alternatively be *A* or *I*; *C* in orthorhombic can alternatively be *A* or *B*. *P* in rhombohedral is often called *R*.)

Addition of symmetry elements to these Bravais lattices give the 230 space groups. Some of these symmetry elements also cause systematic absences in the diffraction pattern. For example, for a two-fold screw axis parallel to **a**, h in the h 0 0 reflections is only even, and for a four-fold screw axis parallel to **a**, h in the h 0 0 reflections is only a multiple of 4. For a glide plane perpendicular to **a** with translation $b/2$ (a b glide), k in the $0kl$ reflections is only even. For more details, *International Tables*, Volume I, or *X-ray Crystallography* by M. J. Buerger (1953 edition, Chapter 4, p. 83) should be consulted.

APPENDIX 2 The Reciprocal Lattice

The relation between the crystal lattice (real space) and the reciprocal lattice (reciprocal space) may be expressed most simply in terms of vectors. If we denote the fundamental translation vectors of the crystal lattice by **a**, **b**, and **c**, and the volume of the unit cell by V_c, and then use the same symbols, starred, for the corresponding quantities of the reciprocal lattice, the relation between the two lattices is:

$$\mathbf{a}^* = \frac{\mathbf{b} \times \mathbf{c}}{V_c}, \qquad \mathbf{b}^* = \frac{\mathbf{c} \times \mathbf{a}}{V_c}, \quad \mathbf{c}^* = \frac{\mathbf{a} \times \mathbf{b}}{V_c} \tag{A2.1}$$

with $V_c = \mathbf{a} \cdot \mathbf{b} \times \mathbf{c} = 1/V_c^*$.

The vectors of the crystal lattice and the reciprocal lattice are thus oriented as follows: any fundamental translation of one lattice is perpendicular to the other two fundamental translations of the second lattice. Thus $\mathbf{a}^*$ is perpendicular to both **b** and **c**, **b** is perpendicular to both $\mathbf{a}^*$ and $\mathbf{c}^*$, and so on. The vectors of the crystal lattice and the reciprocal lattice form an adjoint set in the sense that this term is used in tensor calculus, satisfying the condition that the scalar product of any two corresponding fundamental translation vectors of the two lattices is unity and the scalar product of any two noncorresponding vectors of the two lattices is zero, because, as mentioned above, they are mutually perpendicular. This is expressed by

$$\mathbf{a}_i^*\cdot\mathbf{a}_j = \delta_{ij} \begin{cases} = 1, & i = j \\ = 0, & i \neq j \end{cases} \tag{A2.2}$$

That is,

$$\mathbf{a}\cdot\mathbf{a}^* = \mathbf{b}\cdot\mathbf{b}^* = \mathbf{c}\cdot\mathbf{c}^* = 1$$

and

$$\mathbf{b}\cdot\mathbf{a}^* = \mathbf{c}\cdot\mathbf{a}^* = \mathbf{c}\cdot\mathbf{b}^* = \mathbf{a}\cdot\mathbf{b}^* = \mathbf{a}\cdot\mathbf{c}^* = \mathbf{b}\cdot\mathbf{c}^* = 0$$

As stressed in the section "Diffraction by Regular Arrays" (Chapter 3, especially Figure 3.5), if a structure is arranged on a given lattice, its diffraction pattern is necessarily arranged on a lattice reciprocal to the first. The fact that any fundamental translation of the crystal lattice is perpendicular to the other two fundamental translations of the reciprocal lattice, and the converse, is an example of a quite general relation: every reciprocal lattice vector is perpendicular to some plane in the crystal lattice and, conversely, every crystal lattice vector is perpendicular to some plane in the reciprocal lattice. Furthermore, if the indices of a crystal lattice plane are (hkl) (in the sense defined in the caption of Figure 2.3), the reciprocal lattice vector **H** perpendicular to this plane is the vector from the origin of the reciprocal lattice to the reciprocal lattice point with indices hkl. It is expressed as

$$\mathbf{H} = h\mathbf{a}^* + k\mathbf{b}^* + l\mathbf{c}^* \tag{A2.3}$$

In a monoclinic unit cell,

$$d_{100} = a \sin\beta = \frac{1}{a^*} \tag{A2.4}$$

(See Appendix 3, where this relation is illustrated, and Figure 2.3.) We have the relation, for this reflection, 100,

$$|\mathbf{H}| = |h\mathbf{a}^*| = h/d_{100}. \tag{A2.5}$$

or

$$|\mathbf{H}_{100}| = |\mathbf{a}^*| = 1/d_{100}$$

A comparison with the Bragg equation for 100 with the appropriate values of d and θ,

$$h\lambda = 2d\sin\theta \qquad \text{or} \qquad h/d = \frac{2\sin\theta}{\lambda}$$
$$\lambda = 2d_{100}\sin\theta_{100} \qquad \text{or} \qquad 1/d_{100} = 2\sin\theta_{100}/\lambda \tag{A2.6}$$

indicates that in this case

$$|\mathbf{H}| = \frac{2\sin\theta}{\lambda} \tag{A2.7}$$

This relation holds quite generally.

$$|\mathbf{H}_{100}| = 2 \sin \theta_{100}/\lambda$$
$$|\mathbf{H}_{hkl}| = 2 \sin \theta_{hkl}/\lambda \qquad \text{(A2.8)}$$

The equations relating the real and reciprocal unit cell dimensions are given in *X-ray Crystallography* by M. J. Buerger (1953 edition, Chapter 18, p. 360) and by Stout and Jensen, ref. 2., p. 31. Some of these are listed below:

$$a^* = bc \sin \alpha/V, \qquad b^* = ac \sin \beta/V, \qquad c^* = ab \sin \gamma/V$$

where

$$V = abc\sqrt{(1 - \cos^2 \alpha - \cos^2 \beta - \cos^2 \gamma + 2 \cos \alpha \cos \beta \cos \gamma)}$$

$$V^* = 1/V$$

$$\cos \alpha^* = (\cos \beta \cos \gamma - \cos \alpha)/\sin \beta \sin \gamma$$

$$\cos \beta^* = (\cos \alpha \cos \gamma - \cos \beta)/\sin \alpha \sin \gamma$$

$$\cos \gamma^* = (\cos \alpha \cos \beta - \cos \gamma)/\sin \alpha \sin \beta$$

APPENDIX 3 The Equivalence of Diffraction by a Lattice and the Bragg Equation

(This treatment is adapted from that of Stout and Jensen, ref. 2.)

For simplicity we will consider diffraction by a two-dimensional orthogonal lattice (a rectangular net), but the treatment can be generalized to three dimensions and to the nonorthogonal case. Suppose the lattice has sides a and b for each unit cell and that X rays are incident upon the lattice from a direction such that the incident beams make an angle ψ with the lattice rows in the **a** direction. Consider the scattering in the direction ψ' with respect to the **a** direction. Because a and b are lattice translations, any atom in the structure will be repeated periodically with spacings a and b. Thus atoms may be imagined to be present at the lattice points in Figure A3.1 below (they will normally also be present at other points, lying between these lattice points, but spaced identically in each unit cell). If scattering is to occur in the direction specified by ψ', then the radiation scattered in that direction from every lattice point must be exactly in phase with that from every other lattice point. (If scattering from any two lattice points is somewhat out of phase, that from some other pair of lattice points will be out of phase by a different amount, and the net sum over all lattice points, considering the crystal to be essentially infinite, will consist of equal positive and negative contributions and thus will be zero.)

Consider waves 1 and 2, scattered by atoms separated by a (Figure A3.1). For these waves to be just in phase after scattering, the path difference (PD_1) must be an integral number of wavelengths:

$$PD_1 = p - q = a \cos \psi - a \cos \psi' = h\lambda \tag{A3.1}$$

where h is some integer. Similarly, the path difference for waves 1 and 3, scattered by atoms separated by b, must also be an integral number of wavelengths:

$$PD_2 = r + s = b \sin \psi + b \sin \psi' = k\lambda \tag{A3.2}$$

where k is some integer.

Both of these conditions must hold simultaneously. They are sufficient conditions to ensure that the scattering from all atoms in this two-dimensional net will be in phase in the direction ψ'. In three dimensions, another similar equation, corresponding to the spacing in the third (noncoplanar) direction, must be added. Each of these equations describes a cone. In three dimensions, the *three cones intersect in a line corresponding to the direction of the diffracted beam,* such that the conditions $h\lambda = PD_1$, $k\lambda = PD_2$, and $l\lambda = PD_3$ all are satisfied simultaneously. This is why, when a three-dimensional crystal diffracts, there are very few diffracted beams for any given orientation of the incident beam to the (stationary) crystal. The chance that all three conditions will be satisfied at once is small.

Now let us see how this set of conditions can be related to the Bragg equation. Consider several parallel planes, I, II, and III, each passing through a set of lattice points and making equal angles, θ, with the incident and scattered beams (Figure A3.2). The planes make an angle α with the **a** axis. The angles ψ and ψ' are defined as in Figure A3.1, and so

$$\theta = \psi + \alpha = \psi' - \alpha \tag{A3.3}$$

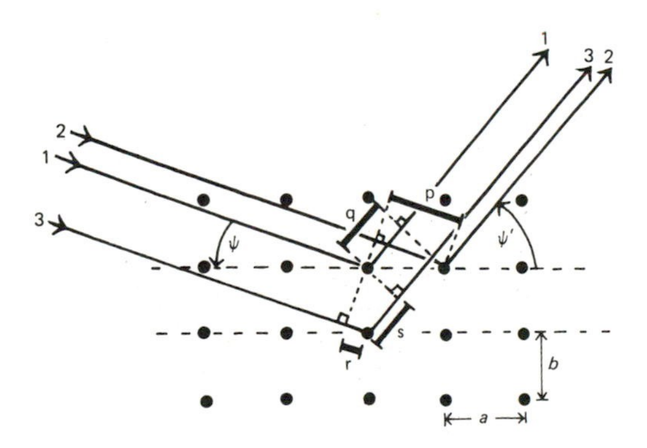

FIGURE A3.1

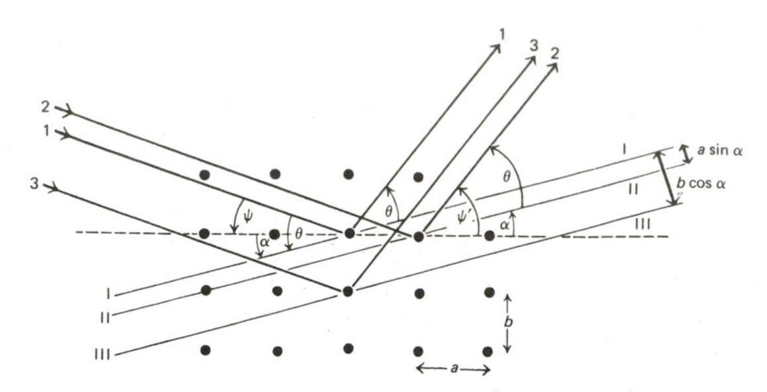

FIGURE A3.2

These are illustrated in Figure A3.2. Substituting for ψ and ψ' from Eq. (A3.3) into Eqs. (A3.1) and (A3.2), we find

$$h\lambda = 2a \sin \alpha \sin \theta \tag{A3.4}$$

$$k\lambda = 2b \cos \alpha \sin \theta \tag{A3.5}$$

or

$$\frac{2 \sin \theta}{\lambda} = \frac{h}{a \sin \alpha} = \frac{k}{b \cos \alpha} \tag{A3.6}$$

Now $a \sin \alpha$ is just the spacing between planes I and II, while $b \cos \alpha$ is just the spacing between planes I and III. If we let d_{hkl} represent the spacing between any two planes in a set of equidistant planes parallel to I, and n be some integer, we can write Eq. (A3.6) generally as

$$\frac{2 \sin \theta}{\lambda} = \frac{n}{d_{hkl}} \tag{A3.7}$$

which is the Bragg equation, $n\lambda = 2d \sin \theta$, Eq. (3.1).

The indices (H K) of the "reflecting planes," I, II and III, are determined, as described in the caption to Figure 2.3, by measuring the intercepts on the axes as fractions of the cell edges. From Figure A3.2 it can be seen that the intercepts along **b** and **a** are in the ratio tan α, whence

$$(b/\mathrm{K})/(a/\mathrm{H}) = \tan \alpha, \tag{A3.8}$$

or

$$\mathrm{H/K} = (a \tan \alpha)/b \tag{A3.9}$$

Equation (A3.6) then shows the relation of H and K to the indices of the reflection (h k),

$$\frac{\mathrm{H}}{\mathrm{K}} = \frac{a \sin \alpha}{b \cos \alpha} = \frac{h}{k} \tag{A3.10}$$

That is, h and k, the indices of the reflection, are proportional to H and K, the indices of the reflecting plane.

APPENDIX 4 The Determination of Unit-Cell Constants and Their Use in Ascertaining the Contents of the Unit Cell

Cell Dimensions

Cell dimensions may be determined, with radiation of a known wavelength, from values of 2θ for reflections of known indices, where 2θ is the deviation of the diffracted beam from the direct beam. The Bragg equation is then used.

Example Monoclinic cell $\alpha = \gamma = 90°$, $\beta = 100.12°$, $\sin\beta = 0.98445$, $\lambda = 1.5418$ Å.

h	k	l	$2\theta(°)$	$\theta(°)$	$\sin\theta$	$n\lambda/2\sin\theta$ (Å)	
20	0	0	85.68	42.84	0.67995	22.675	d_{100}
22	0	0	96.82	48.41	0.74791	22.676	
0	4	0	47.41	23.705	0.40203	7.670	d_{010}
0	0	10	104.14	52.07	0.78876	9.774	d_{001}

Unit cell **b** perpendicular to the plane of the paper. d_{hkl} = spacing between crystal planes hkl (see Figures 2.3 and A4.1).

Cell Contents

Let W = weight in grams of one gram-formula weight of the contents of the unit cell and V = the volume in cm^3 of this weight of the crystal.

Cell volume = 1726 $Å^3$ = 1726×10^{-24} cm^3

Observed density (by flotation) = 1.34 g/cm^3

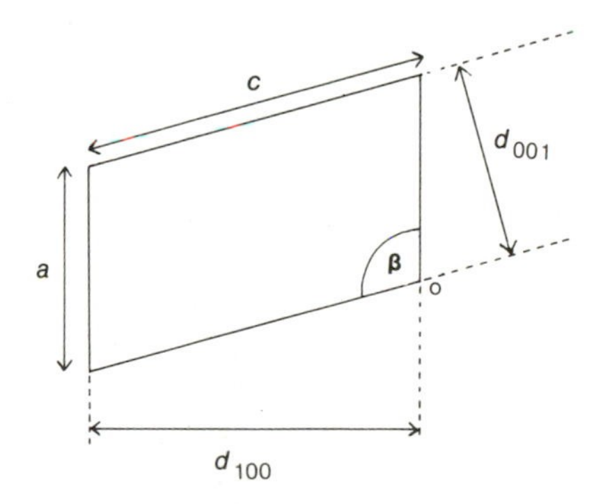

FIGURE A4.1

$N_{Avog.}$ unit cells occupy $1726 \times 10^{-24} \times 6.02 \times 10^{23}$ cm^3 = V = 1039 cm^3
Density = W/V = W/(1726 × 0.602) g/cm^3 = 1.34 g/cm^3
Therefore W = 1.39×10^3
But W also equals (ZM + zm),
where Z is the number of molecules of the compound (molecular weight M) per unit cell, and z is the number of molecules of solvent of crystallization (molecular weight m) per unit cell. In this example M is known to be 340 and m = 18 (for water).

$$(Z \times 340) + (z \times 18) = 1.39 \times 10^3$$

The monoclinic symmetry of the unit cell suggests that Z is 4, or a multiple of 4, leading to the conclusion that Z = 4 and z = 2 (W = 1396) is the correct solution, and that the solution Z = 3 and z = 20 (W = 1380), which is equally probable from the calculated weight alone, is much less likely.

APPENDIX 5 Proof That the Phase Difference on Diffraction Is $2\pi(hx + ky + lz)$

The phase difference for the h 0 0 reflection for diffraction by two atoms one unit cell apart is $(360h)°$ = $2\pi h$ radians. If the atoms are only the fraction x of the cell length apart then the phase difference will be $2\pi hx$ radians. This may be extended to three dimensions to give $2\pi(hx + ky + lz)$ as the phase difference for the hkl reflection for two atoms, one at 0,0,0 and the other at x,y,z.

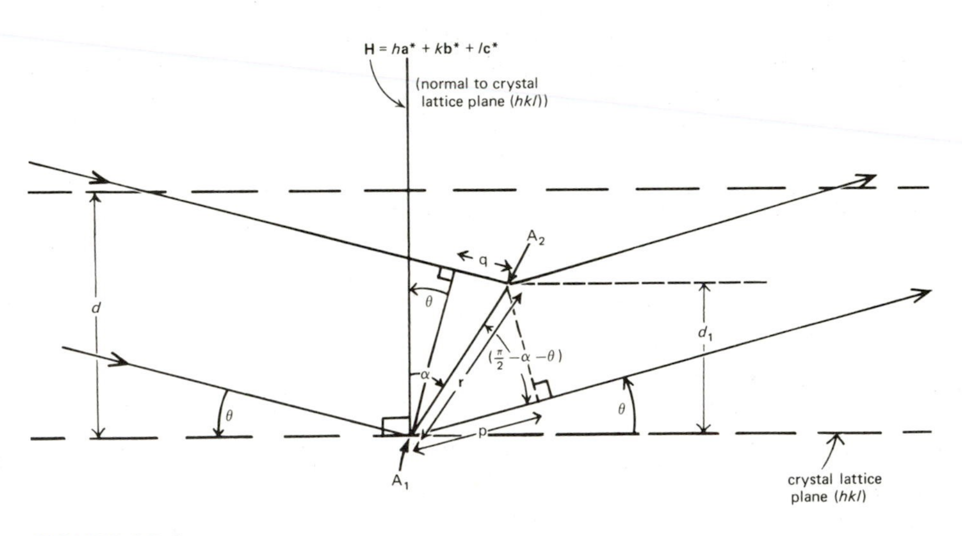

FIGURE A5.1

A proof is given below:

Let A_1 and A_2 be two scattering points (atoms) separated by a vector **r** (Figure A5.1). The phase difference for the beams scattered from these atoms at the angle θ is

$$2\pi \frac{\mathrm{p} - \mathrm{q}}{\lambda} \text{ radians}$$

where

$$\mathrm{p} = |\mathbf{r}| \cos\left(\frac{\pi}{2} - \alpha - \theta\right) = |\mathbf{r}| \sin(\alpha + \theta)$$

$$= |\mathbf{r}| \sin\alpha \cos\theta + |\mathbf{r}| \cos\alpha \sin\theta \qquad \text{(A5.1)}$$

$$\mathrm{q} = |\mathbf{r}| \sin(\alpha - \theta) = |\mathbf{r}| \sin\alpha \cos\theta - |\mathbf{r}| \cos\alpha \sin\theta \qquad \text{(A5.2)}$$

$$\mathrm{p} - \mathrm{q} = 2|\mathbf{r}| \cos\alpha \sin\theta$$

Therefore

$$\frac{2\pi(\mathrm{p} - \mathrm{q})}{\lambda} = 2\pi \cdot \frac{2 \sin\theta}{\lambda} \cdot |\mathbf{r}| \cos\alpha \qquad \text{(A5.3)}$$

But

$$\mathbf{H} = h\mathbf{a}^* + k\mathbf{b}^* + l\mathbf{c}^* \qquad \text{(A5.4)}$$

and is normal to the crystal lattice plane (hkl) (Appendix 2) and, by Eq. (A2.8),

$$|\mathbf{H}| = \frac{2 \sin\theta}{\lambda} \qquad \text{(A5.5)}$$

Since α is the angle between **H** and **r**, where $\mathbf{r} = \mathbf{a}x + \mathbf{b}y + \mathbf{c}z$, then, by Eqs. (A5.3) and (A5.5),

$$\frac{2\pi(\mathrm{p} - \mathrm{q})}{\lambda} = 2\pi |\mathbf{H}| |\mathbf{r}| \cos(\text{angle between } \mathbf{H} \text{ and } \mathbf{r})$$

Therefore, the phase difference on diffraction is

$$\frac{2\pi(\mathrm{p} - \mathrm{q})}{\lambda} = 2\pi\, \mathbf{H} \cdot \mathbf{r} = 2\pi(hx + ky + lz) \qquad \text{(A5.6)}$$

since

$$\mathbf{a}_i^* \cdot \mathbf{a}_j = \delta_{ij} \begin{cases} = 1, & i = j \\ = 0, & i \neq j \end{cases}$$

APPENDIX 6 A Second Example of Direct Methods Structure Solution (see also Figure 8.1)

A monoclinic crystal of a compound, $C_{10}H_{13}N_2O_3P$, was studied. Unit-cell dimensions are a = 12.461, b = 5.497, c = 16.888 Å, β = 97.24°. The space group was $P2_1/n$, centrosymmetric, with atoms at $\pm(x, y, z)$ and $\pm(\frac{1}{2} - x, \frac{1}{2} + y, \frac{1}{2} - z)$. The unit cell contained 4 atoms of phosphorus, 12 oxygen, 8 nitrogen, 40 carbon, and 52 hydrogen atoms. The structure was solved using the direct methods program MULTAN (ref. 31).

A Wilson plot of $\ln(F^2_{obs}/\Sigma f^2)$ vs. $(\sin\theta/\lambda)^2$ gave a line of slope −8.24, intercept 2.8146 corresponding to a temperature factor of 4.12 Å^2 and a scale of 0.0599 where $|F_{absolute}|^2 = 0.0599\,|F_{measured}|^2$. The structure was shown to be centrosymmetric by an examination of E statistics (although in this example the systematic absences showed this also, since there is only one space group corresponding to them).

Average	All data	Theoretical acentric	Theoretical centric
$\lvert E\rvert$	0.811	0.886	0.798
$\lvert E\rvert^2$	1.000	1.000	1.000
$\lvert\lvert E\rvert^2 - 1\rvert$	0.925	0.736	0.968

The 150 largest E values (down to E = 1.739) and the 50 smallest E values ($E \leq 0.056$) were made into a list. The Σ_1 relationships were then computed. It was decided to accept a phase angle if it was computed with a probability greater than 0.95. No reflections fitted this category. Then a Σ_2 listing was made and phases were derived. Three origin-fixing reflections were chosen with phases of 0°: 5,2,−10; 2,1,1, and 9,2,1 [none (even, even, even) and each of a different parity group]. Three other reflections (6,0,−14; 6,1,−15; 0,1,1) were chosen with phases of 0° or 180°, so that 8 sets of phases were generated.

Set #	Phase of 6,0,−14	Phase of 6,1,−15	Phase of 0,1,1	Figures of merit (see footnote below)* ABSFOM	PSIZERO	RESIDUAL
1	0	0	0	1.15	1.52	20.9
2	180	0	0	1.05	1.31	24.7
3	0	180	0	1.14	1.52	21.0 (almost same as set 1)
4	180	180	0	1.03	1.25	26.3
5	0	0	180	1.28	1.42	13.0

Set #	Phase of 6,0,−14	Phase of 6,1,−15	Phase of 0,1,1	Figures of merit (see footnote below)*		
				ABSFOM	PSIZERO	RESIDUAL
6	180	0	180	1.07	1.29	24.4 (almost same as set 2)
7	0	180	180	1.28	1.42	13.0 (same as set 5)
8	180	180	180	1.07	1.29	24.4 (same as set 6)

The computer program MULTAN evaluates (as indicated) three different figures of merit for each solution of the structure, based on different criteria.* The most important is probably the residual, especially if its value is distinctly lower for one solution than for the others, as for set 5 or 7 here, which (after refinement of the phases by MULTAN) were identical. The first electron density map was calculated from set 5, using the entire set of phases

*The following figures of merit are used as a preliminary indicator of the best phase set. For a plausible set of phases one would expect ABSFOM to be large, preferably close to unity, while PSIZERO and RESIDUAL should be relatively small.

PSIZERO is a measure of how well the low-intensity reflections are computed to have low E values. It is computed as:

$$\phi_0 = \sum_{\mathbf{h}} \left| \sum_{\mathbf{h}'} E_{\mathbf{h}'} E_{\mathbf{h}-\mathbf{h}'} \right|$$

where the summation over $\mathbf{h}'$ includes large E values for which phases have been determined and that over $\mathbf{h}$ is for E values that are small or zero.

ABSFOM is a measure of the internal consistency among Σ_2 relationships and is computed (with summations over all reflections) as:

$$\{\Sigma\alpha_{\mathbf{h}} - \Sigma(\alpha_{\mathbf{h}})_{\mathrm{RAN}}\}/\{\Sigma\langle\alpha_{\mathbf{h}}^2\rangle^{1/2} - \Sigma(\alpha_{\mathbf{h}})_{\mathrm{RAN}}\}$$

where $(\alpha_{\mathbf{h}})_{\mathrm{RAN}}$ is the r.m.s. expected value of $\alpha_{\mathbf{h}}$ for random phases, given by

$$(\alpha_{\mathbf{h}})_{\mathrm{RAN}} = \left\{ \sum_{\mathbf{h}'} K_{\mathbf{h},\mathbf{h}'}^2 \right\}^{1/2}$$

where

$$K_{\mathbf{h},\mathbf{h}'} = 2\mathrm{N}^{-1/2} \,|E_{\mathbf{h}} E_{\mathbf{h}'} E_{\mathbf{h}-\mathbf{h}'}|$$

RESIDUAL is analogous to a conventional R value and is computed as:

$$\left\{ \sum_{\mathbf{h}} |\, E_{\mathbf{h}}| - \mathrm{k} |\sum_{\mathbf{h}'} E_{\mathbf{h}'} E_{\mathbf{h}-\mathbf{h}'}| \right\} \Big/ \left\{ \sum_{\mathbf{h}} |E_{\mathbf{h}}| \right\}$$

where the summation over $\mathbf{h}$ is for all E values whose phases are being determined and k is a scaling constant given by

$$\mathrm{k} = \left\{ \sum_{\mathbf{h}} |E_{\mathbf{h}}| \right\} \Big/ \left\{ \sum_{\mathbf{h}} \left| \sum_{\mathbf{h}'} E_{\mathbf{h}} E_{\mathbf{h}-\mathbf{h}'} \right| \right\}$$

generated. High peaks were picked out from this map (this could be done by computer) and distances and angles were then computed. All seemed reasonable and the R value was 0.32 (32 per cent) before any refinement of atomic coordinates or temperature factors. A least-squares refinement of the atomic parameters, location of hydrogen atoms from a difference map and subsequent refinement of hydrogen atom parameters as well gave the structure to high precision. If none of the MULTAN solutions had given a reasonable structure, it would have been necessary to go back to an earlier stage in the direct methods approach. If no good solution could be obtained, it would have been necessary to recheck the original data collection, the space group, and the poorly measured reflections and to look at the Patterson map to see if that would give any indication of where the difficulty might be.

APPENDIX 7 The Patterson Function

The Patterson function, deduced by A. L. Patterson in 1934, is a Fourier series, analogous to Eq. (6.1), in which the coefficients of $|F|$ are replaced by $|F|^2$.

$$P(u,v,w) = \frac{1}{V} \sum\sum\sum_{\text{all } h,k,l} |F|^2 \exp\left[-2\pi i(hu + kv + lw)\right] \quad \text{(A7.1)}$$

This function was an extension from the suggestion of Zernike and Prins in 1926 that, because there may be local order in a liquid, there should be diffraction effects. For example, in a monatomic liquid such as mercury, the nearest neighbors of a given atom should never be at a distance less than two atomic radii, and seldom much more. There is more disorder for second nearest neighbors and, as the distance from the atom under consideration increases, the arrangement becomes random. Zernike and Prins showed that measurements of the diffraction pattern could be used to calculate the *average radial distribution* of matter in a liquid or powdered crystal. The term "radial" is used because the distribution is averaged over all directions and depends only on the distance from its origin. The term "average" implies that the distribution function found represents the average of the distributions of neighbors around each of the atoms in the sample whose diffraction pattern has been used.

These ideas were extended to crystals by Patterson, who recognized the key fact that, because of the high degree of order in the crystal, the averaging over all directions could be eliminated and detailed information about both the *magnitudes* and the *directions* of the interatomic vectors (that is, both radial and angular information) could be obtained.

It is easiest to consider first a one-dimensional case and then extend it to three dimensions. A one-dimensional electron density distribution map,

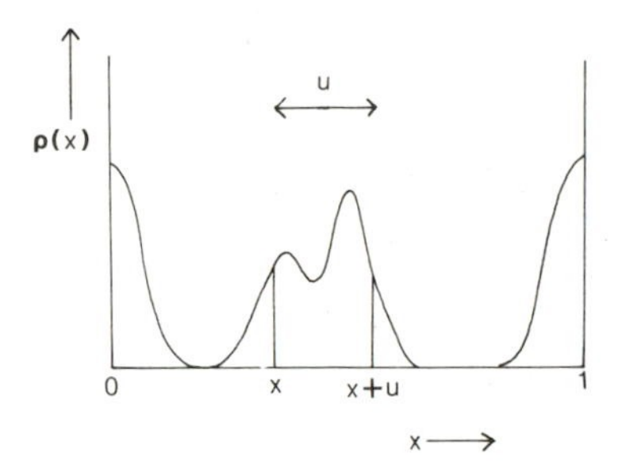

FIGURE A7.1

$\rho(x)$, for a regularly repeating cell, length a, can be expressed by Eq. (A7.2) and is illustrated in Figure A7.1.

$$\rho(x) = \frac{1}{a} \sum_{\text{all } h} F(h) \exp(-2\pi i h x) \tag{A7.2}$$

Consider the distribution of electron density about an arbitrary point, x, in the unit cell. The electron density at a point $+u$ from x is $\rho(x + u)$. Patterson defined the *weighted distribution,* WD, about the point x by allotting to the distribution about x a weight that was equal to $\rho(x)\,dx$, the total amount of scattering material in the interval between x and $x + dx$.

$$\text{WD} = \rho(x + u)\rho(x)\,dx \tag{A7.3}$$

It can be seen that, for a given value of dx, WD is large *only* if both $\rho(x)$ and $\rho(x + u)$ are large, and is thus small if either or both are small.

Values of WD are summed by integrating over all values of x in the cell, keeping u constant, so that the average weighted distribution of density, $P(u)$, is

$$P(u) = a \int_0^1 \rho(x)\rho(x + u)\,dx \tag{A7.4}$$

$$= \frac{1}{a} \int_0^1 \left[\sum_{\text{all } h} F(h) \exp(-2\pi i h x) \right] \times \left[\sum_{\text{all } h'} F(h') \exp[-2\pi i h'(x + u)] \right] dx \tag{A7.5}$$

The properties of the complex exponential are such that the integral in (A7.5) vanishes unless $h = -h'$; consequently, Eq. (A7.5) leads to

$$P(u) = \frac{1}{a} \sum_{\text{all } h} F(h)F(-h) \exp(2\pi i h u) = \frac{1}{a} \sum_{\text{all } h} |F(h)|^2 \exp(2\pi i h u) \tag{A7.6}$$

since $\alpha(\bar{h}) = -\alpha(h)$ and, by Eq. (5.18), $F(h)F(-h) = |F|^2 e^{i\alpha} e^{-i\alpha} = |F|^2$.

This function, $P(u)$, may be visualized by imagining a pair of calipers set to measure a distance u. One point of the calipers is set on each point in the cell in turn; since the unit cell is repeated periodically, the situation at x is repeated at $x - 1, x + 1, x + 2, \ldots$. The sum of all the products of values of $\rho(x)$ at the two ends of the calipers then gives $P(u)$. As the electron density is nearly zero between atoms and is high near atomic centers, the positions of peaks in $P(u)$ correspond to vectors between atoms; in other words, large values of $\rho(x)\rho(x + u)$ give large contributions to $P(u)$.

These equations may be extended to three dimensions, letting V_c = the volume of the unit cell, so that from the definition

$$P(u,v,w) = V_c \int_0^1 \int_0^1 \int_0^1 \rho(x,y,z)\rho(x + u, y + v, z + w)\, dx\, dy\, dz \tag{A7.7}$$

after substitution of values for ρ and integration, most terms are zero, leaving

$$P(u,v,w) = \frac{1}{V_c} \sum_{\substack{\text{all} \\ hkl}}\sum\sum |F(hkl)|^2 \exp[2\pi i(hu + kv + lw)] \tag{A7.8}$$

This equation can easily be reduced to Eq. (9.1),

$$P(u,v,w) = \frac{F^2(000)}{V_c} + \frac{2}{V_c} \sum_{\substack{h\geq 0 \text{ all } k,l \\ \text{excluding } F^2(000)}}\sum\sum |F|^2 \cos 2\pi(hu + ky + lw) \tag{9.1}$$

by noting that $|F|^2(hkl) = |F|^2(\overline{hkl})$ and

$$e^{i\phi} = \cos\phi + i \sin\phi \tag{A7.9}$$

and then grouping reflections in pairs, hkl and $\overline{hkl}$. For every such pair, the value of ϕ for hkl is equal in magnitude and opposite in sign to that for $\overline{hkl}$ for each point u, v, w and thus, since $\cos\phi = \cos(-\phi)$ while $\sin\phi = -\sin(-\phi)$, the sine terms in the expansion of Eq. (A7.8) cancel when summed over each of these pairs of reflections.

To summarize, the importance of the Patterson function is that peaks in it occur at points to which vectors from the origin correspond very closely in direction and magnitude with vectors between atoms in the crystal and that no preliminary assumptions are needed because $|F|^2$ values are independent of phase and can be derived directly from the measured intensities.

APPENDIX 8 Vectors in a Patterson Map

A certain derivative of vitamin B_{12} crystallizes in the space group $P2_12_12_1$ and contains Co, Cl, O, N, C, and H, with atomic numbers 27, 17, 8, 7, 6, and 1, respectively.

(a) Expected approximate relative heights of typical peaks in the Patterson map are

Co–Co	27 × 27	729
Co–Cl	27 × 17	459
Cl–Cl	17 × 17	289
Co–O	27 × 8	216
Co–C	27 × 6	162
O–O	8 × 8	64
H–H	1 × 1	1

(This map will then be dominated by Co–Co and Co–Cl vectors unless there are accidental overlaps of other peaks.)

(b) Derivation of the coordinates of vectors between symmetry-related positions of any atom (for example, Co) in terms of its atomic position parameters:

(i) Atomic positions:
1. x, y, z
2. $\frac{1}{2} - x,\ -y,\ \frac{1}{2} + z$
3. $\frac{1}{2} + x,\ \frac{1}{2} - y,\ -z$
4. $-x,\ \frac{1}{2} + y,\ \frac{1}{2} - z$

(ii) Interatomic vectors between symmetry-related atoms are expected at the following positions, corresponding to the differences in coordinates of the various atomic positions:

$u = 0$	$v = 0$	$w = 0$	(position 1 to position 1)
$u = \frac{1}{2} - 2x$	$v = -2y$	$w = \frac{1}{2}$	(position 2 to position 1)
$u = \frac{1}{2}$	$v = \frac{1}{2} - 2y$	$w = -2z$	(position 3 to position 1)
$u = -2x$	$v = \frac{1}{2}$	$w = \frac{1}{2} - 2z$	(position 4 to position 1)

(The vector between any other positions—for example, 2 to 4 or 4 to 3—is either identical to one of these, or is related by a two-fold axis or by a center of symmetry at the origin. Every Patterson map is centrosymmetric.)

(c) The actual Patterson map for this crystal shows peaks at, among other positions:

u	v	w
0.00	0.00	0.00
±0.20	±0.32	0.50
0.50	±0.18	±0.20
±0.30	0.50	±0.30

(d) A comparison of these observed peak positions with the general expectations in (b) above shows that a consistent set of coordinates for the Co atom is $x = 0.15$, $y = 0.16$, $z = 0.10$. (See *Proceedings of the Royal Society,* A **242** (1957), pp. 228 and 1959; **251,** p. 306.)

APPENDIX 9 Isomorphous Replacement (Centrosymmetric Structure)

$F_T = F_M + F_R$
M = replaceable atom or group of atoms
R = the rest of the structure

If the position of M is known, then F_M is known. If two isomorphous crystals are studied, it is assumed that the position of the remainder of the structure is the same in each. Then, for one crystal:

$$F_T = F_M + F_R$$

while for the second one

$$F'_T = F'_M + F_R$$

The experimentally obtained data* consist of $|F_T|$ and $|F'_T|$ derived from the measured intensities. F_M and F'_M are computed from the positions of M and M′, found by an analysis of the Patterson map. The signs of F_T and F'_T may then be obtained, as illustrated in the following table.

h	k	l	$\|F_T\|$ Rb salt	$\|F_{T'}\|$ K salt	$F_{Rb} - F_K$ (calculated from known metal position)	Sign of F for Rb compd	Sign of F for K compd
0	1	1	16	20	+33	+	−
0	1	3	132	78	−59	−	−
0	2	1	63	70	−11	+	+
0	2	2	56	31	+29	+	+
0	4	0	102	50	+61	+	+
0	4	1	6	12	+17	+	−
0	5	2	9	16	+21	+	−
0	5	3	38	9	−50	−	+

For example, for the 0 1 1 reflection it is known, from computed values of F_{Rb} and F_K, that the difference is approximately +33 between F_T for the rubidium salt (for which the experimental value is ±16) and F_T for the potassium salt (for which the experimental value is ±20). This is only possible if F_{Rb} is +16 and F_K is −20, an experimental difference ($F_{Rb} - F_K$) of

*See *Acta Crystallographica* **16** (1963), 1102.

+36. For the 0 1 3 reflection the calculated value of ($F_{Rb} - F_K$) is −59 and the experimental values are ±132 for F_{Rb} and ±78 for F_K. This difference of −59 indicates that F_{Rb} is −132 and F_K is −78, an experimental difference of −54. Some phases may be ambiguous, in which case they must then be omitted.

APPENDIX 10 Diffraction Data Showing Anomalous Dispersion

The following values of $|F|$ for five pairs of reflections hkl and $\overline{hkl}$ from a crystal of potassium dihydrogen isocitrate* (Figure 10.1d) measured with chromium radiation, show the effect of anomalous scattering as a result of the presence of potassium ions in the structure. With these data the effect was sufficiently large that the absolute configuration of the sample could easily be determined. The $|F_c|$ values given here were calculated for the absolute configuration of the dihydrogen isocitrate ion given in Figure 10.1d; for the enantiomorphous form, the corresponding values for hkl and $\overline{hkl}$ would be reversed.

h	k	l	$\lvert F_o \rvert$	$\lvert F_c \rvert$
1	3	1	19.0	19.2
−1	−3	−1	22.9	23.7
1	3	2	6.4	6.6
−1	−3	−2	11.7	11.7
1	3	3	26.3	25.7
−1	−3	−3	20.7	20.0
4	5	2	7.2	7.0
−4	−5	−2	2.5	2.7
7	1	2	9.2	9.0
−7	−1	−2	13.1	12.9

The effect of anomalous scattering on the electron density calculation was discussed by A. L. Patterson in *Acta Crystallographica* **16** (1963), p. 1255. He showed how to correct the value of F so that the effect is removed, and in so doing, demonstrated that to use $F(\overline{hkl})$ or $F(hkl)$ to compute electron density is not correct when anomalous scattering is appreciable.

$$|F_{\pm}|^2 = A^2 + B^2 + (\delta_1^2 + \delta_2^2)(A_d^2 + B_d^2) + 2\delta_1(AA_d + BB_d) - 2\sigma\,\delta_2\,(AB_d - BA_d) \quad \text{(A10.1)}$$

where A and B are the components of the structure factors for the normally scattering atoms and A_d and B_d are those for the anomalously scattering

*See *Acta Crystallographica* **B24** (1968) 578.

atoms. σ has a value of $+1$ for $F(hkl)$ and -1 for $F(\overline{h}\overline{k}\overline{l})$. $\delta_1 = \Delta f'_d/f_d$ and $\delta_2 = \Delta f''_d/f_d$. As a result two quantities were defined:

$$S = \tfrac{1}{2}\{|F_+|^2 + |F_-|^2\} = A^2 + B^2 + 2\delta_1 (AA_d + BB_d) + (\delta_1^2 + \delta_2^2)(A_d^2 + B_d^2) \quad \text{(A10.2)}$$

$$D = \tfrac{1}{2}\{|F_+|^2 - |F_-|^2\} = -2\,\delta_2 (AB_d - BA_d) \quad \text{(A10.3)}$$

Thus the average of the intensities of Bijvoet-related pairs of reflections (hkl and $\overline{h}\overline{k}\overline{l}$) may be computed by Eq. (A10.2) and the differences may be computed by Eq. (A10.3), provided the structure is known. If the sign of D is wrong, then the structure model has the wrong absolute configuration.

The term that should be used in computing an electron density map is

$$|F|_0 = \{S - 2\,\delta_1 (AA_d + BB_d) - (\delta_1^2 + \delta_2^2)(A_d^2 + B_d^2)\} \quad \text{(A10.4)}$$

Thus it is best, if accurate electron density maps are required, to measure diffraction data far from the absorption edge of any atom in the structure. Data measured near an absorption edge can be used to establish the absolute configuration.

APPENDIX 11 Molecular Geometry

(a) Transformation from fractional coordinates to Cartesian coordinates.

The atomic positions, x, y, z in a unit cell of dimensions $a, b, c, \alpha, \beta, \gamma$, may be expressed in Cartesian coordinates (in units of Å) as follows:

$$X = xa + yb \cos\gamma + zc \cos\beta$$

$$Y = yb \sin\gamma + z\,\{c\,(\cos\alpha - \cos\beta\cos\gamma)/\sin\gamma\}$$

$$Z = zcW/\sin\gamma$$

where

$$W = \sqrt{(1 - \cos^2\alpha - \cos^2\beta - \cos^2\gamma + 2\cos\alpha\cos\beta\cos\gamma)}$$

The orientation of the Cartesian axes relative to the crystallographic axes is:

A parallel to **a**

B in the **a**,**b** plane perpendicular to **a**

C perpendicular to A and B

(b) Interatomic distances.

distance A–B: $$(d_{\text{A-B}}) = \sqrt{(X_\text{A} - X_\text{B})^2 + (Y_\text{A} - Y_\text{B})^2 + (Z_\text{A} - Z_\text{B})^2}$$

$$= \sqrt{\Delta X^2_{\text{A-B}} + \Delta Y^2_{\text{A-B}} + \Delta Z^2_{\text{A-B}}}$$

(c) Interbond angles.

$$\text{angle A–B–C} = \arctan\left(\sqrt{1 - c_{\mathrm{A}}^2}\,/c_{\mathrm{A}}\right)$$

where

$$c_{\mathrm{A}} = -(\Delta X_{\mathrm{A\text{-}B}}\Delta X_{\mathrm{B\text{-}C}} + \Delta Y_{\mathrm{A\text{-}B}}\Delta Y_{\mathrm{B\text{-}C}} + \Delta Z_{\mathrm{A\text{-}B}}\Delta Z_{\mathrm{B\text{-}C}})/d_{\mathrm{B\text{-}C}}d_{\mathrm{A\text{-}B}}$$

(d) Torsion angles.

$$\text{torsion angle A-B-C-D} = \arctan(s_{\mathrm{T}}/c_{\mathrm{T}})$$

where

$$s_T = (\Delta X_{\mathrm{AB}}v_1 + \Delta Y_{\mathrm{AB}}v_2 + \Delta Z_{\mathrm{AB}}v_3)/d_{\mathrm{AB}}$$

and

$$c_{\mathrm{T}} = u_1v_1 + u_2v_2 + u_3v_3$$

where

$$u_1 = (\Delta Y_{\mathrm{AB}}\Delta Z_{\mathrm{BC}} - \Delta Z_{\mathrm{AB}}\Delta Y_{\mathrm{BC}})/(d_{\mathrm{AB}}d_{\mathrm{BC}})$$

$$u_2 = (\Delta Z_{\mathrm{AB}}\Delta X_{\mathrm{BC}} - \Delta X_{\mathrm{AB}}\Delta Z_{\mathrm{BC}})/(d_{\mathrm{AB}}d_{\mathrm{BC}})$$

$$u_3 = (\Delta X_{\mathrm{AB}}\Delta Y_{\mathrm{BC}} - \Delta Y_{\mathrm{AB}}\Delta X_{\mathrm{BC}})/(d_{\mathrm{AB}}d_{\mathrm{BC}})$$

$$v_1 = (\Delta Y_{\mathrm{BC}}\Delta Z_{\mathrm{CD}} - \Delta Z_{\mathrm{BC}}\Delta Y_{\mathrm{CD}})/(d_{\mathrm{BC}}d_{\mathrm{CD}})$$

$$v_2 = (\Delta Z_{\mathrm{BC}}\Delta X_{\mathrm{CD}} - \Delta X_{\mathrm{BC}}\Delta Z_{\mathrm{CD}})/(d_{\mathrm{BC}}d_{\mathrm{CD}})$$

$$v_3 = (\Delta X_{\mathrm{BC}}\Delta Y_{\mathrm{CD}} - \Delta Y_{\mathrm{BC}}\Delta X_{\mathrm{CD}})/(d_{\mathrm{BC}}d_{\mathrm{CD}})$$

A BASIC program for computing these three quantities is given in an Appendix to the text by J. D. Dunitz (ref. 1).

APPENDIX 12 Some Scattering Data for X Rays and Neutrons[a]

Element	Nuclide	X rays: $\sin\theta/\lambda = 0$ (relative to scattering by one electron)	X rays: $\sin\theta/\lambda = 0.5/\text{Å}$ (relative to scattering by one electron)	Neutrons[b] $b/10^{-12}$ cm	Neutrons normalized to ^1H as -1.00
H	^1H	1.0	0.07	−0.38	−1.00
	$^2H{=}D$	1.0	0.07	0.65	1.71
Li	^6Li	3.0	1.0	$0.18+0.025i$	$0.71+0.066i$
	^7Li	3.0	1.0	−0.25	−0.66
C	^{12}C	6.0	1.7	0.66	1.74
	^{13}C	6.0	1.7	0.60	1.58
O	^{16}O	8.0	2.3	0.58	1.53
Na	^{23}Na	11.0	4.3	0.35	0.92
Fe	^{54}Fe	26.0	11.5	0.42	1.11
	^{56}Fe	26.0	11.5	1.01	2.66
	^{57}Fe	26.0	11.5	0.23	0.61
Co	^{59}Co	27.0	12.2	0.25	0.66
Ni	^{58}Ni	28.0	12.9	1.44	3.79
	^{60}Ni	28.0	12.9	0.30	0.79
	^{62}Ni	28.0	12.9	−0.87	−2.29
U	^{238}U	92.0	53.0	0.85	2.24

[a] In the final column on the right we have listed neutron scattering amplitudes arbitrarily normalized to a value of -1.0 (for ^1H) in order to illustrate more clearly the small range of amplitudes observed as compared with that observed for X-ray scattering. For the nuclides considered here, the range of scattering amplitudes for X rays is about 10^2 at $\theta = 0°$ and nearly 10^3 at $\sin\theta/\lambda = 0.05$ Å^{-1}, whereas for neutrons it is smaller than 6, independent of scattering angle.

[b] The neutron scattering factors for ^1H, ^7Li, and ^{62}Ni involve phase shifts of 180° ($\cos 180° = -1$). The anomalous factor for ^6Li involves a phase shift of about 8° ($0.025/0.18 = \tan 8°$) relative to most other nuclei. The quantity b is the neutron coherent scattering amplitude.

Glossary

Absent reflections. Reflections too weak to be observed by the method of measurement used. The fact that they are so weak provides structural information to the crystallographer. *Systematically absent reflections* are those for which a generally applicable description may be made—for example, all reflections for which $h + k$ is odd. Systematic absences (sometimes referred to as systematic extinctions) depend only upon the symmetry of the disposition of the atoms in the crystal and consequently are used in deriving the space group.

Absolute configuration. The structure of a crystal or molecule expressed in an absolute frame of reference. A chiral object or structure is nonsuperimposable on its mirror image and X-ray measurements of anomalous dispersion effects can be used to determine its absolute configuration in the crystalline state. This absolute configuration can often be correlated with chiroptical phenomena, such as the direction of rotation of the plane of polarization of light by a solution of the crystal.

Absolute scale. Structure amplitudes are on an absolute scale when they are expressed relative to the amplitude of scattering by a classical point electron under the same conditions. A scale factor is required to convert structure amplitudes derived from X-ray diffraction intensities measured on an arbitrary scale to absolute values.

Absorption edges. The absorption of X rays by an atom varies with the wavelength, the amount of absorption increasing as the wavelength increases. However, discontinuities occur, at which the absorptivity abruptly drops to a low value and then starts to rise again. These discontinuities, illustrated in Figure 10.2, are absorption edges; they occur at the wavelength at which the incident X-ray quantum has just insufficient energy to excite a bound electron to a higher orbital or to eject it altogether.

Absorption effects. As an X-ray beam passes through a crystal, its intensity is reduced by scattering and by absorption. The extent of reduction by absorption depends on the path length of the beam through the crystal, the nature of the atoms in the crystal, and the wavelength of the incident X-ray beam (see Linear absorption coefficient).

Accuracy Deviation of a result obtained by a particular method from the value accepted as true (cf. Precision).

Amorphous solid. Displays no crystalline nature whatsoever; that is, there is no long-range ordering of atoms. Many substances that appear superficially to be amorphous are, in fact, composed of many tiny crystals.

Amplitude. The maximum numerical value of a periodic function measured from its mean or base value. It is half the peak-to-trough value.

Angle of incidence. The angle between a ray incident on a surface and the normal to the surface at that point of incidence. The Bragg angle, θ, is the complement of this; that is, the angle of incidence is $(90° - \theta)$.

Angle of reflection. The angle between a ray reflected by a surface and the normal to the surface at the point of reflection.

Angstrom unit. $1 \text{ Å} = 10^{-10} \text{ m} = 10^{-8} \text{ cm} = 0.1 \text{ nm}$

Anisotropic. A medium is anisotropic if a certain physical property differs in different directions.

Anisotropic temperature factors. Temperature factors (q.v.) that represent the different apparent amplitudes of vibration for an atom in different directions.

Anomalous dispersion. A discontinuity in the curve of refractive index against wavelength caused by high absorption by the medium at certain wavelengths. This effect occurs at wavelengths in the vicinity of absorption edges (q.v.) of the absorbing substance.

Anomalous scattering. A phase change on scattering of radiation that is strongly absorbed by one or more kinds of atoms present in the crystal. As a result, in crystals that lack symmetry elements of the second kind (rotation-inversion axes and glide planes), reflections from opposite faces (that is, in directions at 180° to one another) are caused to have different intensities. These differences in intensity may be used to determine the absolute configuration of chiral crystals. Also referred to as anomalous dispersion.

Area detector. An electronic device for measuring the intensities of a large number of diffracted intensities at one time. Such a device may, for example, involve a multiwire proportional counter coupled to an electronic device for recording the data in computer-readable form.

Asymmetric unit. The smallest part of a crystal structure from which the complete structure can be obtained from the space group symmetry operations (including translations). The asymmetric unit may consist of only part of a molecule or of several molecules not related by symmetry.

Atomic coordinates. A set of numbers that specifies the position of an atom with respect to a specified coordinate system. Coordinates are generally expressed as dimensionless quantities x, y, z (fractions of unit-cell edges), but sometimes as lengths, either with respect to the axial directions of the crystal or to a Cartesian coordinate system.

Atomic parameters. A set of numbers that specifies the position of an atom in the unit cell and the extent of its vibration. Usually three parameters define position (see Atomic coordinates). One parameter can be used to define isotropic vibration, or six to define simple (harmonic) anisotropic vibration.

Atomic scattering factor. X rays set the electrons of atoms into vibration, causing them to act as sources of secondary radiation. The scattering power of a single atom depends on its electronic structure and the angle of scattering, and is modified if the X rays used are appreciably absorbed by the atom. This scattering is reduced as the vibration of the atom increases (see Temperature factor). These effects are expressed in atomic scattering factors, which represent the scattering power of an atom measured relative to the scattering by a single electron under similar conditions; the scattering power falls off as the scattering angle increases. Atomic scattering factors are computed from theoretical wave functions for free atoms. Atomic scattering factors are modified by anomalous scattering (see discussion in Chapter 10, p. 137 ff.).

Automated diffractometer. An instrument for automatically measuring and recording the intensities of diffracted beams. The mutual orientations of the crystal and of the detector with respect to the source of radiation are computed from some initial data on a few selected reflections. The computer calculates these orientation angles and drives the gears that move the crystal orienter and detector to the desired angular settings.

Avogadro's number. $N = 6.022 \times 10^{23}$ mol^{-1}. This is the number of molecules in a mole of material.

Axial lengths and angles. These are the unit-cell lengths and angles.

Axial ratios. The ratios of the axial lengths, customarily expressed with the value of b equal to unity. These ratios may be deduced from measurements of the angles between faces on a crystal.

Axis of symmetry. When an object can be rotated about an axis passing through it to produce an object indistinguishable from the first, this object is said to possess an axis of symmetry. If such a situation occurs for a rotation of $360°/n$, the object is said to have an n-fold axis of symmetry.

Best plane. The plane through a group of atoms that satisfies the least-squares (q.v.) criterion.

Bijvoet differences. In 1951 Bijvoet and co-workers demonstrated that it is experimentally possible to determine the absolute configuration of an optically active molecule from the effects of anomalous dispersion, using the differences between $|F(hkl)|^2$ and $|F(-h,-k,-l)|^2$.

Birefringence. The separation of a ray of light on passing through a crystal into two unequally refracted, plane-polarized rays (of orthogonal polarizations). This effect occurs in crystals in which the velocity of light is not the same in all directions; that is, the refractive index is anisotropic. Uniaxial crystals have one direction in which double refraction does not occur; biaxial crystals have two.

Body-centered unit cell. A cell having a lattice point at the center ($x = y = z = \frac{1}{2}$) as well as at each corner.

Bragg equation. Each diffracted beam is considered as a "reflection" from a lattice plane. If the angle between the nth order of diffraction of X rays, wavelength λ, and the normal to a set of crystal lattice planes is $(90° - \theta)$, and the perpendicular spacing of the lattice planes is d, then

$$n\lambda = 2d \sin \theta$$

Thus when X rays strike a crystal they will be diffracted, when, and only when, this equation is satisfied. With this equation, Bragg first identified the integers, h, k, and l of the Laue equations with the Miller indices of the lattice planes.

Bravais lattice. One of the 14 possible arrays of points repeated periodically in three-dimensional space such that the arrangement of points about any one of the points is identical in every respect to that about any other point in the array.

Calculated phase. The phase angle, α, computed from the atomic positions of a model structure.

$$A(hkl) = \Sigma f \cos 2\pi(hx + ky + lz)$$

$$B(hkl) = \Sigma f \sin 2\pi(hx + ky + lz)$$

$$|F(hkl)|^2 = [A(hkl)]^2 + [B(hkl)]^2$$

$$\alpha = \tan^{-1} [B(hkl)/A(hkl)]$$

Cartesian coordinates. Points or objects in three dimensions can be located by reference to three orthogonal (mutually perpendicular) axes with equal units along the axes. The location of a point is defined by values of X, Y, and Z as distances from an origin, measured parallel to the a, b, and c axes, respectively. (Named after Descartes, the mathematician and philosopher.)

Cell constants. The unit-cell dimensions of a crystal structure (see Unit cell).

Center of symmetry (or center of inversion). A point through which an inversion operation is performed, converting an object into its enantiomorph (see Inversion).

Centrosymmetric structure. A crystal structure that crystallizes in a space group with a center of symmetry. When the origin is chosen at the center of symmetry, the phase angle for each reflection is either 0° or 180° and a nearly correct trial structure will give an electron density map that needs little refinement, because nearly all of the phases are correct. The electron density map will, of course, contain errors that reflect errors in the measured structure factor amplitudes and the fact that the data set is of limited resolution, even if all phase angles are correct.

Characteristic X-rays. X rays of definite wavelengths, characteristic of a pure substance (generally a metal) and produced when that substance is bombarded by fast electrons. Characteristic X rays are emitted when an electron that has been displaced from an inner shell of the atom being excited (an atom in the target) is replaced by another electron that falls in from an outer shell (see X rays).

Chiral. An object that cannot be superimposed upon its mirror image is said to be chiral (Greek: *cheir* = hand).

Chi-square. The sum of the quotients obtained by dividing the square of the difference between the observed and mean values of a quantity by the square of its e.s.d.

$$\chi^2 = \sum_i \frac{(x_i - \langle x \rangle)^2}{\sigma(x_i)^2}$$

Chi-square test. A test for the mathematical fit of the distribution of chi-square to a standard frequency distribution. This gives the likelihood that an observed distribution arose from fluctuations of random sampling rather than from systematic error.

Cleavage. The property of many crystals of splitting readily in one or more definite directions to give smooth surfaces, always parallel to actual or possible crystal faces.

Coherent scattering. With coherent scattering the fluctuation of the electromagnetic field of an incident X-ray beam causes an electron to oscillate about its nucleus at the same frequency. The oscillating dipole in turn results in a secondary scattered wave with the same wavelength but a phase change of 180°. This is the type of scattering we consider in this book. (See also Incoherent scattering.)

Collimator. A device for producing a parallel beam of radiation.

Complex number. An expression of the form $a + ib$ where a and b, are real numbers and $i = \sqrt{-1}$.

Conformation. A general description of the shape of a molecule. It is generally applied to molecules in which there is a possibility for rotation about bonds. Different rotational positions about bonds are represented by torsion angles (q.v.).

Contact goniometer. A device for measuring angles between faces of a crystal by making direct contact with the crystal faces with two straight edges and then measuring the angle between these straight edges.

Contour map. A map showing electron density by means of contour lines drawn at regular intervals.

Convolution. Consider two functions A (x, y, z) and B (x, y, z). The *convolution* of A and B at the point (u_0, v_0, w_0) is found by multiplying together the values of A (x, y, z) and B $(x + u_0, y + v_0, z + w_0)$ for *each* set of possible values of x, y and z and *summing all these products*. To find the convolution as a general function of (u, v, w), it is then necessary to perform this multiplication and summation for *all* desired values of u, v and w.

Correlation of parameters. Interdependence between mathematical variables. For example, position parameters of an atom refined by least squares in an oblique coordinate system are correlated to an extent dependent upon the cosine of the interaxial angle. Parameters related by symmetry are completely correlated. In a least-squares refinement, temperature factors and occupancy factors are often highly correlated.

Crystal. A solid having a regularly repeating internal arrangement of atoms.

Crystal lattice. Crystals are composed of groups of atoms repeated at regular intervals in three dimensions with the same orientation. For certain purposes it is sufficient to regard each such group of atoms as being replaced by a representative point; the collection of points so formed is the space lattice or lattice of the crystal. The meaning is specific and *the term should not be used to denote the entire atomic arrangement*. Each crystal lattice is a Bravais lattice.

Crystal morphology. (See Morphology.)

Crystal structure. The mutual arrangement of the atoms, molecules, or ions that are packed together on a lattice to form a crystal.

Crystal system. The seven crystal systems, best classified in terms of their symmetry (see Appendix 1), correspond to the seven fundamental shapes for unit cells consistent with the 14 Bravais lattices.

Cubic unit cell. A cell in which there are three-fold rotation axes along all four body-diagonals of the unit cell, and four-fold axes parallel to each crystal axis. There also are two-fold axes. As a result all axial lengths are identical by symmetry and all interaxial angles must be 90° ($a = b = c, \alpha = \beta = \gamma = 90°$).

Deformation density. The promolecule density may be defined as the charge density corresponding to a superposition of free atoms, each centered at its position in the molecule and vibrating as in the model structure. The deformation density is the difference between the experimental electron density and the promolecule density.

Deliquescence. The state of dissolving gradually and becoming a solution by attracting and absorbing moisture from the air.

Densitometer. An instrument for measuring photographic density on a film by means of the fraction of light transmitted at given points on the film.

Density modification. (also called Solvent-flattening). A computational method for improvement of phases, useful when a unit cell contains a high proportion of solvent, which may be 50 per cent or more for macromolecular crystals. An "envelope" defining the approximate boundary of the molecule is determined from the electron density map and all electron density outside this envelope is set to its average value. A new set of phases is then determined by Fourier inversion of this "solvent flattened" map. The usefulness of the method increases as the fraction of solvent in the unit cell increases.

Difference synthesis. A Fourier synthesis (or "map") for which the Fourier coefficients are the differences between the observed and calculated structure factor amplitudes. Such a map, normally calculated during the refinement of a structure, will have peaks where not enough electron density was included in the trial structure and troughs where too much was included. It has proved to be an exceedingly valuable tool both for locating missing atoms and for correcting the positions of those already present.

Differential synthesis. A method of refining parameters of an atom from a mathematical consideration of the slope and curvature of the difference synthesis (q.v.) in the region of the atom.

Diffraction. When radiation passes by the edges of an opaque object or through a narrow slit, the waves appear to be deflected and to produce fringes of parallel light and dark bands. This effect may best be explained as the interference of secondary waves generated in the area of the slit or the opaque object. These secondary waves interfere with one another and the intensity of the beam in a given direction may be determined by a superposition of all the wavelets in that direction. When light passes through a narrow slit all the waves will be in phase in the forward direction. In any other direction, each secondary wave traveling in a given direction will be slightly out of phase with it neighbors by an amount that depends on the wavelength of the light and the angle of deviation from the direct beam. The shorter the wavelength or the larger the angle, the more a wave is out of phase with its neighbor. In X-ray crystallography an analogous experiment is done. The radiation is X rays and the slit is replaced by the electron clouds of atoms in a crystal; these electron clouds scatter the X rays. Because the crystal contains a regularly repeating atomic arrangement, the beams diffracted from one unit cell may be in phase with those from other unit cells and may reinforce each other to produce a strong diffracted beam.

Diffraction grating. A system of close, equidistant parallel lines, usually ruled on a polished surface and used for producing diffraction spectra.

Diffraction pattern. The experimentally measured values of intensities, diffracting angle (direction), and order of diffraction for each diffracted beam obtained when a crystal is placed in a narrow beam of X rays or neutrons (usually monochromatic).

Diffractometer. An instrument for measuring diffraction effects, specifically for measuring the directions and intensities of diffracted beams from crystals (see Automated diffractometer).

Diffuse scattering. Halos or streaks that appear around intense reflections and indicate the presence of disorder in the structure.

Direct phase determination. A method of deriving relative phases of diffracted beams by consideration of relationships among the indices and among the structure factor amplitudes of the stronger reflections. These relationships come from the conditions that the structure is composed of atoms and that the electron density must be positive or zero everywhere. Only certain values for the phases are consistent with these conditions.

Direct space. The crystal lattice is commonly called the direct lattice, as opposed to the reciprocal lattice. Each of the lattices can be thought of as existing in a space defined by its coordinate system—that is, direct (or real) space and reciprocal space.

Discrepancy index, Residual, R. An index that gives a crude (and sometimes misleading) measure of the correctness of a structure and the quality of the data. It is defined as

$$R = \Sigma|(|F_o| - |F_c|)|/\Sigma|F_o|$$

and values of 0.06 to 0.02 are considered good for present-day structure determinations. However, some partially incorrect structures have had R values below 0.10, and many basically correct but imprecise structures have higher R values.

Dislocation. A discontinuity in the otherwise regularly periodic three-dimensional structure of a crystal.

Disordered structure. A crystal structure in which ions or molecules pack in alternate ways in different unit cells. Such disorder may be revealed by the presence of diffuse scattering, either as halos or streaks, around intense reflections.

Dispersion. Variation in the velocity of propagation of a traveling wave with wavelength. The spreading of white radiation by a prism or grating into a colored beam is due to dispersion of light. The variation of velocity with wavelength is usually smooth, but at strongly absorbed wavelengths of the incident radiation the curve is discontinuous, leading to anomalous dispersion.

Displacement parameters. Because *static* displacements of a given atom in a structure that vary in an essentially random fashion from one unit cell to another will simulate vibrations of that atom, the "vibration paramenters" of an atomic temperature factor are better referred to as "displacement parameters." Atomic vibrations are displacements from equilibrium positions with periods that are typically smaller than 10^{-12} seconds.

Distribution of intensities. Intensities from a noncentrosymmetric crystal tend to be clustered more tightly around the mean than do those from a centrosymmetric one. This forms the basis for one test for the presence or absence of a center of symmetry in the crystal.

Domain. Small regions of a crystal containing completely oriented structure. The term is generally used in describing ferromagnetic materials.

Double reflection (also called the "Renninger effect" after its discoverer). X rays that are considered to be reflected by one set of lattice planes may then be reflected by another set of planes which, by chance, are in exactly the right orientation for this. The resultant beam appears in the position in reciprocal space expected for a normal reflection but is sharper in appearance on an X-ray photograph than is an ordinary diffracted beam. The effect causes intensity changes in the reflections involved; it may even cause an ambiguity in the space group determination if a systematically absent reflection gains intensity by it. It can be eliminated for that particular reciprocal lattice point by reorienting the crystal.

Dynamical diffraction. Diffraction theory in which the modification of the primary beam on passage through the crystal is important. The mutual interactions of the incident and scattered beams are taken into account. This is important for perfect crystals and for electron diffraction by crystals.

E-map. A Fourier map, equivalent to an electron density map, with phases derived by "direct methods" and normalized structure amplitudes, $|E(hkl)|$, replacing $|F(hkl)|$ in the Fourier summation. Since the $|E|$ values correspond to sharpened atoms, the peaks on the resulting map are sharper than those computed with $|F|$ values.

E-values. (See Normalized structure factors.)

Efflorescence. The change of the surface of a crystal to a powder as a result of loss of water (or some other solvent) of crystallization on exposure to air.

Electron density. The number of electrons per unit volume (usually per Å^3).

Electron density map. A contoured representation of electron density at various points in a crystal structure. Electron density is expressed in electrons per cubic Å and is highest near atomic centers. The map is calculated using a Fourier synthesis—that is, a summation of waves of known amplitude, frequency, and phase. For three-dimensional maps it is customary to superimpose maps, representing two-dimensional sections, parallel to one another and at different heights in the cell.

Enantiomorph. One of a pair of chiral objects related by mirror-symmetry.

Enantiomorphism. The relationship between enantiomorphs (q.v.).

Epsilon-factor, ϵ. This is a factor used in computing normalized structure factors (q.v.) that takes into account the fact that, depending on which of the 32 crystal classes the crystal belongs to, there will be certain groups of reflections in areas of the reciprocal lattice that will have an average intensity greater than that for the general reflections.

Equi-inclination techniques. In these, such as in the equi-inclination Weissenberg photograph, the incident beam is inclined at a selected angle to the axis of rotation of the crystal instead of being normal to it. The beam exits at the same angle to this axis.

Equivalent positions. The complete set of positions produced by the operation of the symmetry elements of the space group upon any general position.

Equivalent reflections. When a complete set of intensity data has been collected there are eight measurements for each h, k, l, corresponding to combinations of positive or negative values of each. Some of these reflections that are equivalent by the symmetry of the crystal have (within experimental error) identical intensities. For high-symmetry crystals, other reflections may also be equivalent, e.g., hkl, klh, and lhk for cubic crystals.

Estimated standard deviation (e.s.d.). A measure of the precision of a quantity. If the distribution of errors is normal, then there is a 99 per cent chance that a given measurement will differ by less than 2.7 e.s.d. from the mean. A bond length 1.542(7) Å (1.542 Å with an e.s.d. of 0.007 Å) is, by the usual criteria, not considered significantly different from one measured as 1.527(7) Å.

Ewald sphere. (See Sphere of reflection.)

Extinction. Modification of the incident beam as it passes through a single block (perfect) of the crystal. Part of the incident beam may be reflected twice so that it returns to its original direction but is out of phase with the main beam, thus reducing the intensity of the latter (primary extinction). When the crystal is mosaic, part of the beam will be diffracted by one mosaic block and therefore is not available for diffraction by a following block that is accurately aligned with the first. Thus the second block contributes less than expected to the diffracted beam (secondary extinction). The effect of extinction is evidenced by a tendency for $|F_o|$ to be systematically smaller than $|F_c|$ for very intense reflections. The effect can be reduced by dipping the crystal in liquid nitrogen, thereby increasing its mosaicity.

Face-centered cell. A unit cell with a lattice point at the center and corners of each face. If all faces are centered, the designation is F; if only faces perpendicular to the a axis are centered, the description is A; and similarly for B and C.

Figure of merit. A numerical quantity used for indicating comparative effectiveness. In crystallographic studies it is used to indicate an estimate of the average precision in the selection of phase angles, and is particularly used in protein crystallography where phase angles are derived by isomorphous replacement methods.

Film scanner. A device for measuring the intensities of spots on an X-ray diffraction photograph. This is done by a light beam that is caused to scan the photograph systematically. The reduction in intensity of this beam at each point on the film is recorded.

Filter. A semitransparent material that absorbs some of the radiation passing through it. It is possible to choose appropriate filters with different wavelength absorptivities to select a narrow wavelength range. (Electronic filtering is very effective with detectors.)

Form factor. (See Atomic scattering factor.)

Fourier map. A map computed for a periodic function by addition of waves of known amplitude, frequency, and phase. The term is generally used for an electron density or difference electron density map.

Fourier series. A function $f(t)$ that is periodic with period T ($f(t + T) = f(t)$), may be represented by a Fourier series, an infinite series of the form

$$\begin{aligned} f(t) &= a_0/2 + a_1 \cos 2\pi(t/T) + a_2 \cos 2\pi(2t/T) + \cdots \\ &\quad + b_1 \sin 2\pi(t/T) + b_2 \sin 2\pi(2t/T) + \cdots \\ &= a_0/2 + \sum_{n=1}^{\infty} a_n \cos 2\pi\,(nt/T) + \sum_{n=1}^{\infty} b_n \sin 2\pi\,(nt/T) \\ &= \sum_{n=-\infty}^{\infty} c_n e^{2\pi i(nt/T)} \end{aligned}$$

The Fourier theorem states that any periodic function may be resolved into cosine and sine terms involving known constants. Since a crystal has a periodically repeating internal structure, this can be represented, in a mathematically useful way, by a three-dimensional Fourier series, to give a three-dimensional Fourier or electron density map.

Fourier synthesis. The summation of sine and cosine waves to give a periodic function (for example, the computation of an electron density map from waves of known amplitude, $|F|$, phase, and frequency).

Fourier transform. In the pair of equations

$$f(x) = \int_{-\infty}^{\infty} e^{2\pi ixy} g(y)\, dy$$

and

$$g(y) = \int_{-\infty}^{\infty} e^{-2\pi ixy} f(x)\, dx$$

$g(y)$ is the Fourier transform of $f(x)$, and $f(y)$ is the Fourier transform of $g(x)$. In X-ray diffraction the structure factor, F, is related to the electron density, ρ, by

$$F = \int_{-\infty}^{\infty} \rho e^{i\phi}\, dV_c$$

and conversely

$$\rho = \frac{1}{V_c} \sum_{\substack{\text{all} \\ \text{reflections}}} Fe^{-i\phi}$$

where $\phi = 2\pi(hx + ky + lz)$, and V_c = the volume of the unit cell. Summation replaces integration in the latter equation because the diffraction pattern of a crystal is observed only at discrete points. The intensity at a particular point of the diffraction pattern of an object is proportional to $|F|^2$ and thus to the value at that point of the square of the Fourier transform of the structure.

Fractional coordinates. Coordinates of atoms expressed as fractions of the unit cell lengths (see Atomic coordinates).

Fraunhofer diffraction. Diffraction observed with parallel incident radiation, as in the diffraction by slits described in Chapter 3.

Friedel's law. This law states that $|F|^2$ values of centrosymmetrically related reflections are equal (even for an acentric structure): $|F(hkl)|^2 = |F(-h, -k, -l)|^2$. This law only holds under conditions where anomalous scattering can be ignored.

Gaussian distribution. In many kinds of experiments repeated measurements follow a Gaussian or normal error distribution that is bell-shaped and symmetrical.

$$G(x, \sigma) = (2\pi\sigma^2)^{-1/2} \exp\left[-(\tfrac{1}{2})(x^2/\sigma^2)\right]$$

where x is the deviation of a variable from its mean value and σ^2 its variance (square of e.s.d.).

Glide plane. A glide plane is a symmetry element for which the symmetry operation is reflection across the plane combined with translation in a direction parallel to the plane. It is designated by a, b or c if the translation is $a/2$, $b/2$, or $c/2$, by n if the translation is $(a + b)/2$, $(a + c)/2$, or $(b + c)/2$, i.e., half way along one of the face diagonals, and by d if the translation is $(a + b)/4$, $(b + c)/4$ or $(c + a)/4$ or $(a + b + c)/4$.

Goniometer. An instrument for measuring angles (see Contact goniometer and Reflecting goniometer).

Goniometer head. A device for orienting a crystal by means of translational motions and, in some models, movable arcs.

Habit of crystal. The usual appearance of a crystal of a particular substance. It is found that the most rapidly growing faces of a crystal are the smallest and the least well developed. Habit may be strongly dependent upon the conditions of crystallization, especially the solvent used.

Harker–Kasper inequalities. Space group-dependent inequalities among unitary structure factors (q.v.) that allow for the determination of the phases of certain intense reflections in a centrosymmetric crystal. These provided (in 1947) the basis of one of the earliest successful attempts to solve the phase problem by direct methods.

Harker sections. Certain portions of the Patterson map that, depending on the space group, contain a large proportion of the readily interpretable structural information because they contain many vectors between space group-equivalent atoms. An analysis of Harker sections may be useful in structure determination.

Heavy-atom derivative of a protein. The product of soaking a solution of the salt of a metal of high atomic number into a crystal of a protein. If the derivative is to be of use in structure determination, the heavy atom must be substituted in only one or two ordered positions per molecule of protein. Then the method of isomorphous replacement (q.v.) can be used to determine the structure.

Heavy-atom method. A method of deriving phase angles in which the phases calculated from the position of the heavy atom are used to compute the first approximate electron density map, from which further portions of the structure are recognizable as additional peaks in this map. If necessary, successive approximate electron density maps are computed to give the entire structure.

Hexagonal unit cell. A unit cell in which there is a six-fold rotation axis parallel to one axis (arbitrarily chosen as **c**) and also two-fold rotation axes perpendicular to **c**. These symmetry relations dictate that the lengths of a and b are identical, the angle between **a** and **b** is 120°, and the other two angles are 90° ($a = b, \alpha = \beta = 90°, \gamma = 120°$). (Rhombohedral crystals, with three-fold axes, can be referred to hexagonal unit cells, see pp. 80–82 of ref. 1.)

Identity element. A symmetry element whose operation leaves unchanged anything on which it operates.

Imperfect crystal. A mosaic crystal in which primary extinction is negligible is termed "ideally imperfect" (see pp. 290–293 of ref. 1.).

Improper symmetry operation (symmetry operation of second kind). Any symmetry operation (q.v.) that converts a chiral object into its enantiomorph; for example, a center of symmetry, a mirror plane, or a glide plane.

Incoherent scattering. With incoherent scattering of X rays by electrons there is an exchange of energy and momentum on impact, resulting in a small wavelength change for the X rays. This type of scattering is not considered in this book.

Indices. Indices are used to describe the faces of a crystal and the orders of diffraction—that is, to refer to a specific "reflection" (using indices h, k, l) (see Miller indices).

Inequality relationships. (See Harker–Kasper inequalities.)

Integrated intensity. The total intensity measured at the detector as a reflection is scanned.

Intensity distribution. (See Distribution of intensities.)

Interfacial angles, Law of. In all crystals of a given sort, angles between corresponding faces have a constant value. This law, of course, applies to only one particular form of a polymorphous crystalline material. Interfacial angles are measured with a goniometer (contact or reflecting).

Inversion. Each point of an object is converted to an equivalent point by projecting through a common center (called the center of inversion or center of symmetry) and extending an equal distance beyond this center. If the center of symmetry is taken at the origin, every point x, y, z becomes $-x, -y, -z$.

Isomorphism. Similarity of crystal shape, unit-cell dimensions, and structure between substances of similar chemical composition. Ideally, the substances are so closely similar that they can generally form a continuous series of solid solutions.

Isomorphous replacement method. A method of deriving phases from measurements of the intensities of the reflections from two or more isomorphous crystals.

Isotropic. Exhibiting properties that are the same in any direction.

Isotropic temperature factor. A temperature factor that represents an equal amplitude of vibration in all directions. At the start of a least-squares refinement of a structure all atoms are considered to have isotropic temperature factors but in the later stages, anisotropic temperature factors are usually assigned to appropriate atoms.

Kinematical diffraction. Diffraction theory in which it is assumed that the incident beam is not affected, apart from the results of simple diffraction, by its passage through the crystal. This type of diffraction is considered in this book.

Lattice. (See Crystal lattice.)

Laue photograph. Diffraction photograph produced by sending a beam of X rays, with a wide range of wavelengths ("white" X rays), along a principal axis of a stationary crystal. It demonstrates well the diffraction symmetry.

Law of Rational Indices. (See Rational Indices, Law of.)

Layer line. When a crystal is rotated or oscillated about a principal axis, the diffraction spots on a cylindrical film surrounding the crystal are arranged in a series of straight lines called layer lines, which are perpendicular to the axis of rotation (see Figure 4.3c).

Least-squares calculations. A statistical method of obtaining the best fit of a large number of observations to a given equation. This is done by minimizing the sum of the squares of the deviations of the experimentally observed values from their respective calculated ones. The individual terms in the sum are usually weighted to take into account their relative precision. In crystal structure analyses, atomic coordinates and other parameters may be fitted in this way to the observed intensities; ideally, there should be at least 10 measurements for each parameter to be determined. In a similar way the least-squares criterion can be applied to the computation of a plane through a group of atoms and to many other problems.

Libration. A form of motion that may be described as a vibration along an arc rather than along a straight line.

Linear absorption coefficient. The intensity, I, of a beam after passing through a thickness t of an absorbing crystal is given by the equation $I = I_0 \exp(-\mu t)$, where I_0 is the intensity of the incident beam and μ the total linear absorption coefficient for the primary beam (with units of cm^{-1}); μ is a function of wavelength and atomic number. The above equation may be rewritten in terms of a mass absorption coefficient (μ/ρ) (in units of cm^2/g), with ρ the density.

Liquid crystal. A substance, such as para-azoxyanisole, that has observable optical anisotropy as for a crystal but behaves in other ways as a liquid. Thus the term refers to a state of matter with structural order intermediate between those of normal liquids and crystalline solids.

Lorentz factor. A factor that takes into account the time that it takes for a given reflection (represented as a reciprocal lattice point with finite size) to pass through the surface of the sphere of reflection. The value depends on the scattering angle and on the geometry of the measurement of the reflection.

Miller indices. The plane with Miller indices h, k, and l makes intercepts a/h, b/k, and c/l with the unit-cell axes a, b, and c. The "law of rational indices" states that the indices of the faces of a crystal are usually small integers, seldom greater than three. The importance of the Bragg equation is that it identifies the integers h, k, l that specify the "order" of diffraction in the Laue equations with the Miller indices of the lattice planes causing the "reflection."

Minimum function. (See Vector superposition map.)

Mirror plane. A symmetry element for which the corresponding symmetry operation resembles reflection in a mirror, coincident with the plane. It converts a chiral object or structure into its enantiomorph (q.v.).

Molecular replacement method. The use of noncrystallographic symmetry (q.v.), and the constraints on the phases it causes, to solve a protein crystal structure.

Monochromatic. Consisting of radiation of a single wavelength or of a very small range of wavelengths.

Monochromator. An instrument used to select radiation of a single wavelength by use of diffraction from an appropriate crystal, such as a graphite crystal.

Monoclinic unit cell. A unit cell in which there is a two-fold rotation axis parallel to one cell axis (usually chosen as **b**); as a result there are no restrictions on the axial ratios, but the angles made by **b** with **a**, and **b** with **c**, must be 90° ($\alpha = \gamma = 90°$).

Morphology. Study of the shape or form of a material. With crystals, a description of the crystal faces and the angles between them can often be used for identification.

Mosaic spread. The divergence of a scattered X-ray beam that is believed to be caused by the irregularity of orientation of small blocks of unit cells in the crystal. These blocks may have varying sizes, but are very small on a macroscopic scale. The misalignment of these blocks of unit cells is small, of the order of 0.2° to 0.5° for most crystals. This imperfection is convenient because diffracted intensities are more intense from crystals with such a mosaic structure than from "perfect" crystals (see extinction).

Mother liquor. The solution from which the crystals under study were obtained.

Neutron diffraction. Neutrons of wavelengths near 1 Å give useful crystal diffraction patterns because neutrons are scattered by atomic nuclei. Atomic scattering factors for neutrons are not a regular function of atomic number. Neutron diffraction is useful for locating hydrogen atoms (which have a much larger relative effect on neutron than on X-ray scattering). Larger crystals are required than for X-ray studies.

Noncentrosymmetric structure. A crystal structure that crystallizes in a space group with no center of symmetry. The phase angle of each reflection may have any value. An electron density map computed from phases of a trial structure will generally show the features of the trial structure even if it is partially wrong, but some features of the correct structure also will be present in the map.

Noncrystallographic symmetry. Symmetry within the asymmetric unit of a crystal structure. For example, the asymmetric unit of a crystalline protein may contain a dimer whose two subunits may have identical molecular structure but, since they are not related by crystallographic symmetry, may have different environments. This noncrystallographic symmetry in proteins can be used as an aid in structure determination.

Normal equations. Any set of simultaneous equations involving experimental unknowns and derived from a larger number of observational equations in the course of a least-squares adjustment of observations. The number of normal equations is equal to the number of parameters to be determined.

Normalized structure factors. The ratio of the value of the structure amplitude, $|F(hkl)|$, to its root-mean-square expectation value. It is denoted by $|E(hkl)|$, where $|E(hkl)| = |F(hkl)|/(\epsilon\Sigma f(hkl)_j)^{1/2}$ (see Epsilon factor for a definition of ϵ).

Nucleation of crystals. The action of a tiny seed crystal, dust particle, or other "nucleus" in starting a crystallization process, as, for example, in the seeding a cloud for the production of ice crystals.

Observational equation. An equation expressing a measured value as some function of one or more unknown quantities. Observational equations are reduced to normal equations during the course of a least squares refinement.

Occupancy factor. A parameter that defines the partial occupancy of a given site by a particular atom. It is most frequently used to describe disorder in a portion of a molecule, or for describing nonstoichiometric situations—for example, solvent loss when a solvent molecule is being lost to the atmosphere.

Optic axis. The direction in a birefringent crystal along which the ordinary and extraordinary rays travel at the same speed. Uniaxial crystals have one such axis; biaxial crystals have two.

Optical activity. The ability of a substance to rotate the plane of polarization of plane-polarized light.

Order of diffraction. An integer associated with a given interference fringe of a diffraction pattern. It is first order if it arises as a result of a radiation path difference of one wavelength. Similarly, the nth order corresponds to a path difference of n wavelenghts.

Orthogonal system. A system with three mutually perpendicular axes as reference axes.

Orthorhombic unit cell. A unit cell in a lattice in which there are three mutually perpendicular two-fold rotation axes (parallel to the three axes); as a result, while there are no restrictions on axial ratios, all interaxial angles are necessarily equal to 90° ($\alpha = \beta = \gamma = 90°$).

Oscillation photograph. A photograph of the diffraction pattern obtained by oscillating the crystal through a small angular range.

Parity group. A set of structure factors whose three indices are odd or even in an identical way. There are eight parity groups for three indices.

Path difference. This term is used in diffraction to describe the difference in distance that two beams travel when "scattered" from different points. As a result of such path differences, the beams may or may not be in phase.

Patterson function. A map made from the summation of a Fourier series that has the squares of the structure factor amplitudes as coefficients. Because these values of $|F|^2$ can be calculated from the diffraction intensities, the map can be computed directly. Ideally, the positions of the maxima in the map represent the end points of vectors between atoms, all referred to a common origin.

Patterson superposition method. (See Vector superposition map.)

Phase. The difference in position of the crests of two waves of the same wavelength traveling in the same direction. Also, the point to which the crest of a given wave has advanced in relation to a standard position—for example, the starting point. The phase is usually expressed as a fraction of the wavelength in angular measure, with one cycle or period being 360°; that is, if the crests differ by Δx for a wavelength λ, the phase difference is $\Delta x/\lambda$ or $2\pi\ \Delta x/\lambda$ radians or $360°\ \Delta x/\lambda$ degrees (see also Calculated phase).

Phase problem. The problem of determining the phase angle to be associated with each structure factor, so that an electron density map may be calculated from a Fourier series with structure factors (including both amplitude and phase) as coefficients. The measured intensities of diffracted beams give only the squares of the amplitudes; the phases cannot normally be determined experimentally. Phases may, however, be calculated for any postulated structure and combined with the experimentally determined amplitudes to give an electron density map. They may also, in special cases, be measured by the method of isomorphous replacement and by other methods discussed in Part II.

Photometry. Comparison of the intensity of a beam of light when it has passed through a point on a photographic film to that from a source of constant intensity. In this way the intensity distribution on the film can be measured.

Piezoelectric effect. The generation of a small potential difference across certain crystals when they are subjected to stress. The effect is found only for noncentrosymmetric crystals; good examples are quartz and Rochelle salt.

Plane groups. The groups of symmetry elements that produce regularly repeating patterns in two dimensions. There are 17 plane groups (listed in *International Tables*) e.g., 17 symmetry variations of wallpaper.

Plane of symmetry. (See Mirror plane.)

Plane Polarization. The process of affecting electromagnetic radiation so that the electric vectors of the waves are confined to a single plane.

Pleochroism. The property of certain crystals of appearing to have different colors when viewed from different directions under transmitted white light (dichroism if only two colors).

Point group. A group of symmetry operations that leave unmoved at least one point within the object to which they apply. Symmetry elements include simple rotation and rotatory-inversion axes; the latter include the center of symmetry and the mirror plane. Since one point remains invariant, all rotation axes must pass through this point and all mirror planes must contain it. A point group is used to describe isolated objects, such as single molecules or real crystals.

Polarization. (See Plane polarization.)

Polarization factor. A factor that takes into account the reduction in intensity on X-ray scattering due to the state of polarization of the incident beam.

Polymorphism. Property of crystallizing in two or more forms with distinct structures (dimorphism if only two forms).

Powder diffraction. Diffraction by a crystalline powder (a mass of randomly oriented microcrystals), consisting of lines or rings rather than separate diffraction spots. The powder is either glued to a glass fiber, placed on a flat surface (e.g., a microscope slide), or, if it is unstable in air, put in a sealed capillary tube. The diffraction pattern obtained is that expected for a set of randomly oriented crystals.

Precession photograph. A photograph of the diffraction pattern that is an undistorted magnified image of a given layer of the reciprocal lattice. The necessary camera and crystal motion involve the precession of one crystal axis about the direction of the direct beam (see Figure 4.5). The film is continuously maintained in the plane perpendicular to this precessing axis. The photograph resulting from this complicated set of motions is simple to interpret, and the indices (h, k, l) of the diffraction spots may be found by inspection.

Precision. A measure of the experimental uncertainty in a measured quantity, an indication of its reproducibility; cf. Accuracy.

Primitive cell. Designating the unit cell of the smallest possible volume for a given space lattice. The term is used to differentiate this cell from a centered cell or other nonprimitive cells. When a primitive cell is chosen, the symbol P is included in the space-group designation, except in the trigonal system.

Principal axes of thermal ellipsoids. Three mutually perpendicular directions, along two of which the amplitude of vibration of an atom, represented by an ellipsoid, is at a maximum and at a minimum. Each axis is characterized by an amplitude and a direction.

Probability relationships. Equations representing the probability that a phase has a certain value. They form the basis of phase determination in direct methods.

Pyroelectric effect. The development of a small potential difference across certain crystals as the result of a temperature change.

R-factor or R-value. (See Discrepancy index.)

Racemic mixture. A mixture composed of equal amounts of dextrorotatory and levorotatory forms (enantiomorphs) of the same compound. It displays no optical rotatory power.

Rational Indices, Law of A rational number is an integer or the quotient of two integers. The Law of Rational Indices states that all of the faces of a crystal may be described, with reference to three noncollinear axes, by three small whole numbers.

Real space averaging. A computational method for improvement of phases, which may be applied when there are two or more identical chemical units in the crystallographic asymmetric unit. In the trial electron density map, the densities of the identical units are averaged. Then a new set of phases is computed by Fourier inversion of the averaged structure, and with these a new map is synthesized with the observed $|F|$ values. By iteration of this procedure, the electron density is improved.

Reciprocal lattice. The lattice with axes $\mathbf{a}^*$, $\mathbf{b}^*$, $\mathbf{c}^*$, related to the crystal lattice or direct lattice (with axes $\mathbf{a}$, $\mathbf{b}$, $\mathbf{c}$) in such a way that $\mathbf{a}^*$ is perpendicular to $\mathbf{b}$ and $\mathbf{c}$; $\mathbf{b}^*$ is perpendicular to $\mathbf{a}$ and $\mathbf{c}$; and $\mathbf{c}^*$ is perpendicular to $\mathbf{a}$ and $\mathbf{b}$. $\mathbf{a}^*$, $\mathbf{b}^*$, and $\mathbf{c}^*$ are related to $\mathbf{a}$, $\mathbf{b}$, and $\mathbf{c}$ by, e.g.,

$$\mathbf{a}^* = \frac{\mathbf{b} \times \mathbf{c}}{(\mathbf{a} \cdot \mathbf{b} \times \mathbf{c})}$$

Rows of points (zone axes) in the direct lattice are normal to nets (planes) of the reciprocal lattice, and vice versa. The repeat distance between points in a particular row of the reciprocal lattice is inversely proportional to the interplanar spacing between the nets of the crystal lattice that are normal to this row of points (see Figure 3.5). The same relation holds between the spacings in rows of the crystal lattice and the spacings of planes of the reciprocal lattice.

Refinement. A process of improving the parameters of an approximate (trial) structure until the best fit of calculated structure factor amplitudes to those observed is obtained. The process usually requires many successive stages.

Reflecting goniometer. A device for measuring the angle between crystal faces by measuring the angle through which a crystal has to be rotated from a position at which one face reflects a narrow beam of light into a stationary detector to a position at which a second face reflects.

Reflection. Since diffraction by a crystal may be considered as reflection from a lattice plane, this term has come to be used to denote a diffracted beam.

Refraction. The change in direction that occurs when a beam of radiant energy passes from one medium into another in which its velocity is different.

Refractive index. The ratio of the velocity of light *in vacuo* to the velocity in the material under study. When a colorless substance is immersed in a colorless medium of the same refractive index, the substance becomes invisible.

Resolution. The process of distinguishing individual parts of an object when examining it with radiation. Most X-ray structures of small molecules are determined to a "resolution" of 0.8–1.0 Å. At this resolution each atom is fairly distinct. For macromolecules, the resolution is seldom below 1.5 Å and often not even that good. As diffraction data are collected to higher scattering angles, higher resolution is obtained. Sometimes the quality of the crystal limits the resolution that may be obtained experimentally. A limit is also imposed by the wavelength of the radiation.

A second use of this term is for the separation of enantiomorphs.

Rhombohedral unit cell. A unit cell in which there is a three-fold rotation axis along one body diagonal of the unit cell. This symmetry requirement makes all three axial lengths necessarily the same and all three interaxial angles necessarily equal, although the value is not restricted ($a = b = c, \alpha = \beta = \gamma$).

Right-handed coordinate system. A system of three axes, x, y, and z, in which a rotation from x to y, coupled with a translation along z, corresponds to the action of a right-handed screw. If the thumb, index finger and middle finger of the right hand are extended in mutually perpendicular directions, then these digits point to the positive directions of x, y, and z, respectively.

Rotation axis. (See Axis of symmetry.)

Rotation function. A function describing the rotation of a known group of vectors, for example, those corresponding to a known portion of a structure, for comparison with a Patterson map. The orientation of a known part of a structure often may be found through such comparison.

Rotation photograph. A photograph of the diffraction pattern obtained by rotating a crystal continuously about a fixed axis normal to some set of reciprocal lattice planes.

Rotatory-inversion axis. An axis for which the corresponding symmetry operation is a rotation by ($360°/n$) combined with inversion through a point lying on the axis. The special point is a center of symmetry *only* if n is odd.

Salting out. Precipitating, coagulating, or separating a substance from a solution by the addition of a salt.

Scale factor. (See Absolute scale.)

Scattering factor. (See Atomic scattering factor.)

Scintillation counter. A device for measuring the intensity of an X-ray beam. It makes use of the fact that X rays cause certain substances to emit visible light by fluorescence. The intensity of this visible light is proportional to the intensity of the incident X-ray beam and is measured by a photomultiplier. The substance generally used as the X-ray detector is a sodium iodide crystal, activated by a small amount of thallous ion.

Screenless precession photography. A method of measuring the diffraction pattern of a macromolecular crystal by the precession method without the use of a screen to filter out certain reflections. The indexing is done by computer after some preliminary photographs have been taken to determine the crystal orientation with high accuracy. This method is used for data collection for crystalline macromolecules. Small precession angles (1° to 2°) are used so that many sets of photographs with different settings of the dial axis are required.

Screw axis. For a screw axis, designated n_r, the corresponding symmetry operation is a rotation about the axis by ($360°/n$) coupled with a translation parallel to the axis by r/n of the unit-cell length in that direction.

Series-termination error. An effect that results from a limitation in the number of terms in a Fourier series. Ideally an infinite amount of data is required in a Fourier series. In practice, the number of data depends on the reciprocal radius ($\sin \theta/\lambda$) to which the data are collected. Because of truncation of the Fourier series, peaks in the resulting Fourier syntheses are surrounded by series of ripples. These are especially noticeable around a heavy atom. The use of difference syntheses (q.v.) obviates most of the effects of series termination errors.

Sharpened Patterson function. A Patterson map computed with values of $|F|^2$ modified by an exponential or similar function that enhances those reflections with high values of $\sin \theta/\lambda$. The resulting interatomic vectors appear as sharper peaks and the Patterson map may therefore be simpler to interpret.

Sigma Two formula (Σ_2). A formula used in the direct method of structure determination. It relates the phases of three strong reflections to one another.

Small-angle scattering. The study of matter by analysis of the diffraction of X rays with diffraction angles smaller than a few degrees—that is, θ less than 1° for copper radiation. This scattering occurs when the sample is composed of particles with dimensions of the order of several hundred to several thousand Å. Measurement of the intensity distribution gives information on the low resolution structure of the diffracting material; for example, it will give the radius of gyration of the particle.

Solvent-flattening. (See Density modification.)

Space group. A group of symmetry operations consistent with an infinitely extended, regularly repeating pattern. There are 230 such groups, which can be identified (although sometimes with some ambiguity) from the systematic absences in the diffraction pattern. The space groups can be derived by the addition of translational symmetry to the 32 point groups appropriate for structures arranged on lattices. These include simple translations, screw axes, and glide planes. Hence a space group may be considered as the group of operations that converts one molecule or asymmetric unit into an infinitely extending pattern. The 230 space groups are described in *International Tables for X-ray Crystallography Vol.* 1, N. F. M. Henry and K. Lonsdale, eds., Kynoch Press, 1952 and Volume A, T. Hahn, ed., Reidel, 1983 (ref. 20).

Space lattice. A three-dimensional arrangement of points on a lattice (see Bravais lattice).

Sphere of reflection. (Ewald sphere). A construction for considering conditions for diffraction in terms of the reciprocal rather than the real lattice. It is a sphere, of radius $1/\lambda$, with the incident beam along a diameter. The origin of the reciprocal lattice is positioned at the point where the incident beam emerges from the sphere. Whenever a reciprocal lattice point touches the surface of the sphere, the conditions for a diffracted (or reflected) beam are satisifed. Thus, for any orientation of the crystal relative to the incident beam, it is possible to predict which reciprocal lattice points, and thus which planes in the crystal, will be in a "reflecting position" (in the sense used by Bragg).

Stoichiometry. The quantitative relationship of constituents implied by a chemical formula or equation.

Structure factor. The magnitude of the structure factor, $|F|$, is the ratio of the amplitude of the radiation scattered in a particular direction by the contents of one unit cell to that scattered by a classical point electron at the origin of the unit cell under the same conditions. The structure factor has both a magnitude (amplitude) and a phase (relative to the origin of the unit cell). From the intensity of a reflection, we can derive directly the amplitude but not the phase. The structure factor depends on:

1. The nature of the scattering material.
2. The arrangement of the scattering material (including thermal motion).
3. The direction of scattering.

The experimentally measured ("observed") structure factor amplitudes are denoted by $|F_o|$; those calculated for a model of the structure are designated $|F_c|$. The structure factor is the Fourier transform of the unit-cell contents sampled at reciprocal lattice points, h, k, l.

Structure invariant. A reflection or group of reflections whose phase angle(s) do not change when the origin is changed.

Superposition methods. (See Vector superposition map.)

Symbolic addition procedure. One of the direct method procedures in which phases for a few starting reflections are represented by algebraic symbols. The meaning of the symbolic phases is then established during the subsequent analysis.

Symmetry element. A geometrical entity, such as a point, a line, or a plane, with respect to which a particular symmetry operation is performed and the set of symmetry operations associated with this entity, all of which can be generated by at least one of them.

Symmetry operation. When we say that an object is symmetrical, we mean that it may be converted into itself by a symmetry operation or a series of symmetry operations. In crystal structures (assumed infinite in extent), the possible symmetry operations include axes of rotation and rotatory inversion, screw axes, and glide planes as well as lattice translations. Proper operations are translation and rotation while improper operations (see improper symmetry operations) are reflection and inversion.

Synchrotron radiation. Radiation emitted by very high-energy electrons, such as those in an electron storage ring, when their path is bent by a magnetic field. This radiation is characterized by a continuous spectral distribution (which can, however, be "tuned" by appropriate selection), a very high intensity, a pulsed-time structure, and a high degree of polarization.

Systematically absent reflections. (See Absent reflections.)

Tangent formula. A formula used in direct methods of phase determination that allows the development of additional phases.

Temperature factor. An exponential expression by which the scattering of an atom is reduced as a consequence of vibration (or a simulated vibration resulting from static disorder). For isotropic motion the exponential factor is exp $(-B_{iso} \sin^2 \theta/\lambda^2)$, with B_{iso} called, loosely but commonly, the "isotropic temperature factor." It equals $8\pi^2\langle u^2\rangle$, where $\langle u^2\rangle$ is the mean square displacement of the atom from its equilibrium position. For anisotropic motion the exponential expression contains six parameters, the anisotropic vibration or displacement parameters, which describe ellipsoidal rather than isotropic (spherically symmetrical) motion or average static displacements.

Tetragonal unit cell. A unit cell in which there is a four-fold rotation axis parallel to one axis (arbitrarily chosen as **c**); as a result the lengths of a and b are identical and all interaxial angles are 90° ($a = b, \alpha = \beta = \gamma = 90°$).

Thermal vibration. (See Temperature factor.)

Torsion angle. (sometimes called "conformational angle"). The torsion angle (or the angle of twist) about the bond B—C in a series of bonded atoms A—B—C—D is defined as the angle of rotation needed to make the projection of the line B—A coincide with the projection of the line C—D, when viewed along the B—C direction. The positive sense is clockwise. If the torsion angle is 180°, the four atoms lie in a planar zigzag (Z-shaped). If the torsion angle is 60°, one end atom is twisted 60° out of the plane of the other three. Enantiomers have torsion angles of equal absolute value but opposite sign.

Transform. (See Fourier transform.)

Translation. Referring to a motion in which all points of a body move in the same direction—that is, along the same or parallel lines.

Trial-and-error method. A method that involves postulating a structure (that is, assuming locations of the atoms in the unique part of the unit cell), calculating structure factors, F_c, and comparing their magnitudes with the scaled observed values, $|F_o|$. Since the number of trial structures that must be tested increases with the number of parameters, such methods have generally been applied in solving only the simplest structures.

Trial structure. A possible structure for a crystal, which is tested by a comparison of calculated and observed structure factors and by the results of an attempted refinement of the structure.

Triclinic unit cell. A unit cell in which there is no rotational symmetry; as a result there are no restrictions on axial ratios or interaxial angles.

Triple-product sign relationship. The sign relationship $s(h,k,l)s(h',k',l')s(h - h',k - k',l - l') \approx +1$, where $\approx$ means "is probably equal to" and s means "the sign of."

Twin. A composite crystal built from two or more crystal specimens that are grown together in a specific manner so that there is at least one plane and a direction perpendicular to it that are related in the same manner to the crystallographic axes of both parts of the twin.

Unitary structure factor. The ratio of the structure amplitude, $|F(hkl)|$, to its maximum possible value—that is, the value it would have if all atoms scattered exactly in phase. It is denoted by U. $U(hkl) = |F(hkl)|/|\Sigma f(hkl)_j]$, where $f(hkl)_j$ contains a temperature factor expression.

Unit cell. The basic building block of a crystal, repeated infinitely in three dimensions. It is characterized by three vectors, **a**, **b**, and **c**, that form the edges of a parallelepiped and the angles between them, γ, α, and β (γ between **a** and **b**).

Vector. A quantity that requires for its complete description a magnitude, direction, and sense. It is often represented by a line, the length of which specifies the magnitude of the vector, and the orientation of which specifies the direction of the vector. The sense of the vector is then indicated by an arrowhead at one end of the line. Two vectors may be added together by placing the second vector with its origin on the end (arrowhead) of the first. The resultant vector is the directed line from the origin of the first vector to the end of the second. This process of addition can be continued infinitely. The scalar product, $\mathbf{a}\cdot\mathbf{b}$, is $|a||b| \cos \gamma$, where γ is the angle from **a** to **b**. The magnitude of the vector product, $\mathbf{a} \times \mathbf{b}$, is $|a||b| \sin \gamma$.

Vector superposition map. A method of analyzing the Patterson map that involves setting the origin of the Patterson map in turn on the positions of certain atoms whose positions are already known and suitably combining the superposed maps.

Vibration parameters. (See Displacement parameters and Temperature factor.)

Weight of a measurement. A number assigned to express the relative precision of each measurement. In least-squares refinement the weight should be proportional to the reciprocal of the square of the estimated standard deviation of the measurement. Other weighting schemes frequently are used.

Weissenberg photograph. A distorted image of a given layer of the reciprocal lattice. It is an oscillation photograph in which the camera moves in a particular manner as the crystal rotates. A narrow band of diffracted radiation normal to the rotation axis (a layer line) is selected. The position of a spot on the film is related to the angle of rotation of the crystal at the instant of diffraction for that particular spot. Indices can readily be assigned to each spot.

White radiation. Any radiation, such as sunlight, with a continuum of wavelengths. We use the term here for X-radiation with such a continuum of wavelengths.

Wilson plot. A plot of averages of observed intensities in ranges of $\sin^2 \theta/\lambda^2$ with the theoretical values for a unit cell composed of randomly arranged atoms. From this plot a scale factor and an average temperature factor may be derived.

X rays. Electromagnetic radiation of wavelength 0.1–100 Å, produced by bombarding a target (generally a metal such as copper or molybdenum) with fast electrons. As a result, electrons from the innermost shells (K or L) may be ejected from atoms in the target material. When an electron from an outer shell falls back into the vacant shell, energy is emitted in the form of X rays. The spectrum of the emitted radiation has a maximum intensity at a few wavelengths characteristic of the target material.

X-ray camera. A device for holding film in an appropriate manner to intercept an X-ray diffraction pattern.

X-ray tube. The basic parts of an X-ray tube are a source of electrons and a metal anode that emits the X rays, enclosed in a glass envelope under vacuum. Tubes may be classified according to the nature of these parts.

Bibliography

All teachers of X-ray crystallography, including the present authors, have their favorite sources of information. We have chosen those that we consider to have a simple or clarifying approach. Those readers who wish to delve further into the subject should follow the bibliographies given in some of the books listed. We have tried to keep to a minimum the number of different books referred to, thus necessarily omitting some worthy ones. A few books are mentioned for the browser because they seem delightful, but they are not essential for those with limited time or budget.

Almost any book related to crystallography may be obtained from Polycrystal Book Service, P. O. Box 27, Western Springs, IL 60558, U.S.A.

COMPREHENSIVE BOOK LIST

Since the last edition of this book an excellent and comprehensive book list has been published by Dr. John H. Robertson, University of Leeds, U.K. This list contains volumes published between 1970 and 1980 and appears in the *Journal of Applied Crystallography,* **15** (Dec. 1982), pp. 640–676. It is an update of a previous *Crystallographic Booklist,* edited by Helen Megaw, and published for the International Union of Crystallography in 1965. If you cannot obtain a copy (photocopying of this article is freely allowed), then write to Dr. J. N. King, Executive Secretary, International Union of Crystallography, 5 Abbey Square, Chester CH1 2HU, U.K.

GENERAL BOOKS

We recommend the following for a general study of the subject:

1. J. D. Dunitz. *X-ray Analysis and the Structure of Organic Molecules.* Cornell University Press, 1979. The first 300 pages of this 500-page book present a lucid exposition of most of the topics central to modern structural crystallography. The

book was developed from a series of lectures given to an audience of chemists and it is highly recommended to those who want a deeper treatment than that given here. The last part of the book describes some of the ways in which the results of crystal structure analysis have influenced chemistry, particularly organic chemistry, with emphasis on studies emanating from the author's laboratory.

2. G. H. Stout and L. H. Jensen. *X-ray Structure Determination. A Practical Guide.* The Macmillan Company, 1968. A detailed textbook for practitioners of X-ray crystallography; that is, this is a book that describes "how it is done." It is suggested as an ideal companion for the present work, which is an attempt to explain "why it is possible to do it."

3. M. F. C. Ladd and R. A. Palmer. *Structure Determination by X-ray Crystallography.* Plenum, 1977. This is an excellent text with many worked examples. You may find it a very useful practical guide; the diagrams are unusually clear. This book is referred to many times in the listing that follows. For example, the Patterson function for papaverine hydrochloride is illustrated and then it is shown how the structure may be derived by use of a minimum function. The crystal structure of a cyclobutene derivative, "squaric acid," is determined by direct methods. The authors lead the reader by the hand, showing how the "origin set" of signs may be chosen and then giving a Σ_2 listing. The E-map is illustrated, atomic fractional coordinates determined, and the molecular geometry described in detail.

4. B. K. Vainshtein. *Modern Crystallography. I. Symmetry of Crystals, Methods of Structural Crystallography.* Springer-Verlag, 1981. B. K. Vainshtein, V. M. Fridkin, and V. L. Indenboom. *Modern Crystallography. II. Structure of Crystals,* Springer-Verlag, 1982. Volume I contains much information including some on new trends in the development of X-ray crystallography, often described in a little more depth than is common in crystallographic texts. There are some excellent electron-micrographs of crystals in the chapter on microscopic characteristics of crystals. The section on symmetry extends to a discussion of color symmetry. These are also sections on kinematic and dynamic theories of diffraction, scattering by noncrystalline substances (including small-angle scattering), powder diffraction, and electron diffraction. All are highly recommended. Volume II contains accounts of selected structural results from X-ray diffraction studies.

5. P. Luger. *Modern X-ray Analysis on Single Crystals.* W. de Gruyter, 1980. This volume on crystal structure analysis was translated from the German with the assistance of G. A. Jeffrey. The first part deals with the mathematics of the method; then follows a section on X-ray diffraction data collection, with excellent diagrams. Next are sections on symmetry and on diffractometer measurements. The solution of crystal structures and their refinement are combined with some very useful examples and figures.

6. T. L. Blundell and L. N. Johnson. *Protein Crystallography.* Academic Press, 1976. At present this is the definitive book on the subject. The authors discuss X-ray crystallography in general and then go into details of the applications of the principles to protein crystallography. There are most interesting and readable chapters on the crystallization of proteins, the preparation of heavy atom derivatives, and molecular replacement.

The following general books are also highly recommended to the reader. They are listed in reverse chronological order. Some are older, some are new, but each contains information and explanations that help the student and researcher alike.

7. R. Steadman. *Crystallography.* Van Nostrand, 1982. This is a delightful student workbook on structures and lattices, planes and directions, and crystal geometry. Then follow sections on atomic coordinates, powder photographs, the reciprocal lattice, and electron diffraction. The early part about lattices is highly recommended and the first chapter ends with the reminder that "A crystal structure is made of atoms, a crystal lattice is made of points and a crystal system is a set of axes." There are plenty of graphical examples and you will have a good understanding of lattices and planes when you have finished looking through this.

8. L. E. Sutton and M. R. Truter (Senior Reporters). *Molecular Structure by Diffraction Methods. Volume 6. (A Review of the Recent Literature up to September 1977).* A Specialist Periodical Report. The Chemical Society, 1978. This contains discussions of results for many classes of chemical and biochemical compounds, as well as discussions of information retrieval, molecular mechanics calculations, and studies of intermolecular interactions.

9. L. S. Dent Glasser. *Crystallography and Its Applications.* Van Nostrand, 1977. This is a useful small volume with an excellent chapter on the powder method, with illustrative examples. The chapter on optical crystallography is also highly recommended.

10. D. E. Sands. *Introduction to Crystallography.* Benjamin, 1975. This small paperback is devoted equally to discussions of symmetry and of structure determination. It is intended for undergraduate chemists.

11. A. I. Kitaigorodskii. *Molecular Crystals and Molecules.* Academic Press, 1973. The author is very interested in the results of crystal structure analyses and discusses crystal packing with great insight. The chapters on lattice energy, thermodynamic concepts, and conformations of organic molecules are particularly recommended.

12. H. Lipson. *Crystals and X Rays.* Wykeham Science Series 13. Springer-Verlag, 1970. This is a general and lightly written book that will interest those who do not want to go into too much detail. The author has written other texts that delve more deeply, but this is a readable volume.

13. M. M. Woolfson. *An Introduction to Crystallography.* Cambridge University Press, 1970. An excellent text. This is now published as a paperback (1978).

14. H. Lipson and W. Cochran. *The Crystalline State. Volume III. The Determination of Crystal Structures.* Cornell University Press, 1966. A major teaching text for the time. The various stages of structure determination are considered in detail.

15. K. C. Holmes and D. M. Blow. *The Use of X-ray Diffraction in the Study of Protein and Nucleic Acid Structure.* Interscience, Wiley, 1966. This paperback gives a general description of diffraction phenomena and their application to X-ray crystallography. The sections on the Fourier transform and fiber diffraction are particularly recommended.

16. C. W. Bunn. *Chemical Crystallography. An Introduction to Optical and X-ray Methods,* 2nd ed. Oxford at the Clarendon Press, 1961. This book gives clear descriptions by which many older crystallographers were introduced to the art and science of their field.

17. A. Holden and P. Singer. *Crystals and Crystal Growing.* Anchor Books, Doubleday, 1960. This is a delightful paperback, with recipes for growing crystals. It is suitable for high school students and practicing crystallographers alike.

18. R. W. James. *The Crystalline State. Volume II. The Optical Principles of the Diffraction of X Rays.* G. Bell and Sons, 1948. This has been reprinted as a

paperback. It contains detailed descriptions and analyses of many of the physical fundamentals of X-ray diffraction, written by one of the pioneers in the field. The physics of diffraction is dealt with in depth.

19. M. J. Buerger. *X-ray Crystallography.* Kreiger, 1980. This is a reprint of the well-known book of 1942 and it contains, among many useful sections, a most informative practical account of the Weissenberg method.

INFORMATION SOURCES THAT CRYSTALLOGRAPHERS USE

The international organization to which most crystallographers belong, through various national associations, is the International Union of Crystallography, which holds congresses and symposia every 3 years and sponsors three major publications and several compilations.

The tables of data published by the International Union of Crystallography provide not only useful information, but also an explanation of each area of crystallography that is referred to. The reader is urged to browse through *International Tables* (see below), particularly the plane groups in Volume I or Volume A, and some of the details of the experimental methods that are clearly described in Volume III.

(A) INTERNATIONAL TABLES

20. *International Tables for X-ray Crystallography.* N. F. M. Henry and K. Lonsdale (eds.). *Volume I. Symmetry Groups.* Kynoch Press, 1952. This volume is devoted to space group tables, with information on point groups as well. It contains a full listing of all data on plane groups and space groups and various other symmetry properties. Diagrams are included that illustrate the location of equivalent positions of a motif and of the various symmetry elements, the coordinates of equivalent positions with their point symmetry and the number of such positions, and the conditions limiting possible reflections (see Figure 7.2). The structure factor expressions for each space group are given in another part of the volume. One of the best ways to study space groups is to work systematically through Volume I.

Volume II. Mathematical Tables. J. S. Kasper and K. Lonsdale (eds.). Kynoch Press, 1959. This volume presents the general mathematical information required by X-ray crystallographers. It includes data and descriptions of matrix algebra, group theory, vector analysis, tensor analysis, Fourier series, Fourier transforms, the method of least squares, statistics, diffraction, the geometry of direct and reciprocal lattices, and formulas for absorption corrections.

Volume III. Physical and Chemical Tables. C. H. MacGillavry, G. D. Rieck and K. Lonsdale (eds.). Kynoch Press, 1962. This volume contains most of the physical and chemical data necessary in X-ray structure determinations, such as atomic scattering factor tables (including anomalous dispersion corrections), absorption coefficients, X-ray wavelengths, and atomic radii.

Volume IV. Revised and Supplementary Tables to Volumes II and III. J. A. Ibers and W. C. Hamilton (eds.). Kynoch Press, 1974. This volume contains more up-to-date information.

Volume A. Space-group Symmetry. T. Hahn. D. Reidel, 1983. This is a new version of the space group tables in Volume I, devoted to real space. Volume B, in preparation, will deal with reciprocal space. Volume A contains additional comments and shows views for alternative space group settings (such as $P2_1/a$, $P2_1/c$, $P2_1/n$), which correspond to different axial choices in the monoclinic system. Lacking are the relationships between signs of structure factors useful in analyses by "direct methods"; presumably these will be included in Volume B. There are sections at the end entitled "Introduction to space-group symmetry" (pp. 712–731), "crystal lattices" (pp. 734–744), and "symmetry operations" (pp. 788–792) that those with a mathematical bent may find useful.

(B) CRYSTALLOGRAPHIC JOURNALS

21. The two main journals for crystallographers, also published by the International Union of Crystallography are: *Acta Crystallographica* and the *Journal of Applied Crystallography*. The *Journal of Applied Crystallography* was started in 1968 and in that same year, *Acta Crystallographica* was split into two sections: *Section A (Crystal Physics, Diffraction, Theoretical, and General Crystallography)* and *Section B (Structural Crystallography and Crystal Chemistry)*.

In 1983 *Acta Crystallographica* was split into three sections:

Section A: Foundations of Crystallography. This is concerned with basic developments in many areas of crystallography.

Section B: Structural Science. This contains the results of structural studies from disciplines throughout the natural sciences. The articles are generally concerned with assessing structural information in order to advance a branch of science.

Section C: Crystal Structure Determinations. This journal contains results of crystal structure analyses. It is, in part, a continuation of a journal published in Italy for many years and titled *Crystal Structure Communications*.

Structure determinations are also reported in many other journals: *Zeitschrift für Kristallographie, Inorganic Chemistry, Journal of the American Chemical Society, Journal of the Chemical Society, Acta Chemica Scandinavica,* and *Helvetica Chimica Acta,* among others.

(C) STRUCTURE REPORTS

22. A. J. C. Wilson (Vols. 8–18), W. B. Pearson (Vols. 19–35), J. Trotter (Vols. 36–47), and G. Ferguson (Vols. 48 to date) (eds.). *Structure Reports* originally published by A. Oosthoek; now by Reidel. These volumes, one for each year, describe with editorial comment, all the crystal structure determinations published in that year, so that there is, for certain purposes, no need to refer back to the original article. A cumulative index (1913–1973) was published in 1975.

(D) OTHER USEFUL COMPILATIONS

Several other compilations are of interest. A few of them are listed here.

23. J. D. H. Donnay and G. Donnay (eds.). *Crystal Data. Determinative Tables*, 1963. American Crystallographic Association Special Monograph No. 5, 2nd ed. A compendium of data on crystals. It gives unit-cell dimensions, densities, and space groups, and can be very helpful in the identification of diffraction patterns of crys-

tals that have been studied previously. The third edition of Volumes 1 and 2 was published in 1973 with J. D. H. Donnay and H. M. Ondik as editors. These contain organic and inorganic entries through 1966 respectively. Volumes 3 and 4, published in 1979, also contain organic and inorganic entries, respectively, now totalling more than 45,000 entries. This compilation is produced by the International Center for Diffraction Data, Swarthmore, PA 19081.

24. O. Kennard, D. G. Watson, F. A. Allen, and S. M. Weeds (eds.). *Molecular Structures and Dimensions. Bibliography. Organic and Organometallic Crystal Structures,* 1970. This is produced annually and is the ideal library source for information on references to any compounds containing carbon.

25. W. L. Duax and D. A. Norton (eds.). *Atlas of Steroid Structure,* 1975. The "Sears-Roebuck Catalog" of steroid structures, with illustrations of conformations and packing.

(E) COMPUTER DATA BASES

Of particular interest and special importance in the analysis of results are the computer data bases.

26. *Cambridge Crystallographic Data File.* This is produced under the supervision of Dr. Olga Kennard and Dr. D. G. Watson in Cambridge, England, and contains the references, formulas, space groups, cell dimensions, and atomic positions of each organic structure published. The bibliography is also published annually (ref. 24). For more information see F. H. Allen, S. Bellard, M. D. Price, B. A. Cartwright, A. Doubleday, H. Higgs, T. Hummelink, B. J. Hummelink-Peters, O. Kennard, W. D. S. Motherwell, J. R. Rodgers, and D. G. Watson. *Acta Crystallographica,* Section B, **35** (1979), pp. 2331–2339.

27. *Protein Data Bank.* This is maintained at Brookhaven National Laboratory, Upton, Long Island, New York. Atomic coordinates for all available protein and nucleic acid structures are stored here.

28. *Inorganic Crystal Structure Data Base.* This is produced under the direction of Drs. G. Bergerhoff, University of Bonn, Germany (FRG) and I. D. Brown, Institute for Materials Research, McMaster University, Hamilton, Ontario, Canada. It is published as a machine-readable numerical database by the Fachinformationzentrum Energie Physik Mathematik (FIZ) in Karlsruhe. All compounds not containing C–C or C–H bonds (but including metal carbides) are included. For details about the file see: G. Bergerhoff, R. Hundt, R. Sievers and I. D. Brown. *Journal of Chemical Information and Computer Sciences,* **23** (1983), pp. 66–69.

29. *Metals Data File.* Available from Dr. L. D. Calvert, National Research Council, Ottawa, Canada. For more information see: L. D. Calvert, *Acta Crystallographica,* **A37** (1981), pp. C343–C344.

30. *The Powder Diffraction File.* Edited by W. L. Berry and published by the Joint Committee on Powder Diffraction Standards (JCPDS), 1601 Park Lane, Swarthmore, PA 19081.

(F) COMPUTER PROGRAMS OF GENERAL USE

There are many commonly used crystallographic computing packages. Only a few of these are listed here.

31. MULTAN The first general direct-methods program and still the most widely used, MULTAN is updated frequently to incorporate new techniques and approaches. A recent reference is M. Woolfson, in ref. 165 (D. Sayre, ed.), pp. 110–125. Almost any recent volume of *Acta Crystallographica,* Series A, will be found to contain a paper by Michael Woolfson or his co-workers (chiefly Peter Main) on new developments.

32. SHELX A package of crystallographic programs that performs almost every calculation needed in solving and refining a structure, in a fashion that is extremely convenient for the user. A recent reference to it is G. Sheldrick, ref. 165 (D. Sayre, ed.), pp. 506–514. The package is updated at intervals.

33. XRAY and XTAL Systems. The XRAY system is a widely used package of programs for solving and refining structures, developed by James Stewart and his co-workers at the University of Maryland, and originally written to be readily interpreted by all large computers (pidgin FORTRAN). XTAL is a newer computing system, still under development, the result of a group effort by individuals in many different laboratories. Crystallographic computing systems are discussed in a general way by J. Stewart, ref. 165 (D. Sayre, ed.), pp. 497–505.

34. ORTEP This is doubtless the most widely used crystallographic program, the "Oak Ridge Thermal Ellipsoid Program," developed two decades ago by Carroll Johnson and used for illustrations in innumerable journals and books (including this one).

There is also an enormous variety of specialized programs available. Such programs are listed in publications in *Acta Crystallographica.*

CRYSTALS

(A) PREPARATION OF CRYSTALS

35. Holden and Singer, 1960. Reference 17.

36. A. McPherson. *Preparation and Analysis of Protein Crystals.* Wiley Interscience, 1982. The entire book is a gold mine of information for the would-be crystal grower.

37. Stout and Jensen, 1968. Reference 2, Ch. 4, pp. 62–82.

38. Blundell and Johnson, 1976. Reference 6, Ch. 3, pp. 59–82.

39. Holmes and Blow, 1966. Reference 15, pp. 174–179.

40. R. F. Strickland-Constable. *Kinetics and Mechanism of Crystallization,* 1968. Academic Press, pp. 7–38. A discussion of the principles of crystal growth from the melt, from vapor and from solution.

(B) CRYSTAL MORPHOLOGY AND OPTICAL CRYSTALLOGRAPHY

41. P. Groth. *Elemente der Physikalischen und Chemischen Krystallographie.* von R. Ohlenbourg, 1921. Contains 4 tables, 962 figures, and 25 stereoviews of crystals together with a text on optical crystallography.

42. Vainshtein, 1981. Reference 4, Ch. 1, pp. 1–26, and Ch. 3, pp. 180–206. These include some electron micrographs of crystals and discussions of crystals, holohedry, and hemihedry and goniometry.

43. J. W. Mullin. *Crystallization.* Butterworths, 1961. Chapter 1, pp. 1–20. The symmetry and habits of some crystals are described.

44. C. J. Schneer (ed.). *Crystal Form and Structure.* Benchmark Papers in Geology, No. 34. Dowden, Hutchinson and Ross, Inc., 1977, pp. 1–31. Contain a description of early ideas on internal periodicity in crystals such as those of Kepler, Hauy, and Bravais, with excerpts from their publications.

45. N. H. Hartshorne and A. Stuart. *Crystals and the Polarizing Microscope,* 2nd ed. Edward Arnold, 1950, Ch. 4, pp. 89–159. Optical properties of crystals are described including an especially good section on birefringence (double refraction).

46. L. V. Azaroff. *Introduction to Solids.* McGraw-Hill, 1960, Ch. 3, pp. 55–72: On the closest packing of spheres. Chapter 7, pp. 140–165: On the formation of crystals. This chapter contains a description of the development of faces on a crystal, and mechanisms of growth. Twinning is also discussed.

47. Bunn, 1961. Reference 16, Ch. 2, pp. 24–58. The shapes of crystals and methods of indexing their faces.

48. Ladd and Palmer, 1977. Reference 3, Ch. 1, pp. 1–48: Description of stereograms, Laue symmetry, and a diagram of a contact goniometer. Chapter 3, pp. 101–112: Discussion of polarized light and birefringence.

49. Dent Glasser, 1977. Reference 9, Ch. 2, pp. 26–44. This is a very useful discussion of optical crystallography.

50. E. E. Wahlstrom. *Optical Crystallography,* 5th ed. Wiley, 1979.

51. W. L. Bond. *Crystal Technology.* Wiley, 1976. This book is concerned with crystal orienting and shaping—that is, cutting plates from raw crystals at precise orientations using optical properties and X-ray diffraction.

52. E. A. Wood. *Crystals and Light: An Introduction to Optical Crystallography,* 1964. Reprinted by Dover in 1977.

53. W. A. Bentley and W. J. Humphreys. *Snow Crystals.* Dover, 1962.
(First published by McGraw-Hill in 1931.) Photographs of 2000 snow crystals taken by W. A. Bentley with a microscope over a period of 50 years in Vermont. Collected in 1931 by the American Meteorological Society, with a text by W. J. Humphreys. There is a section on the "Crystallography of the Snow Crystal." This book shows the beauty and the great diversity of forms that snow crystals can have.

(C) UNIT-CELL DIMENSIONS

54. N. F. M. Henry, H. Lipson, and W. A. Wooster. *The Interpretation of X-ray Diffraction Photographs.* Macmillan, 1951, Ch. 13, pp. 189–197. Discussion of the accurate determination of cell dimensions and the problems encountered in obtaining cell dimensions of the highest accuracy.

55. R. W. M. D'Eye and E. Wait. *X-ray Powder Photography in Inorganic Chemistry,* 1960. Academic Press, Ch. 5, pp. 101–120. The derivation of accurate cell dimensions by film methods free from systematic errors such as film shrinkage, absorption of the X-ray beam by the specimen, and refraction of X rays by the crystal.

56. Stout and Jensen, 1968. Reference 2. Ch. 5, pp. 94–97: The measurement of unit-cell dimensions from rotation photographs. Pages 117–122: Measurements from Weissenberg photographs. Page 133: Measurements from precession photographs. Page 192: Measurements from diffractometers.

(D) MATHEMATICS

57. E. Prince. *Mathematical Techniques in Crystallography and Materials Science.* Springer-Verlag, 1982. In this volume are collected many of the mathematical expressions that crystallographers use. There are chapters on matrices (including linear transformations and rotations of axes), group theory and space groups, vectors (including the reciprocal lattice and orientation matrices), tensors (including anisotropic temperature factors and rigid-body motion), precision and constrained crystal structure refinement.

58. D. E. Sands. *Vectors and Tensors in Crystallography.* Addison-Wesley, 1982. This is a clear and extremely helpful exposition of the use of vectors and tensors in crystallography and in any other field in which vector or tensor analysis in rectilinear systems is important. There are many valuable exercises to illustrate the many different applications discussed.

59. W. A. Wooster. *Tensors and Group Theory for the Physical Properties of Crystals.* Oxford University Press, 1973.

60. Luger, 1980. Reference 5, Ch. 1, pp. 1–51. Theoretical basis of crystallography.

61. *International Tables.* Reference 20, Volumes II and IV. The discussion of matrices and of vector and tensor analysis by Patterson in Volume II is masterfully lucid.

DIFFRACTION

(A) X RAYS

62. L. V. Azaroff. *Elements of X-ray Crystallography.* McGraw-Hill, 1968, Ch. 6, pp. 86–95: A discussion of coherent and incoherent scattering and the polarization of X rays. Chapter 6, pp. 95–103: The theory of absorption of X rays is discussed, including the dependence of absorption on the wavelength of the radiation, the existence of absorption edges and the application of absorption edges in the selection of X-ray filters: Chapter 6, pp. 103–130. Details of the production of X rays and their properties.

63. *A Guide to the Safe Use of X-ray Crystallographic and Spectrometric Equipment,* 1978. Association of University Radiation Protection Officers, U.K. (Obtainable from F. Griffiths. UWIST. Cardiff. Great Britain.) This is included for the practitioner as a reminder that X-radiation is not good for you.

(B) PHYSICS OF DIFFRACTION

64. F. A. Jenkins and H. E. White. *Fundamentals of Optics,* 4th ed. McGraw-Hill, 1976. This has excellent chapters, updated too, on the superposition of waves (Ch. 12, pp. 238–257), the interference of two beams of light (Ch. 13, pp. 259–266), Fraunhofer diffraction by a single opening (Ch. 15, pp. 315–322), the diffraction grating (Ch. 17, pp. 355–365). You will recognize some of the photographs that we have reproduced in the present volume.

65. R. P. Feynman. R. B. Leighton, and M. Sands. *The Feynman Lectures on Physics.* Addison-Wesley, 1963, Volume 1, pp. 29-1 to 31-11 and 37-1 to 38-10. Discussions of interference, diffraction, and refraction. The fact that in diffraction a small hole may be regarded as a set of uniformly distributed sources is discussed.

66. A. H. Compton and S. K. Allison. *X-rays in Theory and Experiment,* 2nd ed. Van Nostrand, 1954.

67. Woolfson, 1970. Reference 13, Ch. 2, pp. 34–65. Discusses diffraction including incoherent scattering in pp. 45–47. Also the reciprocal lattice (Ch. 2, pp. 65–71) and extinction (Ch. 2, pp. 182–189).

68. J. Cowley. *Diffraction Physics,* 2nd ed. North-Holland, 1975.

69. G. Harburn, C. A. Taylor, and T. R. Welberry, *Atlas of Optical Transforms.* G. Bell and Sons, Ltd. 1975. This very instructive volume contains optical diffraction patterns of a variety of objects and is arranged so that the object is diagrammed on the left-hand page and a photograph of its diffraction pattern is given on the facing right-hand page. We recommend it highly.

(C) DIFFRACTION BY CRYSTALS

70. J. D. Dunitz, 1979. Reference 1, Ch. 6, pp. 281–298. Description of the meaning of integrated intensity, Lorentz and polarization factors, absorption, extinction, and double reflections.

71. Vainshtein, 1981. Reference 4, Volume 1, pp. 235–272.

72. James, 1948. Reference 18, reprinted, Ch. 1, pp. 1–14: A mathematical account, using vectors, of the geometry of diffraction by a simple lattice: Chapter 3, pp. 93–134: Atomic scattering factors. Chapter 8, pp. 413–457: Contains a description of dynamical theory.

73. J. P. Glusker (ed.). *Structural Crystallography in Chemistry and Biology.* Volume 4 in the Series of Benchmark Papers in Physics and Chemistry. Hutchinson Ross, 1981. Dynamical theory, including a reprint of a general article by Ewald, pp. 40–47.

74. Blundell and Johnson, 1976. Reference 6, Ch. 5, pp. 117–118. Discusses resolution, an important parameter in macromolecular structure determination.

(D) DIFFRACTION AND THE CRYSTAL LATTICE

75. Dunitz, 1979. Reference 1, Ch. 1, pp. 33–37. The reciprocal lattice.

76. Woolfson, 1970. Reference 13, Ch. 2, pp. 65–71. The reciprocal lattice.

77. James, 1948. Reference 18, Appendix 2, pp. 598–607.

(E) MEASUREMENT OF THE DIFFRACTION PATTERN

78. Ladd and Palmer, 1977. Reference 3, Ch. 3, pp. 122–124: Laue method. Chapter 3, pp. 124–134: Oscillation method. Chapter 3, pp. 134–137: Weissenberg method.

79. Dunitz, 1979. Reference 1, Ch. 6, pp. 266–298. Experimental aspects of X-ray analysis.

80. Dent Glasser, 1977. Reference 9, Ch. 4, pp. 69–104. Photographic methods.

81. Luger, 1980. Reference 5, Ch. 2, pp. 52–110: Covers the various methods of collecting X-ray diffraction data. Chapter 4, pp. 176–204: Diffractometer measurements.

82. Holmes and Blow, 1966. Reference 15. This gives some "still" photographs of protein crystals.

83. U. W. Arndt and B. T. M. Willis. *Single Crystal Diffractometry.* Cambridge University Press, 1966. A detailed account of the use of single crystal diffractometers with X-rays or neutrons.

84. Stout and Jensen, 1968. Reference 2, Ch. 6, pp. 165–177, on film methods and pp. 177–194, on counter methods for intensity measurements. Practical aspects of intensity measurements are dealt with in detail.

85. Blundell and Johnson, 1976. Reference 6, Ch. 9, pp. 240–309. This describes protein data collection, including the various types of cameras and detecting devices that can be used. Precession method on pp. 274–283, diffractometry on pp. 284–309.

86. M. J. Buerger. *The Precession Method.* Wiley, 1964.

87. Henry, Lipson and Wooster, 1951. Reference 54, Weissenberg photographs, Ch. 7, pp. 87–100; Laue photographs, Ch. 6, pp. 71–86; oscillation photographs, Ch. 5, pp. 48–70.

88. R. Rudman. *Low Temperature X-ray Diffraction.* Plenum, 1976.

89. S. Block and G. Piermarini. *High Pressure Crystallography.* In McLachlan and Glusker, 1983, Ref. 210, Section D, Ch. 14, pp. 265–267.

90. P. Coppens. The evaluation of absorption and extinction in single-crystal structure analyses, in Ahmed, 1969, Ref. 169.

91. N. W. Alcock. The analytical method for absorption correction, pp. 271–278, in Ahmed, 1969, Ref. 169.

92. James, 1948. Reference 18, Ch. 2, pp. 27–92: Intensities of reflection. Chapter 6, pp. 268–341: Extinction.

93. Azaroff, 1968. Reference 61, Ch. 6. pp. 114–120: Monochromators. Chapter 9, pp. 207–226: Extinction and absorption. Chapter 13, pp. 207–226: Diffractometers. Chapter 14, pp. 390–415: Laue photography. Chapter 15, pp. 416–431: Rotation method. Chapter 16, pp. 433–446: Weissenberg methods. Chapter 16, pp. 446–459: Precession methods.

94. *Transactions of the American Crystallographic Association. Proceedings of the Symposium on New Crystallographic Detectors and the Workshop on Crystallographic Detectors,* Volume 18, 1982. Contains articles on area detectors. The chapter on "Multiwire Proportional Counters for use in Area X-ray Diffractometers" by R. Hamlin, pp. 95–123 is helpful. For a diagram of a multiwire proportional counter see Figure 3, p. 100.

95. U. W. Arndt and A. J. Wonacott. *The Rotation Method in Crystallography.* Data Collection from Macromolecular Crystals. North-Holland, 1977. This is a book for the expert or would-be expert in the subject.

FOURIER SERIES

96. Dunitz, 1979. Reference 1, Ch. 1, pp. 59–65. Fourier transforms are explained, pp. 29–31; convolutions, pp. 38–41; the crystal as a Fourier series, pp. 59–65.

97. J. Waser. "Pictorial Representation of the Fourier Method of X-ray Crystallography," *Journal of Chemical Education* **45**(1968), pp. 446–451. Highly recommended. Waser starts with a consideration of periodic functions and then of some one-dimensional crystals, showing the effects of addition of sinusoidal waves, particularly the effects of the application of different relative phases to these waves.

98. Lipson and Cochran, 1966. Reference 14, Ch. 4: Calculation of structure factors, pp. 66–82. Chapter 5: Summation of Fourier series, pp. 83–108. Details of the calculation of structure factors and of Fourier series are given.

99. Ladd and Palmer, 1977. Reference 3, Ch. 6, pp. 201–218. The superposition of waves.

100. Woolfson, 1970. Reference 13, Ch. 4, pp. 84–123.

101. R. A. Jacobson. *The Fast Fourier Transform in Crystallography.* In McLachlan and Glusker, 1983, Ref. 210, Section D, Ch. 7, pp. 223–234.

102. C. A. Taylor and H. Lipson. *Optical Transforms. Their Preparation and Application to X-ray Diffraction Problems.* G. Bell and Sons, 1964. This entire book should at least be browsed through. Plate 26, used in the present work as Figure 3.7, is one of the best illustrations of "what it is all about."

103. J. Amoros and M. Amoros. *Molecular Crystals, Their Transforms and Diffuse Scattering.* Wiley, 1968. Figure 14, p. 92 shows how the diffraction pattern of a two-dimensionsal lattice is the diffraction pattern of a small disc sampled at reciprocal lattice points. Figure 16, p. 96 shows the diffraction pattern of an anthracene crystal.

104. Holmes and Blow, 1966. Reference 15: Fourier transform and its inverse, pp. 117–132. This explains, in a very clear way, not only Fourier transforms but also convolutions.

105. Harburn, Taylor, and Welberry, 1975. Reference 69.

106. James, 1948. Reference 18, Ch. 7, pp. 342–412. Fourier series. Also appendixes 4 and 5, pp. 611–631.

SYMMETRY

107. *International Tables.* Reference 20, Volume I (1952) and also Volume A (1983).

108. F. C. Phillips. *An Introduction to Crystallography.* Longmans, Green and Co, 1946. Crystal symmetry, from a study of the seven crystal systems to the space groups, is discussed with clarity.

109. H. H. Jaffe and M. Orchin. *Symmetry in Chemistry.* Wiley, 1965.

110. F. A. Cotton. *Chemical Applications of Group Theory.* Wiley, 1963.

111. Dunitz, 1979. Reference 1, Ch. 2, pp. 73–98. This is a discussion of space groups with detailed examples of the most common space groups $P\bar{1}$, $P2_1/c$, and $P2_12_12_1$.

112. P. Luger, 1980. Reference 5, Ch. 3: Crystal symmetry, pp. 111–175.

113. Prince, 1982. Reference 57, Chs. 2 and 3, pp. 20–48.

114. Ladd and Palmer, 1977. Reference 3, Ch. 2, pp. 73–98, especially an analysis of the space group symbol on pp. 97–98. Systematic absences (space group extinctions), Ch. 4, pp. 160–179.

115. S. B. Hendricks. *Some statistical aspects of crystal symmetry.* In McLachlan and Glusker, 1983, Ref. 210, Section E, Ch. 3, pp. 296–301.

116. C. P. Brock and E. C. Lingafelter, 1980. "Common misconceptions about crystal lattices and crystal symmetry," *Journal of Chemical Education*, **57**, p. 552.

117. A. Holden, 1965. *The Nature of Solids*. Columbia University Press, Ch. 5, symmetry, pp. 49–69. A very readable account illustrating the idea of symmetry.

STRUCTURE ANALYSIS

(A) TRIAL-AND-ERROR

This approach is now only of historical interest in practice, but these accounts give many valuable insights into "why it works."

118. Lipson and Cochran, 1966. Reference 14, Ch. 6, trial-and-error methods, pp. 109–143. Some examples of methods include spatial considerations, evidence from physical properties, and the use of structure-factor graphs.

119. Lipson, 1970. Reference 12, Ch. 6, pp. 78–94. How some simple structures were determined.

(B) DIRECT METHODS

120. M. F. C. Ladd and R. A. Palmer. *Theory and Practice of Direct Methods in Crystallography*. Plenum, 1980.

121. Dunitz, 1979. Reference 1, intensity statistics in Ch. 2, pp. 99–103, and some detailed examples of direct methods in Ch. 3, pp. 148–182, including a specific example (Table 3.2, p. 152) and an illuminating diagram that illustrates triple product relationships in terms of reciprocal lattice points.

122. Woolfson, 1970. Reference 13, Ch. 8, pp. 294–316. Intensity statistics, Ch. 7, pp. 234–247.

123. Woolfson, M. M. *Direct Methods in Crystallography*. Oxford at the Clarendon Press, 1961. An excellent early monograph, which includes some numerical examples. It predicts great advances in the use of "direct methods" in the following decade, and these advances did, indeed, take place, many as a result of the work of the author.

124. Glusker, 1981. Reference 73, pp. 117–153. Includes reprints of several early papers on direct methods and Wilson's paper on the "probability distribution of X-ray intensities."

(C) PATTERSON FUNCTION

125. P. P. Ewald (ed.). *Fifty Years of X-ray Diffraction*. A. Oesthoek, 1962. "A. L. Patterson: Experiences in Crystallography—1924 to 1961." pp. 612–622. This is a brief autobiographical sketch, describing the perseverance of a man who was convinced that "something was to be learned about structural analysis from Fourier theory." It conveys some of the excitement of the realization that this was possible through the summation that now bears his name and that has done so much for the science of X-ray structure analysis.

126. Glusker, 1981. Reference 73, pp. 78–91. The Patterson function and Harker sections with copies of the original papers.

127. Dunitz, 1979. Reference 1, Ch. 1, pp. 65–72. This contains an analysis of the Patterson map of an irregular pentagon of atoms.

128. Ladd and Palmer, 1977. Reference 3: Sharpened Patterson maps with an example. Also a minimum function is computed with an example. Highly recommended—Ch. 6, pp. 218–245.

129. H. F. Judson. *The Eighth Day of Creation: Makers of the Revolution in Biology*. Simon and Schuster, 1979. A first-class description of the X-ray crystallographic, biological, chemical, and other studies that led to the deduction of the double helical structure of DNA. Try pp. 70–187 and 493–567, but we are sure you will not stop there. Read about the Patterson fuction pp. 544–545 and "A Postcard from Mount Fourier," an appendix on the Fourier series. Also read about Pauling (pp. 71–76), and Astbury (pp. 71–79).

130. Buerger, 1942. Reference 19.

(D) HEAVY-ATOM METHOD

131. Dunitz, 1979. Reference 1, Ch. 3, pp. 117–123. Including a discussion of spurious symmetry in electron density maps.

132. Blundell and Johnson, 1976. Reference 6. Heavy-atom derivatives of proteins. Ch. 8, pp. 183–239. This chapter describes how to prepare heavy-atom derivatives of proteins.

(E) ISOMORPHOUS REPLACEMENT METHOD

133. Dunitz, 1979. Reference 1, Ch. 3, pp. 123–129. This goes into the details of the determination of phase angles for proteins.

134. Glusker, 1981. Reference 73. pp. 92–116. With copies of the paper on the alums and the application to hemoglobin.

135. Blundell and Johnson, 1976. Reference 6: Shows precession photographs (Figure 6.2, pp. 154, 155) of a native protein and its heavy-atom derivative, and a description of the phase determination, including the use of anomalous dispersion data.

136. Ladd and Palmer, 1977. Reference 3, Ch. 6, pp. 254–262. Very clear diagrams of phase determination for proteins, with some data for ribonuclease, showing its Patterson map and phase determination.

(F) ANOMALOUS SCATTERING

137. Dunitz, 1979. Reference 1, Ch. 3, pp. 129–148. The Bijvoet experiment, the experiment of Coster, Knol, and Prins on zinc blende and different kinds of Patterson functions using anomalous scattering data are described clearly.

138. Glusker, 1981. Reference 73, pp. 154–167 includes the abstract of the zinc blende paper translated into English (the entire paper in German is reprinted in Early Papers, ref. 212), plus an epilogue by Prins and copies of the Bijvoet papers on sodium rubidium tartrate.

139. S. Ramaseshan and S. C. Abrahams (eds.). *Anomalous Scattering.* Munksgaard, 1975. This volume gives many of the details of the use of anomalous scattering.

140. Blundell and Johnson, 1976. Reference 6, Ch. 7, pp. 165–182. An informative discussion of anomalous dispersion and its use in phase determination for proteins, and the way to choose the correct absolute configuration for heavy-atom-phased maps.

141. W. Klyne and J. Buckingham. *Atlas of Stereochemistry: The Absolute Configuration of Organic Molecules,* 2nd ed. Volume 1 (to the end of 1971) and Volume 2 (from 1971 to 1976), Chapman and Hall, 1978.

142. D. H. Templeton, L. K. Templeton, J. C. Phillips, and K. O. Hodgson. "Anomalous scattering of X-rays by cesium and cobalt measured with synchrotron radiation," *Acta Crystallographica,* **A36** (1980), pp. 436–442.

143. J. C. Phillips and K. O. Hodgson. "The use of anomalous scattering effects to phase diffraction patterns from macromolecules," *Acta Crystallographica.* **A36** (1980), pp. 856–864.

144. W. A. Hendrickson and M. M. Teeter, "Structure of the hydrophobic protein crambin determined directly from the anomalous scattering of sulfur," *Nature* (London) **290** (1981), pp. 107–113.

145. James, 1948. Reference 18, Ch. 4, pp. 135–192. Also Appendix 3. Gives a list of anomalous scattering values pp. 608–610.

146. *International Tables. Volume IV.* Reference 20: A list of anomalous scattering factors.

REFINEMENT

(A) DIFFERENCE FOURIERS

147. Lipson and Cochran, 1966. Reference 14, Ch. 12: Difference synthesis pp. 331–340. The difference electron density map is described, with illustrations and examples, together with the mathematics involved in the analysis of the map.

148. Glusker, 1981. Reference 73, pp. 168–180, includes a reprint of the 1951 article by W. Cochran on difference Fouriers. This shows hydrogen atoms in a difference map long before it was commonly believed that such detail could be obtained from X-ray data.

149. Dunitz, 1979. Reference 1. Read pp. 185–190, 219–220, 391–392. An account of difference maps and what they mean.

(B) THE METHOD OF LEAST SQUARES

150. Dunitz, 1979. Reference 1, Ch. 4, pp. 190–224. Gives the details of the computations, and discussions of correlations of errors, constrained refinement, least-squares weights, and wrong structures.

151. Luger, 1980. Reference 5, Ch. 6, pp. 268–278. Discusses the theory and some of the computer programs used.

152. Glusker, 1981. Reference 73, pp. 168–171 and 181–184. Contains the original paper by Hughes in which he suggests the use of least squares.

153. Stout and Jensen, 1968. Reference 2, Ch. 16: The method of least squares, pp. 385–397. Discussions of the normal equations that must be set up, of the matrix notation used to describe them simply, of weighting functions and of other practical details.

154. Lipson and Cochran, 1966. Reference 14, Ch. 12: Method of least squares, pp. 340–351. A brief outline of the method is given, with a numerical example, together with a discussion of its application to structure refinement and the interpretation of results.

155. D. W. J. Cruickshank. Least-squares refinement of atomic parameters, pp. 187–197 in Ahmed, 1969, Ref. 169.

(C) ERRORS

156. Dunitz, 1979, Reference 1, Ch. 5, pp. 261–265. Describes explicitly the computation of estimated standard deviations of bond lengths, interbond angles, and torsion angles.

157. Woolfson, 1970. Reference 13, Ch. 9, pp. 323–352.

158. Stout and Jensen, 1968. Reference 2, Ch. 17: Random and systematic errors, pp. 399–414. A discussion of ways of determining the standard deviations of such quantities as the electron density and the atomic parameters from the experimental measurements. Possible causes of systematic errors are also considered.

159. D. R. Lide, Jr. and M. A. Paul. *Critical Evaluation of Chemical and Physical Structural Information.* National Academy of Sciences, 1974. In this most interesting and informative volume, "well-behaved" and "problem" structures are discussed in the context of methods other then X-ray diffraction for structure determination.

(D) DEFORMATION DENSITY MAPS

160. P. Coppens and M. B. Hall. *Electron Distributions and the Chemical Bond.* Plenum, 1982. This book represents the proceedings of a symposium held at the meeting of the American Chemical Society in Atlanta in 1981. Some most informative deformation density maps are illustrated throughout the articles.

161. P. Coppens. *Journal of Chemical Education,* 1984. Volume 61, pp. 761–765.

162. P. Becker (ed.). *Electron and Magnetization Densities in Molecules and Crystals.* NATO Advanced Study Institutes, Series B: Physics. Plenum, 1980.

163. F. L. Hirshfeld (ed.). *Electron Density Mapping in Molecules and Crystals.* Weizmann Scientific Press of Israel, 1977.

164. Dunitz, 1979. Reference 1, Ch. 8, pp. 406–417. Some excellent examples are given.

CRYSTALLOGRAPHIC COMPUTING

These books contain papers presented at various international "schools" (short intensive courses) by recognized leaders in the field. Each was timely when published and many of the papers in the earlier volumes are still valuable. They include much on direct methods and diffractometer control as well as crystallographic refinement.

165. D. Sayre (ed.). *Computational Crystallography.* Clarendon Press, Oxford, 1982. Proceedings of a Summer School in Ottawa in 1981.

166. R. Diamond, S. Ramaseshan, and K. Venkatesan (eds). *Computing in Crystallography: Proceedings of the 1980 Winter School, Bangalore.* Indian Academy of Sciences, 1980.

167. H. Schenk, R. Olthof-Hazekamp, H. van Koningsveld, and G. C. Bassi (eds.). *Computing in Crystallography: Proceedings of an International Summer School on Crystallographic Computing held in Twente, the Netherlands,* 24 July–1 August, 1978. Delft University Press, 1978.

168. S. Ramaseshan, M. F. Richardson, and A. J. C. Wilson. *Crystallographic Statistics.* Indian Academy of Sciences, 1982.

169. F. R. Ahmed (ed.). *Crystallographic Computing. Proceedings of the 1969 International Summer School on Crystallographic Computing.* Munksgaard, 1970.

GEOMETRY AND ATOMIC VIBRATIONS

(A) MOLECULAR GEOMETRY

170. Dunitz, 1979. Reference 1, Ch. 5, pp. 225–243 and 495–501. This is an excellent discussion of the use of atomic position coordinates in the calculation of interatomic distances, interbond angles, torsion angles, and least-squares planes through parts of the molecule. Some BASIC computer programs are provided.

(B) THERMAL MOTION ANALYSIS

171. Dunitz, 1979. Reference 1, Ch. 1, pp. 43–49; Ch. 5, pp. 244–261. Includes an analysis of rigid-body motion and internal torsional motion. This is a general, and therefore very useful, account.

172. James, 1948. Reference 18, Ch. 5. Temperature effects, pp. 193–267.

173. W. Cochran. *The Dynamics of Atoms in Crystals.* Arnold, 1973 (Crane, Russak and Co., Inc., 1973). Phase transitions and lattice dynamics, leading to a study of the specific heats of crystals. Optical and dielectric properties and thermal conductivity.

174. B. T. M. Willis and A. W. Pryor. *Thermal Vibrations in Crystallography.* Cambridge University Press, 1975. This valuable volume considers lattice dynamics, the effect of thermal vibrations on the intensities of the Bragg reflections, and thermal diffuse scattering of X rays and neutrons.

175. J. L. Amoros and M. Amoros. Reference 103: A most interesting volume on diffuse scattering, the result of disorder, with excellent diagrams. In particular it emphasizes the diffraction patterns of benzene and anthracene. See Figure 16, p. 96 and Figure 20, p. 157 and pp. 164–166.

176. C. K. Johnson. *An Introduction to Thermal-motion Analysis.* In (Ahmed, 1969, Ref. 169), pp. 207–219. *The effect of thermal motion on interatomic distances and angles,* pp. 220–226.

177. Lipson and Cochran, 1966. Reference 14, Ch. 11. Effects of thermal vibration, pp. 300–310. A discussion of the ellipsoidal anisotropic temperature factor,

together with a numerical example to show how the mean square displacements of atoms in the directions of the principal axes of the ellipsoid may be determined from the general exponential expression.

ADDITIONAL TOPICS

(A) NEUTRON DIFFRACTION

178. G. E. Bacon. *Neutron Diffraction,* 2nd ed. Oxford University Press, 1962. A useful and interesting book to read or page through, both from the point of view of the study of neutron diffraction itself and also because analogies and comparisons with X-ray diffraction are constantly made.

179. Vainshtein, 1981. Reference 4, Vol. 1, Ch. 4, pp. 350–363.

180. Dent Glasser, 1977. Reference 9, Ch. 8, pp. 180–183.

(B) ELECTRON DIFFRACTION

181. Vainshtein, 1981. Reference 4, Vol. 1, Ch. 4, pp. 336–350.

182. Dent Glasser, 1977. Reference 9, Ch. 8, pp. 183–191.

(C) POWDER DIFFRACTION

183. W. Parrish. *History of the X-ray Powder Method in the USA.* In McLachlan and Glusker, 1983, Ref. 210, Section D, Ch. 1, pp. 201–214.

184. L. V. Azaroff and M. J. Buerger. *The Powder Method in X-ray Crystallography.* McGraw-Hill, 1958.

185. Dent Glasser, 1977. Reference 9, Ch. 6, pp. 125–155.

186. H. P. Klug and L. E. Alexander. *X-ray Diffraction Procedures for Polycrystalline and Amorphous Materials,* 2nd ed. Wiley, 1974.

187. H. Lipson and H. Steeple. *Interpretation of X-ray Powder Diffraction Patterns.* Macmillan, 1970.

(D) SMALL-ANGLE SCATTERING

188. A. Guinier. *X-ray Diffraction in Crystals, Imperfect Crystals and Amorphous Bodies.* (Transl. P. and D.S-M. Lorrain.) Freeman, 1963. Chapter 10, pp. 319–350.

189. *Transactions of the American Crystallographic Association. Proceedings of the Symposium on Small-Angle Scattering at the University of Missouri, Columbia, Missouri, March 15, 1983.* Volume 19, 1983.

(E) LIQUID CRYSTALS

190. J. D. Litster (ed.). *Physics Today* **35**(No. 5), May (1982). *Special Issue: Liquid Crystals.* pp. 25–74. This is a series of articles on liquid crystals; each is clearly written and forms a good basis for anyone interested in this topic.

(F) LESS-THAN-CRYSTALLINE MATERIALS

191. J. M. Cowley, J. B. Cohen, M. B. Salamon, and B. J. Wuensch. *Modulated Structures—1979.* American Institute of Physics, 1979. The proceedings of a conference. The term "modulated structures" is used in descriptions of a wide range of phenomena involving periodic or partially periodic perturbations of a basic crystal structure, that is, "any periodic, or partially periodic perturbation of a crystal structure with a repetition distance appreciably greater than the basic unit cell dimensions."

192. H. R. Wilson. *Diffraction of X-rays by Proteins, Nucleic Acids and Viruses.* E. Arnold, 1966. A monograph in which the application of X-ray diffraction methods to the study of proteins, nucleic acids, nucleoproteins, and viruses is discussed. The section on diffraction by helical molecules is excellent and informative. Figure 2.25, pp. 42–43 is particularly recommended.

193. B. K. Vainshtein. *Diffraction of X-rays by Chain Molecules.* Elsevier, 1966. Chapters 1–4, diffraction by chain molecules, pp. 1–202. The chapter on diffraction by helical structures is particularly recommended. An excellent text on the subject.

194. Holmes and Blow, 1966. Reference 15: Fibers, pp. 203–234. Optical transforms of continuous helices are shown.

HISTORY

195. J. G. Burke. *Origins of the Science of Crystals.* University of California Press, 1966. The detailed study of crystals in the seventeenth and eighteenth centuries laid the foundation of crystallography and the geometry of molecular packing. This history, particularly the contribution of Hauy, is described.

196. Ewald (ed.), 1962. Reference 125. A commemorative book in honor of the fiftieth anniversary of Max von Laue's discovery of diffraction of X rays by crystals. Included are many fascinating personal reminiscences by famous crystallographers.

197. W. H. And W. L. Bragg. *X-rays and Crystal Structure,* 4th ed. G. Bell and Sons, 1924. Chapter 1, pp. 1–5. This is an early account, first published in 1915, of the experiment of von Laue.

198. Glusker, 1981. Reference 73. A description of early investigations, including a translation by J. J. Stezowski of the Laue, Friedrich, and Knipping paper, pp. 10–39.

199. G. Dodson, J. P. Glusker, and D. Sayre (eds.). *Structural Studies on Molecules of Biological Interest: A Volume in Honour of Dorothy Hodgkin.* Oxford Univeristy Press, 1981.

200. G. M. Caroe. *William Henry Bragg, 1862–1942.* Cambridge University Press, 1978. This is a biography written by the daughter of W. H. and the sister of W. L. Bragg.

201. P. Goodman (ed.). *Fifty Years of Electron Diffraction.* Reidel, 1981. A description of the discovery and development of electron diffraction. The publication of this volume was commissioned by the International Union of Crystallography.

202. D. C. Hodgkin. *Kathleen Lonsdale: A Biographical Memoir.* Royal Society of London, 1975.

203. D. C. Hodgkin. *J. D. Bernal.* Royal Society of London, 1980.

204. D. C. Phillips. *W. L. Bragg.* Royal Society of London, 1979.

205. J. D. Watson. *The Double Helix.* McClelland and Stewart, Ltd., 1968.

206. R. Olby. *The Path to the Double Helix.* University of Washington Press, Seattle, 1974. Describes the work of Astbury, pp. 59–70, Bernal, pp. 254–263, Pauling and the α-helix, pp. 267–295, and the double helical structure of DNA, pp. 297–434, including many quotations from notes and letters and copies of some laboratory notes.

207. Judson, 1979. Reference 129.

208. W. L. Bragg (edited by D. C. Phillips and H. Lipson). *The Development of X-ray Analysis.* Bell, 1975.

209. Schneer, 1977. Reference 44, pp. 1–31. Contain a description of early ideas on internal periodicity in crystals such as those of Kepler, Hauy, and Bravais with excerpts from their publications. Pages 54–71 contain ideas on the packing of ions—for example, rock salt. These ideas were proven by X-ray diffraction. Reprints of the articles discussed are included.

210. D. McLachlan, Jr. and J. P. Glusker (eds.). *Crystallography in North America.* American Crystallographic Association, 1983. This considers the history of the method up to the present. An article by C. Frondel entitled "An Overview of Crystallography in North America," pp. 1–24, tells of the crystallographic studies before the advent of X-ray diffraction.

211. S. H. Mauskopf. *Crystals and Compounds. Molecular Structure and Composition in Nineteenth Century French Science.* Transactions of the American Philosophical Society, 1976. Volume 66, part 3, (82 pages).

212. J. M. Bijvoet, W. G. Burgers, and G. Hagg. *Early Papers on Diffraction of X-rays by Crystals.* Volume I (1969), Volume II (1972). A. Oesthoek. An invaluable source of reprints of many of the important early papers in X-ray crystallography.

Index

Boldface page numbers represent the principal discussion of the topic and italicized page numbers refer to the Glossary.